IMPACTS OF CLIMATE CHANGE ON TRANSPORTATION AND INFRASTRUCTURE – A GULF COAST STUDY

CLIMATE CHANGE AND ITS CAUSES, EFFECTS AND PREDICTION

Global Climate Change
Horace M. Karling (Editor)
2001. ISBN: 1-56072-999-6

Global Climate Change Revisited
Harace B. Karling (Editor)
2007. ISBN: 1-59454-039-X

Climate Change Research Progress
Lawrence N. Peretz (Editor)
2008. ISBN: 1-60021-998-5

Climate Change: Financial Risks
United States Government Accountability Office
2008. ISBN: 978-1-60456-488-4

Post-Kyoto: Designing the Next International Climate Change Protocol
Matthew Clarke
2008. ISBN: 978-1-60456-840-0
2008. ISBN: 978-1-61668-101-2 (E-book)

The Effects of Climate Change on Agriculture, Land Resources, Biodiversity in the United States
Peter Backlund, Anthony Janeto and David Schimel
2009. ISBN: 978-1-60456-989-6

Economics of Policy Options to Address Climate Change
Gregory N. Bartos
2009. ISBN: 978-1-60692-116-6

Thresholds of Climate Change in Ecosystems
Vicente Orostegui (Editor)
2009. ISBN: 978-1-60741-487-2

Climate Variability, Modeling Tools and Agricultural Decision-Making
Angel Utset (Editor)
2009. ISBN: 978-1-60692-703-8
2009. ISBN: 978-1-60876-791-5 (E-book)

Carbon Offsets: Examining their Role in Greenhouse Gas Reduction
Karen T. Morningstar (Editor)
2010. ISBN: 978-1-60741-444-5

Emissions Trading: Lessons Learned from the European Union and Kyoto Protocol Climate Change Programs
Ervin Nagy and Gisella Varga (Editors)
2009. ISBN: 978-1-60741-194-9

Constructing Climate Change Legislation: Background and Issues
Gerald P. Overhauser (Editor)
2009. ISBN: 978-1-60692-986-5

Designing Greenhouse Gas Reduction and Regulatory Systems
Sonja Enden (Editor)
2009. ISBM: 978-1-60741-195-6

Global Climate Change: International Perspectives and Responses
Elias D'Angelo (Editor)
2009. ISBN: 978-1-60741-233-5

Disputing Global Warming
Anton Horvath and Boris Molnar (Editors)
2009. ISBN: 978-1-60741-235-9
2009. ISBN: 978-1-60876-503-4 (E-book)

Focus on Climate Change and Health
Viroj Wiwanitkit
2009. ISBN: 978-1-60741-247-2

Land Use and Climate Change
Suresh C. Rai
2009. ISBN: 978-1-60741-362-2

Impacts of Climate Change on Transportation and Infrastructure - A Gulf Coast Study
Iason Pavlopoulos (Editor)
2009. ISBN: 978-1-60741-424-7

Coastal Sensitivity to Sea Level Rise - Focusing on the Mid-Atlantic Region
Melvin C. Urajner (Editor)
2009. ISBN: 978-1-60741-440-7

Lightning in the Tropics: From a Source of Fire to a Monitoring System of Climatic Changes
Osmar Pinto, Jr.
2009. ISBN: 978-1-60741-764-4

Carbon Capture and Greenhouse Gases
Imrus Juhász and Gyorgy Halász (Editors)
2010. ISBN: 978-1-60692-089-3

The Science of Climate Change and Policy Implications
Kyle S. Hartzell (Editor)
2010. ISBN: 978-1-60741-448-3

Effects of Climate Change on Energy Production and Use in the U.S.
Alrik M. Solberg (Editor)
2010. ISBN: 978-1-60741-426-1

Policy Option Issues for CO2 Emissions
Nikolaus Vogler (Editor)
2010. ISBN: 978-1-60741-381-3

The Role of Agriculture in Carbon Capture and Climate Change
Arvid Bjurstrom (Editor)
2010. ISBN: 978-1-60741-445-2

Carbon Sequestration: Methods, Modeling and Impacts
Elke Hoch and Siegbert Grunwald (Editors)
2010. ISBN: 978-1-60741-498-8

Climate Change Litigation and Law
Jean-François Masson (Editor)
2010. ISBN: 978-1-60876-089-3

Carbon Tax and Cap-and-Trade Tools: Market-Based Approaches for Controlling Greenhouse Gases
Nelson E. Burney (Editor)
2010. ISBN: 978-1-60876-137-1

Global Change and Forestry: Economic and Policy Impacts and Responses
Jianbang Gan, Stephen Grado and Ian A. Munn (Editors)
2010. ISBN: 978-1-60876-262-0

Climate Change Measures and Trade Considerations
George R. Fried (Editor)
2010. ISBN: 978-1-60876-756-4

Assessing Climate Change Impacts on the United States
Alex B. McNeill (Editor)
2010. ISBN: 978-1-60876-160-9
2010. ISBN: 978-1-60876-160-9 (E-book)

Responding to Impacts of Climate Change on Water Resources
Zachary E. Quinn (Editor)
2010. ISBN: 978-1-60741-992-1
2010. ISBN: 978-1-61728-095-5 (E-book)

Climate Change: Fundamental Issues and Policy Tools
Elise M. Farrugia (Editor)
2010. ISBN: 978-1-60741-997-6
2010. ISBN: 978-1-61728-052-8 (E-book)

Effects of Climate Change on Aquatic Invasive Species
Sofia A. Contreras (Editor)
2010. ISBN: 978-1-61728-005-4
2010. ISBN: 978-1-61728-251-5 (E-book)

CLIMATE CHANGE AND ITS CAUSES, EFFECTS AND PREDICTION

IMPACTS OF CLIMATE CHANGE ON TRANSPORTATION AND INFRASTRUCTURE – A GULF COAST STUDY

IASON PAVLOPOULOS
EDITOR

Nova Science Publishers, Inc.

New York

For permission to use material from this book please contact us:
Telephone 631-231-7269; Fax 631-231-8175
Web Site: http://www.novapublishers.com

LIBRARY OF CONGRESS CATALOGING-IN-PUBLICATION DATA

Impacts of climate change on transportation and infrastructure : a Gulf Coast study / editor, Iason Pavlopoulos.
p. cm.
Includes bibliographical references.
ISBN 978-1-60741-424-7 (hardcover)
1. Roads--Design and construction--Climatic factors--Gulf Coast (U.S.) 2. Climatic changes--Gulf Coast (U.S.) 3. Gulf Coast (U.S.)--Climate. 4. Storms--Gulf Coast (U.S.) 5. Floods--Gulf Coast (U.S.) 6. Sustainable engineering--Gulf Coast (U.S.) 7. Transportation--Environmental aspects--Gulf Coast (U.S.) 8. Infrastructure (Economics)--Gulf Coast (U.S.) I. Pavlopoulos, Iason.
TE23.5.I46 2009
388.0976--dc22

2009044385

Published by Nova Science Publishers, Inc. ✦ New York

CONTENTS

Contents		**ix**
Preface		**xi**
Executive Summary	*Lead Authors: Joanne R. Potter, Michael J. Savonis and Virginia R. Burkett*	**xiii**
Chapter 1	**Why Study Climate Change Impacts on Transportation?** *Lead Authors: Robert C. Hyman, Joanne R. Potter Michael J. Savonis, Virginia R. Burkett and Jessica E. Tump*	**1**
Chapter 2	**Why Study the Gulf Coast?** *Lead Authors: Virginia R. Burkett, Robert C. Hyman, Ron Hagelman, Stephen B. Hartley and Matthew Sheppard* *Contributing Authors: Thomas W. Doyle, Daniel M. Beagan, Alan Meyers, David T. Hunt, Michael K. Maynard, Russell H. Henk, Edward J. Seymour, Leslie E. Olson, Joanne R. Potter and Nanda N. Srinivasan*	**49**
Chapter 3	**How Is the Gulf Coast Climate Changing?** *Lead Authors: Barry D. Keim, Thomas W. Doyle and Virginia R. Burkett* *Contributing Authors: Claudia Tebaldi, Ivor Van Heerden, S. Ahmet Binselam, Michael F. Wehner, Tamara G. Houston and Daniel M. Beagan*	**95**

Chapter 4 **What Are the Implications of Climate Change and Variability for Gulf Coast Transportation?** **155**
Lead Authors: Robert S. Kafalenos and Kenneth J. Leonard

Contributing Authors: Daniel M. Beagan, Virginia R. Burkett, Barry D. Keim, Alan Meyers, David T. Hunt, Robert C. Hyman, Michael K. Maynard, Barbara Fritsche, Russell H. Henk, Edward J. Seymour, Leslie E. Olson, Joanne R. Potter and Michael J. Savonis

Chapter 5 **How Can Transportation Professionals Incorporate Climate Change in Transportation Decisions?** **233**
Lead Authors: Kenneth J. Leonard, John H. Suhrbier and Eric Lindquist

Contributing Authors: Michael J. Savonis, Joanne R. Potter and Wesley R. Dean

Chapter 6 **What Are the Key Conclusions of This Study?** **263**
Lead Authors: Michael J. Savonis, Virginia R. Burkett and Joanne R. Potter

Contributing Authors: Thomas W. Doyle, Ron Hagelman, Stephen B. Hartley, Robert C. Hyman, Robert S. Kafalenos, Barry D. Keim, Kenneth J. Leonard, Matthew Sheppard, Claudia Tebaldi and Jessica E. Tump

Appendix A **Gulf Coast Study GIS Datasets** **273**

Appendix B **Additional Data on Social and Economic Setting** **275**

Appendix C **Additional Rail Data** **283**

Appendix D **Water Balance Model Procedures** **287**

Appendix E **HURASIM Model Description** **291**

Appendix F **Projecting Future Sea Level Rise with the SLRRP Model** **295**

List of Selected Acronyms **299**

Glossary of Terms **301**

Index **319**

PREFACE

Climate affects the design, construction, safety, operations, and maintenance of transportation infrastructure and systems. The prospect of a changing climate raises critical questions regarding how alterations in temperature, precipitation, storm events, and other aspects of the climate could affect the nation's roads, airports, rail, transit systems, pipelines, ports, and waterways. This book addresses these concerns for the region of the U.S. Central Gulf Coast between Galveston, Texas and Mobile, Alabama. This region contains multimodal transportation infrastructure that is critical to regional and national transportation services. Furthermore, this book reviews the climate changes that are anticipated during the next 50 to 100 years for the central Gulf Coast including warming temperatures, changes in precipitation patterns, and increased storm intensity. The authors hope that an increased awareness of these factors will lead to changes in today's transportation decisions and planning processes, leading to a more robust, resilient, and cost-effective transportation network in the coming decades.

EXECUTIVE SUMMARY*

Lead Authors: Joanne R. Potter, Michael J. Savonis and Virginia R. Burkett

ABSTRACT

Climate affects the design, construction, safety, operations, and maintenance of transportation infrastructure and systems. The prospect of a changing climate raises critical questions regarding how alterations in temperature, precipitation, storm events, and other aspects of the climate could affect the nation's roads, airports, rail, transit systems, pipelines, ports, and waterways. Phase I of this regional assessment of climate change and its potential impacts on transportation systems addresses these questions for the region of the U.S. central Gulf Coast between Galveston, Texas and Mobile, Alabama. This region contains multimodal transportation infrastructure that is critical to regional and national transportation services.

Historical trends and future climate scenarios were used to establish a context for examining the potential effects of climate change on all major transportation modes within the region. Climate changes anticipated during the next 50 to 100 years for the central Gulf Coast include warming temperatures, changes in precipitation patterns, and increased storm intensity. The warming of the oceans and decline of polar ice sheets is expected to accelerate the rate of sea level rise globally. The effects of sea level rise in most central Gulf Coast counties will be exacerbated by the sinking of the land surface, which is accounted for in this assessment.

The significance of these climate factors for transportation systems was assessed. Warming temperatures are likely to increase the costs of transportation construction, maintenance, and operations. More frequent extreme precipitation events may disrupt transportation networks with flooding and visibility problems. Relative sea level rise will make much of the existing infrastructure more prone to frequent or permanent inundation – 27 percent of the major roads, 9 percent of the rail lines, and 72 percent of the ports are built on land at or below 122 cm (4 feet) in elevation. Increased storm intensity may lead to increased service disruption and infrastructure damage: More than half of the area's major highways (64 percent of Interstates; 57 percent of arterials), almost half of the rail miles, 29 airports, and virtually all of the ports are below 7 m (23 feet) in elevation and

* This is an edited, reformatted and augmented version of a U. S. Climate Change Science Program and Subcommittee on Global Change Research publication dated March 2008.

subject to flooding and possible damage due to hurricane storm surge. Consideration of these factors in today's transportation decisions and planning processes should lead to a more robust, resilient, and cost-effective transportation network in the coming decades.

The changing climate raises critical questions for the transportation sector in the United States. As global temperatures increase, sea levels rise, and weather patterns change, the stewards of our Nation's infrastructure are challenged to consider how these changes may affect the country's roads, airports, rail, transit systems, and ports. The U.S. transportation network – built and maintained through substantial public and private investment – is vital to the Nation's economy and the quality of our communities. Yet little research has been conducted to identify what risks this system faces from climate change, or what steps managers and policy makers can take today to ensure the safety and resilience of our vital transportation system.

This study: *The Impacts of Climate Change and Variability on Transportation Systems and Infrastructure: Gulf Coast Study, Phase I* has investigated these questions through a case study of a segment of the U.S. central Gulf Coast. The research, sponsored by the U.S. Department of Transportation (DOT) in partnership with the U.S. Geological Survey (USGS), has been conducted under the auspices of the U.S. Climate Change Science Program (CCSP). The study is 1 of 21 "synthesis and assessment" products planned and sponsored by CCSP. The interdisciplinary research team included experts in climate and meteorology; hydrology and natural systems; transportation; and decision support.

A case study approach was selected for this research as an approach that would generate useful information for local and regional decision makers, while helping to develop research methodologies for application in other locations. In defining the study area, the DOT sought to design a project that would increase the knowledge base regarding the risks and sensitivities of all modes of transportation infrastructure to climate variability and change, the significance of these risks, and the range of adaptation strategies that can be considered to ensure a robust and reliable transportation network. The availability of reliable data, interest of local agencies and stakeholders, and transferability of findings were also important criteria in selecting the study area. This study focuses on those climate factors which are relevant to the Gulf Coast; in other areas different aspects of climate change may be significant. The modeled climate projections and the specific implications of these scenarios for transportation facilities are specific to the Gulf Coast study area However, the methods presented in this report can be applied to any region.

This report presents the findings of the first phase of a three phase research effort. The ultimate goal of this research is to provide knowledge and tools that will enable transportation planners and managers to better understand the risks, adaptation strategies, and tradeoffs involved in planning, investment, design, and operational decisions. The objective of Phase I was to conduct a preliminary assessment of the risks and vulnerabilities of transportation in the region, after collecting and integrating the range of data needed to characterize the region – its physiography and hydrology, land use and land cover, past and projected climate, current population and trends, and transportation infrastructure. Subsequent phases will conduct more detailed analyses. Phase II will conduct an in-depth assessment of risks to transportation in a selected location, reporting on implications for long-range plans and impacts on safety, operations, and maintenance. This phase will also develop a risk assessment methodology and identify techniques to incorporate environmental and climate data in transportation decisions.

Phase III will identify and analyze adaptation and response strategies and develop tools to assess these strategies, while enumerating future research needs.

The Gulf Coast Study Area

The Gulf Coast study area includes 48 contiguous coastal counties in four States, running from Houston/Galveston, TX, to Mobile, AL. This region is home to almost 10 million people living in a range of urban and rural settings and contains critical transportation infrastructure that provides vital service to its constituent States and the Nation as a whole. It is also highly vulnerable to sea level rise and storm impacts. A variety of physical datasets were compiled for review and use by the project research team. Most of the spatial data is organized in GIS formats or "layers" that can be integrated to assess the vulnerability and risks of the transportation infrastructure in the study area and inform the development of adaptation strategies.

Physical and Natural Environment

The coastal geography of the region is highly dynamic due to a unique combination of geomorphic, tectonic, marine, and atmospheric forcings that shape both the shoreline and interior land forms. Due largely to its sedimentary history, the region is low lying; the great majority of the study area lies below 30 m in elevation. Due to its low relief, much of the central Gulf Coast region is prone to flooding during heavy rainfall events, hurricanes, and lesser tropical storms. Land subsidence is a major factor in the region, as sediments naturally compact over time. Specific rates of subsidence vary across the region, influenced by both the geomorphology of specific locations as well as by human activities. Most of the coastline also is highly vulnerable to erosion and wetland loss, particularly in association with tropical storms and frontal passages. It is estimated that 56,000 ha (217 mi^2) of land were lost in Louisiana alone during Hurricane Katrina. Further, many Gulf Coast barrier islands are retreating and diminishing in size. The Chandeleur Islands, which serve as a first line of defense for the New Orleans region, lost roughly 85 percent of their surface area during Hurricane Katrina. As barrier islands and mainland shorelines erode and submerge, onshore facilities in low-lying coastal areas become more susceptible to inundation and destruction.

The Gulf Coast Transportation Network

The central Gulf Coast study area's transportation infrastructure is a robust network of multiple modes – critical both to the movement of passengers and goods within the region and to national and international transport as well:

- The region has 17,000 mi (27,000 km) of major highways – about 2 percent of the Nation's major highways – that carry 83.5 billion vehicle miles of travel annually.

The area is served by 13 major transit agencies; over 136 providers offer a range of public transit services to Gulf Coast communities.

- Roughly two-thirds of all U.S. oil imports are transported through this region, and pipelines traversing the region transport over 90 percent of domestic Outer Continental Shelf oil and gas. Approximately one-half of all the natural gas used in the United States passes through or by the Henry Hub gas distribution point in Louisiana.
- The study area is home to the largest concentration of public and private freight handling ports in the United States, measured on a tonnage basis. These facilities handle a huge share – around 40 percent – of the Nation's waterborne tonnage. Four of the top five tonnage ports in the United States are located in the region: South Louisiana, Houston, Beaumont, and New Orleans. The study area also has four major container ports.
- Overall, more than half of the tonnage (54 percent) moving through study area ports is petroleum and petroleum products. Additionally, New Orleans provides the ocean gateway for much of the U.S. interior's agricultural production.
- The region sits at the center of transcontinental trucking and rail routes and contains one of only four major points in the United States where railcars are exchanged between the dominant eastern and western railroads.
- The study area also hosts the Nation's leading and third-leading inland waterway systems (the Mississippi River and the Gulf Intracoastal) based on tonnage. The inland waterways traversing this region provide 20 States with access to the Gulf of Mexico.
- The region hosts 61 publicly owned, public-use airports, including 11 commercial service facilities. Over 3.4 million aircraft takeoffs and landings take place at these airports annually, led by the major facilities at George Bush Intercontinental (IAH), William P. Hobby, and Louis Armstrong New Orleans International. IAH also is the leading airport in the study area for cargo, ranking 17th in the Nation for cargo tonnage.

Given the scale and strategic importance of the region's transportation infrastructure, it is critical to consider the potential vulnerabilities to the network that may be presented by climate change. A better understanding of these risks will help inform transportation managers as they plan future investments.

THE GULF COAST CLIMATE IS CHANGING

The research team's assessment of historical and potential future changes in the Gulf Coast study region draws on publications, analyses of instrumental records, and models that simulate how climate may change in the future. The scenarios of future climate referenced in this report were generated by the National Center for Atmospheric Research (NCAR) by using an ensemble of 21 different atmosphere-ocean coupled general circulation models (GCM) for the Gulf Coast region. Model results, climatic trends during the past century, and climate theory all suggest that extrapolation of the 20th century temperature record would likely underestimate the range of change that could occur in the next few decades. While there is still considerable uncertainty about the *rates* of change that can be expected, there is a fairly strong consensus regarding the direction of change for most of the climate variables that affect transportation in the Gulf Coast region. Key findings for the study region include:

- **Rising relative sea levels –** Relative sea level in the study area is likely to increase at least 0.3 meter (1 foot) across the region and possibly as much as 2 meters (6 to 7 feet) in some parts of the study area. Relative sea level rise (RSLR) is the combined effect of the projected increase in the volume of the world's oceans (eustatic sea level change), which results from increases in temperature and melting of ice, and the projected changes in land surface elevation at a given location due to subsidence of the land surface. The highest rate of relative sea level rise will very likely be in the central and western parts of the study area (Louisiana and East Texas), where subsidence rates are highest. The analysis of a "middle range" of potential sea level rise of 0.6 to 1.2 meters (2 to 4 feet) indicates that a vast portion of the Gulf Coast from Houston to Mobile may be inundated over the next 50 to 100 years. The projected rate of relative sea level rise for the region is consistent with historical trends, other published region-specific analyses, and the IPCC 4th Assessment Report findings, which assumes no major changes in ice sheet dynamics.

- **Storm activity –** Hurricanes are more likely to form and increase in their destructive potential as the sea surface temperature of the Atlantic and Gulf of Mexico increase. The literature indicates that the intensity of major storms could possibly increase by 10 percent or more. This indicates that Category 3 storms and higher may return more frequently to the central Gulf Coast and thus cause more disruptions. Rising relative sea level will exacerbate exposure to storm surge and flooding. Depending on the trajectory and scale of individual storms, facilities at or below 9 meters (30 feet) could be subject to direct storm surge impacts.

- **Warming temperatures –** All GCMs available from the Intergovernmental Panel on Climate Change (IPCC) for use in this study indicate an increase in average annual Gulf Coast temperature through the end of this century. Based on GCM runs under three different emission scenarios developed by the IPCC Special Report on Emissions Scenarios (SRES) (the low-emissions B1, the high-emissions A2, and the mid-range A1B scenarios), the average temperature in the Gulf Coast region appears likely to increase by at least 1.5°C ± 1°C (2.7°F ± 1.8°F) during the next 50 years. Extreme high temperatures are also expected to increase – with the number of days above 32.2°C (90°F) very likely to increase significantly across the study area. Within 50 years the probability of experiencing 21 days a year with temperatures of 37.8°C (100°F) or above is greater than 50 percent.

- **Changes in precipitation patterns –** Some analyses, including the GCM results from this study, indicate that average precipitation will increase in this region while others indicate a decline of average precipitation during the next 50 to 100 years. In either case, it is expected that average runoff could decline, due to increasing temperatures and resulting higher evapotranspiration rates. While *average* annual rainfall may increase or decrease slightly, the *intensity* of individual rainfall events is likely to increase during the 21st century.

In the near term, the direction and scale of these modeled outcomes are consistent regardless of the assumptions used for level of greenhouse gas emissions: Model outputs are relatively similar across a range of IPCC SRES emission scenarios for the next four decades. However, long-range projections (modeled to 100 years) do vary across scenarios, with the magnitude of impacts indicated being more severe under higher-emission assumptions.

Climate Change Has Implications for Gulf Coast Transportation

The four key climate drivers in the region: rising temperatures, changing precipitation patterns, rising relative sea levels, and increasing storm intensity, present clear risks to transportation infrastructure in the study area. These factors can be incorporated into today's transportation decisions to help prepare for and adapt to changing environmental conditions.

- **Warming temperatures may require changes in materials, maintenance, and operations.** The combined effects of an increase in mean and extreme high temperatures across the study region are likely to affect the construction, maintenance, and operations of transportation infrastructure and vehicles. Higher temperatures may also suggest areas for materials and technology innovation to develop new, more heat- tolerant materials. Some types of infrastructure deteriorate more quickly at temperatures above 32.2°C (90°F). As the number of very hot days increases, different materials may be required. Further, restrictions on work crews may lengthen construction times. Rail lines may be affected by more frequent rail buckling due to an increase in daily high temperatures. Ports, maintenance facilities, and terminals are expected to require increased refrigeration and cooling. Finally, higher temperatures affect aircraft performance and the runway lengths that are required. However, advances in aircraft technology are expected to offset the potential effects of the temperature increases analyzed in this report, so that current runway lengths are likely to be sufficient. The effects of increases in average temperatures and in the number of very hot days will have to be addressed in designing and planning for vehicles, facilities, and operations.

- **Changes in precipitation patterns may increase short-term flooding.** The analysis of future annual precipitation change based on results of climate model runs is inconclusive: some models indicate an increase in average precipitation and some indicate a decrease. In either case, the hotter climate may reduce soil moisture and average run-off, possibly necessitating changes in right-of-way land management. The potential of changes in heavy rainfall may have more significant consequences for transportation; more frequent extreme precipitation events may result in more frequent flooding, stressing the capacity of existing drainage systems. The potential of extreme rainfall events and more frequent and prolonged flooding may disrupt traffic management, increase highway incidents, and impact airline schedules – putting additional strain on a heavily used and increasingly congested system. Further, prolonged flooding – inundation in excess of one week – can damage pavement substructure.

- **Relative sea level rise may inundate existing infrastructure.** To assess the impact of relative sea level rise (RSLR), the implications of rises equal to 61 cm and 122 cm (2 and 4 ft) were examined. As discussed above, actual RSLR may be higher or somewhat lower than these levels. Under these scenarios, substantial portions of the transportation infrastructure in the region are at risk: 27 percent of the major roads, 9 percent of the rail lines, and 72 percent of the ports are at or below 122 cm (4 ft) in elevation, although portions of the infrastructure are guarded by protective structures such as levees and dikes. While protective structures will continue to be an important strategy in the area, rising sea levels significantly increase the challenge to transportation managers in ensuring reliable transportation services. Inundation of even small segments of the intermodal system can render much larger portions impassable, disrupting connectivity and access to the wider transportation network.

- **Increased storm intensity may lead to greater service disruption and infrastructure damage.** This study examined the potential for flooding and damage associated with storm surge levels of 5.5 m and 7.0 m (18 ft and 23 ft). These modeled outputs are comparable to potential surge levels during severe storms in the region: Simulated storm surge from model runs across the central Gulf Coast demonstrated a 6.7- to 7.3-m (22- to 24-ft) potential surge for major hurricanes. These levels may be conservative; surge levels during Hurricane Katrina (rated a Category 3 at landfall) exceeded these heights in some locations. The specific location and strength of storm surges are of course determined by the scale and trajectory of individual tropical storms, which are difficult to predict. However, substantial portions of the region's infrastructure are located at elevations below the thresholds examined, and recent storms have demonstrated that major hurricanes can produce flooding miles inland from the location of initial landfall. With storm surge at 7 m (23 ft), more than half of the area's major highways (64 percent of Interstates; 57 percent of arterials), almost half of the rail miles, 29 airports, and virtually all of the ports are subject to flooding.

Other damage due to severe storms is likely, as evidenced by the damage caused by Hurricanes Katrina and Rita in 2005. Damage from the force of storm surge, high winds, debris, and other effects of hurricanes can be catastrophic, depending on where a specific hurricane strikes. This study did not examine in detail these effects; the cumulative direct and indirect impacts of major storms need to be further analyzed. However, given the expectation of increasing intensity of hurricanes in the region, consideration should be given to designing new or replacement infrastructure to withstand more energy-intensive, high-category storms.

CLIMATE CHANGE CONSIDERATIONS NEED TO BE INCORPORATED IN TRANSPORTATION DECISIONS

This preliminary assessment raises clear cause for concern regarding the vulnerability of transportation infrastructure and services in the central Gulf Coast due to climate and coastal changes. The effects of potential climate changes, particularly when combined with other factors such as subsidence, are likely to be significant. These changes threaten to cause both major and minor disruptions to the smooth provision of transport service through the study area. As transportation agencies work to meet the challenges of congestion, safety, and environmental stewardship – as well as maintaining transportation infrastructure in good repair – addressing the risks posed by a changing climate can help ensure that the substantial investments in the region's infrastructure are protected in the coming decades by appropriate adaptation strategies.

While several of the impacts of climate change identified above are significant, transportation planners and managers can incorporate effective adaptation strategies into transportation decisions today. Some level of adaptation will be required in the near term to address the effects of climate change processes that are underway. Concentrations of greenhouse gases already in the atmosphere will further force climate changes for the next three to four decades. The scale of adaptation required over the longer term – through this century – will be shaped in part by future emissions levels, as projections of lower- emission scenarios demonstrate lesser impacts.

Transportation Planning Processes

Transportation decisions are made by a number of different entities, both public and private, and transportation infrastructure is financed through a range of government and private investments. Within the study area, four State departments of transportation (DOTs) – for Texas, Louisiana, Mississippi, and Alabama – and 10 Metropolitan Planning Organizations (MPOs) lead surface transportation planning, in close coordination with local governments. To use Federal funding, these agencies must adhere to Federal requirements for surface transportation planning and investment. These laws are contained in Titles 23 and 49 of the United States Code (USC) and were most recently amended in August 2005 by the *Safe, Accountable, Flexible, Efficient Transportation Equity Act: A Legacy for Users* (SAFETEA-LU), the latest six-year authorization of Federal funding for surface transportation.

In surface transportation management, separate but coordinated long-range transportation plans are cooperatively developed on a statewide basis by each State DOT and for each urbanized area by an MPO. The long-range transportation plan is developed with a minimum of a 20-year forecast period, with many areas using a 30-year timeframe. These plans provide a long-range vision of the future of the transportation system, considering all passenger and freight modes and the intermodal system as a whole. The planning and investment process is highly collaborative; transportation agencies need to work in partnership with natural resource agencies, communities, businesses, and others as they chart a course for the transportation network that will meet multiple goals, supporting mobility, economic development, community, safety, security, and environmental objectives.

While climate and environmental projections inherently have a degree of uncertainty, this is not unusual to transportation. Transportation decision makers are well accustomed to planning and designing systems under conditions of uncertainty on a range of factors – such as future travel demand, vehicle emissions, revenue forecasts, and seismic risks. In each case, decision makers exercise best judgment using the best information available at the time. In an ongoing iterative process, plans may be revised or refined as additional information becomes available. Incorporating climate information and projections is an extension of this well developed process.

Similarly, environmental considerations have long played a role in the planning and development of transportation projects. As awareness of the complex interactions among environmental factors and transportation systems has grown, the transportation community has assumed increasing responsibilities for environmental stewardship. Integration of climate factors into transportation decisions continues this trend. However, interviews with a number of transportation managers in the region confirmed that most agencies do not consider climate change projections per se in their long-range plans, infrastructure design, or siting decisions. This appears to be changing, spurred in part by the devastating effects of Hurricanes Katrina and Rita. The damage caused by these storms highlighted the need to incorporate more information and model data related to climate change and other longterm shifts in environmental conditions as transportation plans are developed and implemented.

New Approaches to Incorporate Climate Information

The incorporation of climate factors into transportation decisions may require new approaches.

- **Planning timeframes –** The timeframes generally used for the Federal transportation planning process – 20 to 30 years – are short compared to the multidecadal period over which climate changes and other environmental processes occur. The longevity of transportation infrastructure – which can last beyond a century – argues for a long timeframe to examine potential impacts from climate change and other elements of the natural environment. While the current timeframe is realistic for investment planning, agencies need to consider incorporating longer-term climate change effects into their visioning and scenario planning processes that inform their long-range plans.

- **Risk assessment approach –** Given the complexities of climate modeling and the inherent uncertainties regarding the magnitude and timing of impacts of climate factors, the deterministic methods currently used to support decisions cannot fully address the range of potential environmental conditions that transportation managers need to consider. Adopting an iterative risk management approach would provide transportation decision makers, public officials, and the public a more robust picture of the risks to – and level of resilience of – various components of the transportation network.

A conceptual framework and taxonomy for consideration of climate factors was developed. This approach incorporates four key factors that are critical to understanding how climate change may impact transportation:

- *Exposure:* What is the magnitude of stress associated with a climate factor (sea level rise, temperature change, severe storms, precipitation) and the probability that this stress will affect a transportation segment or facility?
- *Vulnerability:* Based on the structural strength and integrity of the infrastructure, what is the potential for damage and disruption in transportation services from this exposure?
- *Resilience:* What is the current capacity of a system to absorb disturbances and retain transportation performance?
- *Adaptation:* What response(s) can be taken to increase resilience at both the facility (e.g., a specific bridge) and system levels?

Adaptation Strategies

Ultimately, the purpose of a risk assessment approach is to enhance the resilience of the transportation network. Analysis of these factors can help transportation decision makers identify those facilities most at risk and adopt adaptation strategies to improve the resilience of facilities or systems. Structures can be hardened, raised, or even relocated as need be, and – where critical to safety and mobility – expanded redundant systems may be considered as well.

What adaptation strategies are employed, and for which components of the system, will be determined considering the significance of specific parts of the network to the mobility and safety of those served, the effects on overall system performance, the cost of implementation, and public perceptions and priorities. Generally speaking, as the importance of maintaining uninterrupted performance increases, the appropriate level of investment in adaptation for high-risk facilities should increase as well. This study does not make recommendations about specific facilities or adaptation strategies, but rather seeks to contribute to the information available so that States and local communities can make more informed decisions.

Future Research Would Benefit Decision Makers

The analysis of how a changing climate might affect transportation is in its infancy. While there is sufficient information today to begin to assess risks and implement adaptation strategies, further development of data and analysis would help planners, engineers, operators, and maintenance personnel as they create an even more robust and resilient transportation system, ultimately at lower cost. Key research opportunities include:

- **Integrated climate data and projections** – It would be useful to the transportation community if climatologists could continue to develop more specific data on future impacts. Higher resolution of climate models for regional and subregional studies would support the integration of region-specific data with transportation infrastructure information. More information about the likelihood and extent of extreme events, including temperature extremes, storms with associated surges and winds, and precipitation events, could be utilized by transportation planners.

- **Risk analysis tools** – In addition to more specific climate data, transportation planners also need new methodological tools to address the uncertainties that are inherent in projections of climate phenomena. Such methods are likely to be based on probability and statistics as much as on engineering and materials science. The approaches taken to address risk in earthquake-prone areas may provide a model for developing such tools.

- **Region-based analysis** – The impacts that a changing climate might have on an area depends on where the region is and its natural environment. Replication of this study in other areas of the country could help determine the possible impacts of climate change on transportation infrastructure and services in those locations. Transportation in northern climates will face much different challenges than those in the south. Coastal areas will similarly face different challenges than interior portions of the country. Further, additional analysis on demographic responses to climate change, land use interactions, and secondary and national economic impacts would help elucidate what impacts climate will have on the people and the Nation as a whole, should critical transportation services in the region be lost.

- **Interdisciplinary research** – This study has demonstrated the value of cross-disciplinary research that engages both the transportation and climate research communities. Continued collaboration will benefit both disciplines in building methodologies and conducting analyses to inform the Nation's efforts to address the implications of climate change.

In: Impacts of Climate Change on Transportation and... ISBN: 978-1-60741-424-7
Editor: Iason Pavlopoulos

Chapter 1

WHY STUDY CLIMATE CHANGE IMPACTS ON TRANSPORTATION?

Lead Authors: Robert C. Hyman, Joanne R. Potter, Michael J. Savonis, Virginia R. Burkett and Jessica E. Tump

Transportation is such an integral part of daily life in the United States that few pause to consider its importance. Yet the Nation's strong intermodal network of highways, public transit, rail, marine, and aviation is central to our ability to work, go to school, enjoy leisure time, maintain our homes, and stay in touch with friends and family. U.S. businesses depend on reliable transportation services to receive materials and transport products to their customers; a robust transportation network is essential to the economy. In short, a sound transportation system is vital to the Nation's social and economic future. Transportation professionals – including planners, designers, engineers, financial specialists, ecologists, safety experts, and others – work hard to ensure that U.S. communities have access to safe and dependable transportation services.

Given the ongoing importance of the Nation's transportation system, it is appropriate to consider what effect climate change may have on this essential network. Through a regional case study of the central Gulf Coast, this report begins to examine the potential implications of climate change on transportation infrastructure, operations, and services. Investments in transportation are substantial and result in infrastructure that lasts for decades. Transportation plans and designs should, therefore, be carefully considered and well informed by a range of factors, including consideration of climate variability and change. Climate also affects the safety, operations, and maintenance of transportation infrastructure and systems. This research investigates the potential impacts of climate variability and change on transportation, and it assesses how planners and managers may incorporate this information into their decisions to ensure a reliable and robust future transportation network. This report does not contain recommendations about specific facilities or adaptation strategies, but rather seeks to contribute to the information available so that States and local communities can make more informed decisions when planning for the future.

Four key questions guide this investigation:

1. How important are the anticipated changes in climate?
2. Can we anticipate them with confidence?
3. What information is useful to transportation decisions?
4. How can decision makers address uncertainty?

The answers to these questions require first developing an understanding of how the climate is changing and the range of potential climate effects and then considering the relevance of these changes to transportation.

To set the context for this regional case study, this chapter first provides in section 1.1 an overview of how climate change is occurring globally, based on current scientific research. Section 1.2 introduces the questions these changes raise for the transportation sector and the research required to support effective responses to climate change. Section 1.3 provides a synthesis of the state of existing research regarding the impacts of climate change on transportation, discussing the focus of current investigations – both in terms of specific climate factors and individual transportation modes, major findings, and what entities are sponsoring and conducting this research. Section 1.4 draws conclusions from this literature review to identify what is known – and what research questions remain – on this multifaceted topic. Section 1.5 then discusses how the U.S. Department of Transportation (DOT) selected the Gulf Coast region for its first case study of the potential impacts of climate change on transportation and describes the objectives and organization of the research effort.

1.1. THE CLIMATE IS CHANGING

The natural "greenhouse" effect is an essential component of the planet's climate process. Naturally occurring greenhouse gases – carbon dioxide, methane, and nitrous oxide – effectively prevent part of the heat radiated by the Earth's surface from otherwise escaping to space. In the absence of these greenhouse gases, the Earth's temperature would be too cold to support life as we know it today.

However, atmospheric concentrations of greenhouse gases have increased markedly since the industrial age began. The concentration of carbon dioxide (CO_2) in the atmosphere has been increasing due to the combustion of fossil fuels and, to a lesser extent, land use changes. Direct atmospheric measurements made over the past 50 years have documented the steady growth in carbon dioxide concentrations. In addition, analysis of ice bubbles trapped in ice cores show that atmospheric carbon dioxide has increased by roughly one- third since 1750. Atmospheric concentration of CO_2 was 379 parts per million (ppm) in 2005, compared to a preindustrial level of 280 ppm (IPCC, 2007). Other heat-trapping gases – methane and nitrous oxide – also are increasing as a result of human activities. Finally, once in the atmosphere these greenhouse gases have a relatively long life time, on the order of decades to centuries, which means that the atmospheric warming taking place today will continue.

Temperature has increased and is projected to continue to do so. Temperatures have been rising over the last century, with more rapid increases since 1970 than earlier. According to the International Panel on Climate Change (IPCC) Working Group I Fourth Assessment Report (AR4), average global temperatures increased 0.74°C (1.33°F) during the past 100 years, with most of that increase – 0.65°C (1.17°F) experienced in the last 50 years. Recent years have set record highs; 11 of the past 12 years were the warmest years on record since 1850. While some of this change may be due to natural variability, human activities have contributed to the Earth's warming. The IPCC report finds with very high confidence that the globally averaged net effect of human activities since 1750 has been one of warming. The last major challenge to whether the planet was warming or not was resolved in April 2006 with publication of "Temperature Trends in the Lower Atmosphere" (U.S. Climate Change Science Program, Synthesis, and Assessment Product 1.1, 2006). This study reconciled the remaining analytical issues regarding differences between surface and satellite temperature readings.

The climate models used to estimate temperature changes agree that it will be warmer in the future. According to the IPCC report, global average warming is expected to be about 0.4°C (0.72°F) during the next 20 years. Even if the concentrations of all greenhouse gases and aerosols had been stabilized at 2000 levels, warming of 0.2°C (0.36°F) would be expected during this period (IPCC, 2007). Over the longer term, the IPCC models project average global temperature increases ranging from 1.1°C (1.98°F) to 6.4°C (11.5°F) by the end of the 21st century, although climate responses in specific regions will vary. These projections are the result of reviewing a robust set of global climate models under a variety of future scenarios – using a range of assumptions for future economic activity and energy use – for the Earth as a whole.

The average increase in temperature may not be as important to the transportation community as the changes in extreme temperature, which also are expected to increase. Over the last 50 years, the frequency of cold days and nights has declined, while hot days, hot nights, and heat waves have become more frequent. The number of days with temperature above 32°C (90°F) and 38°C (100°F) has been increasing since 1970, as has the intensity and length of periods of drought. The IPCC report finds that it is virtually certain that the next century will witness warmer and more frequent hot days and nights over most land areas (IPCC, 2007).

Precipitation patterns are changing, and more frequent intense precipitation events are expected. Over the past century precipitation amounts have increased in several regions – including the eastern parts of North and South America – while drying has been observed in other regions in Africa and Asia. During the 21st century, the IPCC (2007) anticipates that increases in the amount of precipitation are *very likely* in high latitudes, while decreases are likely in most subtropical land regions, continuing observed patterns in recent trends. While total average levels of precipitation will vary by region, the incidence of extreme precipitation events is expected to increase.

According to NOAA analyses, the magnitude of the highest precipitation events has been increasing since 1970. A Simple Daily Intensity Index that examines the total precipitation for the United States divided by the number of days with precipitation clearly demonstrates an increase in average intensity from 1970 to 2005. These observed increases in extreme precipitation are not only in keeping with observational analyses but also with model projections for the future. The IPCC AR4 (2007) concludes that heavy precipitation events will continue to become more frequent during the coming decades.

Sea level is rising, and the rate of change is likely to accelerate. As the Earth warms, two changes are occurring that are causing sea levels to increase: glacial melting and thermal expansion of the oceans. Sea level rise is perhaps the best documented and most accepted impact of climate change. The IPCC reports that – on a global level – the total 20th century rise is estimated to be 0.17 m (0.56 ft) and that global sea level rose at an average rate of 1.8 mm (0.07 inches) per year between 1961 and 2003. Excluding rapid changes in ice flow, the IPCC model-based projections for global sea level rise over the next century across multiple scenarios range from 0.18 to 0.59 m (0.59 to 1.94 ft). Should the melting of the land-based polar ice caps accelerate, sea level could rise much higher.

The intensity of severe storms is expected to increase. It is likely that future tropical cyclones (typhoons and hurricanes) will become more intense, with larger peak wind speeds and heavier precipitation (IPCC, 2007). (There is insufficient evidence to identify changing trends for other storm phenomenon, such as tornadoes, hail, and lightning [IPCC, 2007]; these types of storm activity are not addressed by this report.) There are several aspects of tropical storms that are relevant to transportation: precipitation, winds, and wind-induced storm surge. All three tend to get much worse during strong storms. Strong storms tend to have longer periods of intense precipitation, and wind damage increases exponentially with wind speed. The primary concern with hurricanes is for strong storms of Categories 3, 4, and 5. These storms have considerably more destructive energy. For example, a Category 5 storm may have winds only twice as fast as a Category 1 storm, but its kinetic energy is over four times that of a Category 1 storm.

Chapter 3.0 of this report provides a detailed discussion of how the climate is changing in the central Gulf Coast study area.

1.2. How Will Changes in Climate Affect Transportation?

That the climate is changing leads to a number of intriguing and critically important questions for transportation. For the transportation community – the planners, engineers, builders, operators, and stewards of our Nation's roads, airports, rail, transit systems, and ports – the primary question is how such changes will affect infrastructure and associated services and the trillions of dollars of investment these facilities represent. Transportation services are vital to our economy and quality of life. Individuals use transportation not only to get to and from work but for a wide variety of personal travel. Further, as producers seek to reduce warehousing costs through "just-in-time" delivery, transportation systems increasingly are functioning in effect as mobile warehouses. This places new stresses on service providers to make sure that economic goods are delivered on time. As the number of vehicles – and miles traveled – continues to grow, congestion on our roadways is an increasing concern.

Nationally, we invest about $110 billion annually in highways and transit alone. Federal investment in passenger rail approaches $2 billion a year. Add to this the considerable investment made by the private sector in freight rail, airports, and ports, and it is clear that the value that we place on these systems is enormous. Any disruption to the goods and services provided through the U.S. transportation network can have immediate impacts ranging from the annoying, such as flight delays due to severe weather, to the catastrophic, such as the chaos wrought by Hurricanes Katrina and Rita.

The question of how a changing climate might affect transportation infrastructure and services led the U.S. DOT, under the auspices of its Center for Climate Change and Environmental Forecasting (hereafter "the Center"), to hold a first-ever workshop on October 1-2, 2002. Cosponsored by the Environmental Protection Agency, the Department of Energy, and the U.S. Climate Change Science Program, the workshop brought together noted climate scientists, top transportation executives and practitioners, and experts in assessment research, environment, planning, and energy. This interdisciplinary group was charged to explore the potential impacts of climate change for transportation and to delineate the research necessary to better understand these implications. In preparation, the Center commissioned a series of white papers on overviews of climate change, regional case studies, potential system impacts, and environment and planning. The workshop participants identified significant gaps in the knowledge and processes necessary to fully incorporate climate science information into transportation decisions and developed a framework to pursue future research in this multifaceted area of investigation. The two-day session deepened practitioners' understanding of the significance of climate change for transportation and led to a firm commitment by the U.S. DOT to pursue needed research. The current Gulf Coast Study was designed to begin to address the research needs identified at this important forum.

1.2.1. What Are the Challenges to Research?

Several research challenges must be met to successfully incorporate climate information into transportation decisions. Framing this new area of research is a complex undertaking that requires a new style of interdisciplinary work among scientists, planners, engineers, and policy makers.

- **Articulating data and information needs** – First, transportation practitioners need to be able to articulate the types of climate data and model projections that will be relevant to transportation decisions: *What information could lead a public or private transportation agency to change a transportation investment plan, road location, or facility design?* Determining what climate information is useful includes identifying the appropriate regional scale and timeframe for climate scenarios, as well as the types of climate factors that could result in a revised decision. Generating this practical information may require scientists to analyze and portray existing data in different ways in order to be useful to transportation decisions.

- **Identifying most relevant climate information** – At the same time, climate scientists need to be able to explain to transportation and planning professionals what information is available today that may be relevant to transportation decisions. The pace of climate science is advancing rapidly, and new and increasingly reliable climate findings are being released regularly. The sheer volume of significant climate information poses a major challenge to the scientific community: How can scientists effectively translate the findings of basic research into information that can be understood by other professions – and the general public – and be applied to the choices transportation managers need to make?

- **Integrating multiple environmental factors** – Further, climate factors need to be considered, not in isolation, but as part of a broader set of social and ecological factors that provide the context for thoughtful and informed transportation decisions. This will require that natural scientists and geospatial specialists work with transportation planners to integrate climate information into maps and data addressing other environmental factors. Incorporating new types of information – including longer- range climate scenario projections – may require the transportation community to adopt new approaches to planning and visioning exercises that engage a broader range of stakeholders and subject matter experts.

- **Incorporating uncertainty** – An additional challenge is learning how to incorporate uncertainty in transportation decisions – how to assess risk and vulnerability of the transportation system and individual facilities given a range of potential future climate conditions. While transportation practitioners historically have planned and designed to meet established standards – for weight loads, flood levels, temperature extremes, etc. – today's transportation planner needs to consider the most effective strategies to ensure a robust transportation system across a broader range of possible futures, potentially encompassing longer timeframes and a wider variety of impacts. This challenge may require new approaches to design and investment that use probabilistic, rather than deterministic, analysis.

To begin to explore these complex research questions, the team conducted a review of existing literature regarding climate change impacts on transportation to determine the state of science.

1.3. State of Science Regarding Climate Change Impacts on Transportation

What is the state of knowledge about climate change impacts on transportation? The research team undertook a review of the literature to assess the depth and breadth of existing research that specifically examines changes in climate and the resulting implications for transportation infrastructure and services.

Although there is a large body of research concerning climate change and how transportation contributes to greenhouse gas emissions, less work has been done concerning the impacts of climate change on transportation. A review of existing literature indicates that the impacts of climate change on transportation is an emerging area of research and one that is growing steadily more sophisticated. As a new field, the level of analysis given to the variety of subtopics within this broad area of research has been uneven; some aspects of climate change impacts on transportation have received much greater scrutiny than others depending on the particular concerns of individual authors and research sponsors.

1.3.1. Overview of State of Practice

Although there are relevant studies going back at least two decades, the pace of investigation has accelerated in more recent years. Several studies were conducted in this field in the late 1980s and early 1990s as international agreements on climate change were first under serious discussion (Marine Board, 1987; Hyman, 1989; Black, 1990; Irwin and Johnson, 1990). However, citations from this period are relatively infrequent, and as recently as 1998, the U.S. Federal Highway Administration (1998) found relatively little literature on this topic. Since then, the citations show growing recognition of climate impacts on transportation as an issue; research on this topic was highlighted in the United States' Third National Communication (U.S. Department of State, 2002). In fact, the majority of references cited are from the new millennium (table 1.1).

In addition to the growing number of research efforts, the analytic rigor of studies – particularly in the use of climate information – has progressed as well. While early discussions tend to be exploratory in nature, recent work has incorporated more sophisticated climate information and model outputs, addressed issues of uncertainty, and begun to examine the implications of climate factors on specific regions and infrastructure. This trend is likely to continue as awareness of the issues grows within the transportation community and decision makers seek improved information and tools to assess risks and adaptation strategies.

The literature encompasses a wide variety of studies conducted for different time periods, sponsored by a range of organizations, and undertaken for different purposes. General characteristics of the literature reviewed are described below:

- **Key climate factors examined** – The major climate factors most often discussed in the literature in terms of transportation impacts are temperature, precipitation, and sea level rise. Some articles explicitly dealt with storm activity or storm surge. (These climate factors are also analyzed as significant drivers in the Gulf Coast Study.) Many northern studies also examined permafrost thawing and navigation issues relating to ice cover on seaways and inland waterways.

- **Modal focus** – Information on modes is uneven. The majority of articles dealt with highways and marine transport; other modes such as rail, aviation, and transit were not as well represented. Relatively few articles addressed pipelines or emergency management issues in the context of climate change.

- **Geographic focus** – Much of the work done in this field has a national or regional focus; only the IPCC (1996 and 2001) has considered the topic at a truly global level. The Arctic Climate Impacts Assessment (Instanes et al., 2005) is a rare example of transnational regional study, in that it focused on impacts throughout the Arctic nations.

  In addition, some studies focused on specific urban areas (Kirsten et al., 2004; Suarez, 2005; Greater London Authority, 2005).

- **Climate zones examined** – The literature does not examine all climate zones equally or in proportion to the amount of transportation infrastructure present. In particular, transportation in Arctic climates received substantial study, as warming impacts already are being observed in those regions. Many other studies looked at temperate climates, as in the United States or Europe. Australian studies were among the few that examined desert climates or hot climates. In addition, most of the literature focused on the industrialized world.

- **Timeframe examined** – Most studies examined time horizons of 50- to 100-years into the future, consistent with the timescale of projections and scenarios often used in the climate literature. Though this is well beyond the 20- to 30-year planning horizons typically used in transportation planning, it was noted in the literature that some infrastructure (such as bridges) is designed with life expectancies of 100 years or more (Eddowess et al., 2003; Wooler, 2004; Norwell, 2004). Other researchers eschewed timescales and instead chose specific thresholds to consider. For instance, Marine Board (1987) chose to examine the impacts of 0.5-, 1.0-, and 1.5-m (1.6-, 3.3-, and 4.9- foot) rises in sea levels, without specifying a projected year for when these might take place. Finally, several Arctic studies focused on changes *presently* occurring, as in Grondin's (2005) study of the effect of thawing permafrost on airfields and roads in Nunavik due to increasingly warmer winters.

1.3.2. Major Sponsors Conducting Related Research

Studies on the impacts of climate change on transportation have been conducted by a variety of researchers and organizations, including governmental agencies, academic researchers, and the private sector, reflecting the range of stakeholders with an interest in the topic. These studies incorporate a variety of approaches and can be found as standalone assessments of transportation impacts or as one aspect of a broader examination of climate impacts.

Two very significant impact assessment efforts have dealt with this issue in a limited fashion. The IPCC's multivolume assessment reports (IPCC, 1996; IPCC, 2001) discussed the topic in general terms, particularly noting the vulnerability of transportation infrastructure in coastal zones and permafrost regions to climate impacts, with the 2001 report broadly discussing some transportation operations impacts and more detail on Europe-specific concerns, such as impacts to aviation operations and river navigation.

Similarly, the U.S. National Assessment, which represents one of the broadest examinations of climate impacts to date in the U.S., did not include transportation as a sector of interest (National Assessment Synthesis Team, 2000). However, some of the regional studies conducted under the umbrella of the national assessment process did examine transportation impacts, most notably the Metro East Coast and Alaska studies (Zimmerman, 2002a; Weller et al., 1999). The 2002 U.S. DOT report, *The Potential Impacts of Climate Change on Transportation: Summary and Discussion Papers*, contains 15 discussion papers addressing potential climate impacts on various modes of transportation across the Nation and a summary of priority research needs. The importance of weather and climate and its potential impacts on the Nation's transportation system was studied in *Weather Information for Surface Transportation: A National Needs Assessment Report* (OFCM, 2002). The report established national needs and requirements for weather information associated with decision-making for surface transportation operation modes including highway, transit, rail, marine, pipeline, and airport ground operations. It was issued as part of the cross-agency Weather Information for Surface Transportation (WIST) initiative, supported by the Federal Committee for Meteorological Services and Supporting Research (FCMSSR) and the agencies it represents.

The United Kingdom (U.K.) Climate Impacts Programme, an initiative similar to the U.S. National Assessment, specifically included impacts on the transportation sector in the overall assessment and in each of the regional reports prepared under its umbrella. The Canadian and Australian governments also have commissioned studies to examine transportation impacts of special interest to them – Canada with permafrost concerns and interest in the opening of the Northwest Passage; Australia with dry land salinity impacts due to its unusual soil and climatic conditions (Andrey and Mills, 2003; Norwell, 2004). References to research on this topic also were seen for New Zealand, Finland, and the Netherlands (Kinsella and McGuire, 2005; Ministry of Housing, Spatial Planning, and the Environment, 2001). A small number of city agencies also have commissioned studies examining impacts to their own transportation networks, such as in Seattle and London (Soo Hoo, 2005; Greater London Authority, 2005).

Many studies also were identified in engineering and transportation journals, ranging from transportation-specific publications such as the National Academy of Science Transportation Research Board's (TRB) *Transportation Research Review* to more general sources such as *Civil Engineering – ASCE* or the *Journal of Cold Regions Engineering*, and even some transportation trade journals (Barrett, 2004). A small number of private sector reports, all from the U.K., were identified, including one study from a ports company and two from the insurance industry (ABP Marine Environmental Research, Ltd., 2004; Dlugolecki, 2004; Climate Risk Management and Metroeconomica, 2005).

Finally, though many nongovernmental organizations (NGO) are engaged in research and policy advocacy related to climate change, we found few NGOs producing literature on climate impacts on transportation. For instance, the Union of Concerned Scientists (UCS) and the Pew Center on Global Climate Change have both published multiple reports on impacts and adaptation (see the UCS regional impact studies[1] and Easterling, 2004), yet transportation implications have received little direct attention in these reports.

1.3.3. State of Technical Analysis

The level of technical analysis in current research regarding their use of climate data and modeling varies, depending both on when the study was done and the magnitude of the study. Early studies, for instance, focused on CO_2-doubling scenarios (i.e., examining an equilibrium state at an unspecified point in the future), because standardized emissions and climate change scenarios had not yet been developed for researchers to use (Hyman, 1989; Black, 1990; Irwin and Johnson, 1990). Later studies took advantage of the climate projections developed by the IPCC process or by other large modeling efforts, such as the United States and United Kingdom national assessments. Several studies demonstrated advanced approaches to climate modeling, making use of multiple climate models and regional models to generate projections of climate variables (Instanes et al., 2005; Kinsella and McGuire, 2005; National Assessment Synthesis Team, 2000; Entek UK Limited, 2004). Other studies took more simplified approaches, using global temperature or sea level rise projections as the basis for examining potential impacts. A few studies did not use climate modeling at all, instead relying on historical trend data (Sato and Robeson, 2006; ABP Marine Environmental Research Ltd., 2004).

In many cases, climate variables produced by global or regional climate models were used as inputs into secondary effects models relevant for specific transportation questions. For example, Cheng (2005) used permafrost models to assess the impact of rising temperatures on road and rail structures in Tibet. Lonergan et al. (1993) integrated climate projections into snowfall and ice cover models for northern Canada to understand climate impacts on freight shipments via ice roads and waterways.

On the whole, relatively few studies attempted to quantify the estimated costs, benefits, or effects on performance resulting from climate change; more commonly, they identified potential impacts without a quantitative assessment. Some examples of the kinds of quantitative analyses performed include:

- Hyman et al. (1989) estimated that it would cost more than $200 million (in 1989 dollars) to elevate affected Miami streets to compensate for rising groundwater levels due to sea level rise and that increases in winter temperatures and decreases in snowfall would reduce Cleveland's snow and ice control budget by 95 percent (about $4.4 million, or nearly 2 percent of the city's operating budget).
- Kirshen et al. (2004) estimated an 80 percent increase in traveler delays due to increased incidence of flooding in the Boston area. They also tested overall monetary and environmental costs for three adaptive strategies, finding that aggressive adaptation strategies proved less costly in the long run than doing nothing.
- Kinsella and McGuire (2005) estimated the approximate cost of retrofitting or redesigning New Zealand's road bridges to accommodate increased precipitation (and higher stream flows). They found that although designing for climate change increased initial costs by about 10 percent, over the life of the structure the incremental cost was small (less than 1 percent) due to the decreased probability of climate-related damage.
- Olsen (2005) conducted a Monte Carlo simulation of total annual losses to shippers on the Mississippi River from having to switch to more expensive modes of transport when barge travel is restricted due to low or high water flows. He found that future losses could range from $1.5 million to $41 million per year, compared to an historical average of $12 million per year.
- Associated British Insurers used insurance catastrophe models to examine the financial implications of climate change through its effects on severe storms (Climate Risk Management and Metroeconomica, 2005), estimating that climate change could increase the annual costs of flooding in the United Kingdom almost 15-fold by the 2080s under high-emissions scenarios.

Studies also have been done on the cost of severe storms on transportation networks, which will provide useful data for future studies relating them to climate change. For instance, Grenzeback and Lukmann (2006) summarize some costs to the transportation network resulting from Hurricane Katrina. Although they do not attempt a full accounting of these costs, they note that infrastructure restoration costs will run into the billions of dollars – replacement of the I-10 Twin Span Bridge between New Orleans and Slidell, LA, alone will cost $1 billion and of the CSX rail line another $250 million.

1.3.4. Impacts, Assessment, and Adaptation

A review of the literature indicates that the potential impacts of climate changes on transportation are geographically widespread, modally diverse, and may affect both transportation infrastructure and operations. Indeed, numerous transportation impacts were discussed in the literature. However, the degree to which a study discussed an impact varied; some studies addressed impacts at length, while others gave an impact only a passing mention. A complete list of impacts and adaptations addressed in the literature, along with references, can be found in table 1.1.

Four major categories of climate change factors are addressed most frequently in the literature. These closely parallel the major factors addressed later in this report's study of the Gulf Coast region. These climate factors and their major impacts are:

1. **Increasing temperatures,** which can damage infrastructure, reduce water levels on inland waterways, reduce ice cover in the Arctic, and melt permafrost foundations;

2. **Increasing precipitation,** which can degrade infrastructure and soil conditions;

3. **Rising sea levels,** which can inundate coastal infrastructure; and

4. **Changes in storm activity,** which can damage infrastructure and operations due to increased storm intensity, though winter snowstorms may decrease in frequency.

A summary of the literature findings regarding these impacts, and their corresponding adaptation measures, is presented below. This is followed by a brief discussion of the indirect or secondary impacts on the economy, environment, population, and security of a region.

1.3.5. Direct Climate Impacts on Transportation Addressed in Existing Literature

Increasing Temperatures

Increasing temperatures have the potential to affect multiple modes of transportation, primarily impacting surface transportation. The transportation impacts mentioned most often in the literature included pavement damage; rail buckling; less lift and fuel efficiency for aircraft; and the implications of lower inland water levels, thawing permafrost, reduced ice cover on seaways, and an increase in vegetation. These are discussed in greater detail below:

- **Pavement damage –** The quality of highway pavement was identified as a potential issue for temperate climates, where more extreme summer temperatures and/or more frequent freeze/thaw cycles may be experienced. Extremely hot days, over an extended period of time, could lead to the rutting of highway pavement and the more rapid breakdown of asphalt seal binders, resulting in cracking, potholing, and bleeding. This, in turn, could damage the structural integrity of the road and/or cause the pavement to become more slippery when wet. Adaptation measures mentioned included more frequent maintenance, milling out ruts, and the laying of more heat resistant asphalt.

- **Rail buckling –** Railroads could encounter rail buckling more frequently in temperate climates that experience extremely hot temperatures. If unnoticed, rail buckling can result in derailment of trains. Peterson (2008) noted, "Lower speeds and shorter trains, to shorten braking distance, and lighter loads to reduce track stress are operational impacts." Adaptation measures included better monitoring of rail temperatures and ultimately more maintenance of the track, replacing it when needed.

- **Vegetation growth –** The growing season for deciduous trees that shed their leaves may be extended, causing more slipperiness on railroads and roads and visual obstructions. Possible adaptation measures included better management of the leaf foliage and planting more low-maintenance vegetation along transportation corridors to act as buffers (Wooler, 2004).

- **Reductions in aircraft lift and efficiency –** Higher temperatures would reduce air density, decreasing both lift and the engine efficiency of aircraft. As a result, longer runways and/or more powerful airplanes would be required. However, one analyst projected that technical advances would minimize the need for runway redesign as aircraft become more powerful and efficient (Wooler, 2004).

- **Reduced water levels –** Changes in water levels were discussed in relation to marine transport. Inland waterways such as the Great Lakes and Mississippi River could experience lower water levels due to increased temperatures and evaporation; these lower water levels would mean that ships and barges would not be able to carry as much weight. Adaptation measures included reducing cargo loads, designing vessels to require less draft, or dredging the water body to make it deeper.

- **Reduced ice cover –** Reduced ice cover was generally considered a positive impact of increasing temperatures in the literature. For example, a study conducted by John D. Lindeberg and George M. Albercook, which was included in the *Report of the Great Lakes Regional Assessment Group for the U.S. Global Change Research Program,* stated, "the costs of additional dredging [due to lower water levels] could be partially mitigated by the benefits of additional shipping days on the [Great] Lakes caused by less persistent ice cover" (Sousounis, 2000, p. 41). Additionally, arctic sea passages could open; for example, the *Arctic Climate Impact Assessment* noted, "projected reductions in sea-ice extent are likely to improve access along the

Northern Sea Route and the Northwest Passage" (Instanes et al., 2005, p. 934). However, negative environmental and security impacts also may result from reduced ice cover as well from as the increased level of shipping. These are discussed below in the subsection on indirect impacts (Section 1.3.6.).

- **Thawing permafrost** – The implications of thawing permafrost for Arctic infrastructure receive considerable attention in the literature. Permafrost is the foundation upon which much of the Arctic's infrastructure is built. The literature consistently noted that as the permafrost thaws the infrastructure will become unstable – an effect being experienced today. Roads, railways, and airstrips are all vulnerable to the thawing of permafrost. Adaptation measures vary depending on the amount of permafrost that underlies any given piece of infrastructure. The literature suggested that some assets will only need rehabilitation, other assets will need to be relocated, and different construction methods will need to be used, including the possibility of installing cooling mechanisms. According to the Arctic Research Commission, "roads, railways, and airstrips placed on ice-rich continuous permafrost will generally require relocation to well-drained natural foundations or replacement with substantially different construction methods" (U.S. Arctic Research Commission Permafrost Task Force, 2003, p. 29).

- **Other** – Other impacts of increasing temperatures included a reduction in ice loads on structures (such as bridges and piers), which could eventually allow them to be designed for less stress, and a lengthening of construction seasons due to fewer colder days in traditionally cold climates.

Increasing Precipitation

Increases in precipitation will likely affect infrastructure in both cold and warm climates, although in different ways. Increases in the frequency and intensity of the precipitation could impact roads, airstrips, bikeways/walkways, and rail beds. The literature suggested most of the impact would be felt in the more rapid deterioration of infrastructure. According to a report released by Natural Resources Canada (2004, p. 138), "accelerated deterioration of these structures may occur where precipitation events and freeze-thaw cycles become more frequent, particularly in areas that experience acid rain." Other impacts of increased flooding include subsidence and heave of embankments (ultimately resulting in landslides), and deterioration in water quality due to run-off and sedimentation. Adaptation measures included monitoring infrastructure conditions, preparing for service delays or cancellations, and replacing surfaces when necessary (Warren, 2004). Although mentioned less frequently, some attention was given in the literature to bridge scour from increased stream flow. Bridge scour could cause abutments to move and damage bridges.

Rising Sea Levels

Sea level rise could impact coastal areas. While incremental sea level rise impacts may not be as immediate or severe as the storm activity, the impacts could nevertheless affect all modes of transportation. Low-level roads and airports are at risk of inundation, and ports may see higher tides. Titus (2002, p. 139) concluded "the most important impact of sea level rise on transportation concerns roads. In many low-lying communities, roads are lower than the surrounding lands, so that land can drain into the streets. As a result, the streets are the first to flood." Adaptation measures include more frequent maintenance, relocation, and the construction of flood-defense mechanisms (such as dikes) (Titus, 2002). Although mentioned less often in the literature, deeper water caused by sea level rise could permit greater ship drafts in ports and harbors.

Changes in Storm Activity

Storm activity was discussed as an issue for all climates, impacting both inland areas and coastal areas. Impacts most frequently mentioned in the literature include storm surges that could potentially cause damage to coastal areas and a decrease in winter snowstorms (with more winter precipitation falling as rain). These are discussed in greater detail below:

- **Increased storm activity or intensity –** In coastal areas, increased storm activity or intensity could lead to an increase in storm surge flooding and severe damage to infrastructure, including roads, rails, and airports. These effects could be exacerbated by a rise in sea level. In addition, coastal urban areas, like New York City, could potentially see storm surges that flood the subway system. As Zimmerman (2002a, p. 94) noted, "transportation systems are traditionally sited in low-lying areas already prone to flooding." She went on to state that, "New York City alone has over 500 miles of coastline, much of which is transgressed [sic] by transportation infrastructure – roadways, rail lines, and ventilation shafts, entrances and exits for tunnels and transit systems, many are at elevations at risk of being flooded even by traditional natural hazards" (p. 94). Adaptation measures included construction of barriers to protect against storm surges, relocating infrastructure, and preparing for alternative traffic routes (Zimmerman, 2002a).
  Other impacts related to storm activity included an increase in wind speed and an increase in lightning. Increased wind speeds could damage signage and overhead cables. Increased lightning strikes could cause electrical disturbances disrupting electronic transportation infrastructure, like signaling.

- **Reduced snowfall –** A decrease in winter snowstorms could potentially relieve areas that typically see large amounts of snow from some of the cost of maintaining winter roads. Natural Resources Canada concluded, "empirical relationships between weather variables and winter maintenance activities indicate that less snowfall is associated with reduced winter maintenance requirements. Thus, if populated areas were to receive less snowfall and/or experience fewer days with snow; this could result in substantial savings for road authorities" (Warren et al., 2004, pp. 13 8-139).

1.3.6. Indirect Climate Impacts on Transportation Addressed in Existing Literature

Four secondary, or indirect, impacts were addressed to some degree in the literature: economic, environmental, demographic, and security impacts.

Economic

The economic impact of climate change received considerable attention. Some studies made an attempt to approximate the cost of replacing infrastructure or to place a monetary figure on loss of specific aspects of system performance, such as traffic disruptions. For example, Suarez et al. (2005, p. 240), when discussing the effects flooding could have on the Boston Metro area, stated, "over the period 2000 to 2100, the results indicate that delays and trips lost (i.e., canceled trips) increased by 80 percent and 82 percent under the climate change scenario. While this is a significant increment in percentage terms, the magnitude of the increase is not enough to justify a great deal of infrastructure improvements."

The economic implications of impacts on freight were particularly studied. Three climate factors were analyzed in most depth: changing inland water levels, specifically on the Great Lakes; thawing permafrost and warmer temperatures in traditionally colder climates; and the potential opening of the Northwest Sea Passage through the Canadian Arctic as a result of sea ice melt. These are discussed in greater detail below:

- **Changing inland waterway levels –** Quinn analyzed the economic impacts of lower water levels in the Great Lakes, which would require ships to lighten their loads because of lower water levels. According to Quinn (2002, p. 120), "a 1,000-foot bulk carrier loses 270 tons of capacity per inch of lost draft." If lower water levels occur on a regular basis, Great Lakes shippers are likely to see less profit and will run the risk of the freight being transported by competing modes (e.g., rail or truck). A few analyses considered the impacts of rising inland water levels (Olsen, 2005).

- **Increasing temperatures in northern regions –** Other analysts assessed the economic impacts of warming temperatures on trucking in northern regions. Typically, trucks are allowed to carry more weight when the underlying roadbeds are frozen, and some Arctic regions are served by ice roads over the tundra in winter. If temperatures increase and northern roads thaw before their usual season, truckloads may have to be reduced during the traditionally higher weight-limit trucking season. This impact already is occurring in some regions of the United States and Canada. As a result, a few highway authorities are adjusting their weight restrictions based on conditions, rather than linking them to a given date (Clayton et al., 2005).

- **Opening of the Northwest Passage –** The literature indicated that the reduction of waterway ice cover and the eventual opening of an Arctic Northwest Passage have by far the largest economic consequences of all the impacts. The passage could provide an alternative to the Panama Canal and stimulate economic development in the Arctic region (Johnston, 2002).

Environmental

A small number of environmental impacts have been addressed in the literature to date, focusing on the effects of specific adaptation responses to changing climate and weather conditions. These included the potential of increased dredging of inland waterways, reduced use of winter road maintenance substances, and the environmental impact increased shipping could have on the Arctic.

- **Dredging –** Dredging of waterways – in response to falling water levels – could have unintended, harmful environmental impacts. According to the Great Lakes Regional Assessment, "in a number of areas the dredged material is highly contaminated, so dredging would stir up once buried toxins and create a problem with spoil disposal" (Sousounis, 2000, p. 30).

- **Increased shipping in the Arctic –** The transportation benefits of the Northwest Passage could be offset by the negative environmental impacts associated with its use, particularly oil spills (Struck, 2006). Johnston (2002, p. 153) noted that there is "serious concern on the part of many Inuit and other residents that regular commercial shipping will, sooner or later, cause serious harm to the Arctic ecology."

- **Reduced winter maintenance –** Some positive environmental impacts also were mentioned, particularly in relation to milder winter weather in northern regions. For example, according to Warren et al. (2004, p. 139) "less salt corrosion of vehicles and reduced salt loadings in waterways, due to reduced salt use" during winter months could positively impact the environment. According to Natural Resources Canada, "experts are optimistic that a warmer climate is likely to reduce the amount of chemicals used, thus reducing costs for the airline industry, as well as environmental damage caused by the chemicals" (Warren et al., 2004, p. 139).

Demographic

Demographic shifts were rarely addressed in the literature. A few reports raised the potential for shifts in travel destinations and mode choices. For instance, in a U.K. Climate Impacts Programme Report on the West Midlands it was noted: "higher temperatures and reduced summer cloud cover could increase the number of leisure journeys by road. There could be a possible substitution from foreign holidays if the climate of the West Midlands becomes more attractive relative to other destinations, reducing demand at Birmingham International Airport" (Entek UK Limited, 2004, p. 24). In addition, the Arctic regions, located near the Northwest Passage, could see an influx of population (Entek UK Limited, 2004).

Security

Security was identified as an issue in relation to the Northwest Passage. Given the enormous changes the development of the Northwest Passage would precipitate, it is no surprise that global diplomacy, safety, and security is of concern. Johnston (2002, p. 152) stated, "even if the remoteness of the Northwest Passage seems to make it an unlikely target for terrorists, security concerns will centrally have to be factored in to any major undertaking in the Arctic or elsewhere that would be perceived by enemies as an important component of the North American economy." If the Northwest Passage does become practical for shipping, security, ownership, maintenance, and safety of the waterway will become an issue. Indeed, the U.S. Navy already had begun thinking about the implications of an ice-free Arctic during a symposium held in April 2001 (Office of Naval Research, 2001). Sovereignty issues also will need to be resolved to clarify whether the passage will be considered international or Canadian waters (Johnston, 2002).

1.3.7. Decision Making Processes and Tools

Until recently, studies typically concluded with recommendations for additional analysis of uncertainty, thresholds, and prioritization of actions. Recent work has begun to respond to this need, but the field still has a long way to go. Some reports have begun to make suggestions for institutional changes necessary to integrate climate impacts into the decision making processes for transportation planning and investment. Studies have suggested some approaches to more adequately dealing with uncertainty. Finally, several studies have attempted to develop methodologies that can integrate potential climate impacts into risk prioritization processes, decision trees, and other decision support tools.

The following sections discuss institutional changes that were identified in the literature, evaluate the manner in which uncertainty and probability was addressed, and present four case studies highlighting different methodologies used in risk analysis and impact assessment.

Institutional Changes

On the whole, analysis and recommendations concerning needed changes in standard design practice or institutional changes are beginning to emerge but are at a nascent stage. A few recent studies illustrate this point:

- **Urban-scale planning –** Two recent studies developed recommendations for London and Seattle. The Greater London Authority (2005) urged transportation decision makers to incorporate climate into routine risk management procedures, build adaptation measures into new infrastructure when appropriate, and make certain that whatever measures are taken are flexible and easily adaptable to future climatic changes. However, the report gave little direction on how they should go about this; suggestions about how and when officials should incorporate these adjustments were not well defined. Likewise, a 2005 Seattle study, authored by the city auditor, recommended that the Seattle Department of Transportation "identify, prioritize, and quantify the potential effects of climate change impacts; and plan appropriate responses to changes in the region's climate" (Soo Hoo et al., 2005, p. 12). A specific institutional recommendation made was the synchronization of sea level rise assumptions among Seattle's various city agencies (for instance, in the assumptions made for construction of seawalls) (Soo Hoo et al., 2005).

- **Arctic maritime regulatory regime –** For the Arctic, several studies identified the need for a new regulatory system to govern ships in Arctic waters. Johnston (2002) recommended a new "transit management regime" be developed for the Northwest Passage to clarify Canadian and international responsibilities and jurisdiction over maritime passage, and the Arctic Marine Transport Workshop (Brigham, 2004) suggested the development of harmonized safety and environmental measures for the larger Arctic region.

- **General planning considerations** – Several other reports recommended that as a first step a process be developed for including climate impacts in planning. For instance, the Northern Ireland assessment recommended that a formalized policy on climate impacts be developed within three years (Smyth et al., 2002), and Associated British Ports indicated that it planned to periodically re-examine potential impacts to ports in order to see if their assessment changes with new information (ABP Marine Environmental Research Ltd., 2004). Interestingly, Norwell (2004) noted that planning for sea level rise already has been incorporated into planning documents in several Australian States.

In general, the mismatch between typical planning horizons and the longer-term timeframe over which climate impacts occur appears to be a barrier to incorporating climate change factors in decision making. For example, Kinsella and McGuire (2005) concluded that for infrastructure with replacement horizons of less than 25 years, there was no need to consider longer-term climate effects in the present day, as the infrastructure would turn over before it became a problem.

Uncertainty and Probability

The literature indicates that only recently have analysts begun to address the issue of transportation risk assessment and decision making under uncertainty. Even now, the analytical sophistication of studies that attempt to address these concerns is in its infancy. The studies consistently showed awareness of the uncertainty of climate projections, quoting ranges for potential climate changes. However, probabilistic approaches were not implemented in the literature reviewed and were rarely discussed. Nor was there a focus on the development of "robust" strategies that can bear up under multiple possible futures or other strategies designed specifically to deal with decision making under uncertainty. Dewar and Wachs (forthcoming) note that this is a gap in transportation planning more generally and not simply in the matter of climate change. They call for a paradigmatic shift in transportation planning approaches.

Several studies did discuss possible approaches to the issue of uncertainty and decision making, without applying them to specific cases. For example, Meyer (forthcoming) noted that, "in recent years, many engineering design analyses have been incorporating more probabilistic approaches into their design procedures that account for uncertainty in both service life and in environmental factors." He continued, "In considering wind speeds, for example, probabilities of different wind speeds occurring based on an underlying distribution of historical occurrences are used to define a design wind speed. Other analysis approaches are incorporating risk management techniques into the tradeoff between design criteria that will make a structure more reliable and the economic costs to society if the structure fails." Furthermore, Dewar and Wachs (forthcoming) discuss a wide variety of conceptual decision making tools that could be considered when designing frameworks to understand how to incorporate climate uncertainty into transportation infrastructure decisions.

Approaches to Risk Analysis and Impact Assessment

Among those studies that attempted to implement a risk analysis or impact assessment framework for a particular transportation system, a number of different approaches were taken. For instance, Associated British Ports demonstrates an approach to risk evaluation that relies on expert elicitation to make a judgment on risk levels for U.K. ports (ABP Marine Environmental Research Ltd., 2004). Risk was broken into four themes: (1) flooding; (2) insurance; (3) physical damage; and (4) disruption. Port managers were asked to evaluate the risk level of each impact by indicating whether they thought it was a: (1) very low risk; (2) low risk; (3) moderate risk; (4) high risk; or (5) very high risk. Using this methodology, the study concluded that storm surge events represent the biggest threat to U.K. ports.

For the U.K. rail network, Eddowess et al. (2003) developed a framework for prioritizing risks that integrates the probability that a particular climate effect would impact the rail industry ("risk likelihood") with the scale of the impact, if it did occur ("risk impact"). The "risk likelihood" essentially combined an assessment of the present-day vulnerability to specific climate factors with projections of how they might change under global climate change scenarios, while the "risk impact" took into account the severity of a given impact, the amount of infrastructure affected, and the ability to adapt to the change. Their study did not, however, explicitly specify thresholds for when a given level of adaptation was worth implementing.

Transit New Zealand developed a methodology for determining thresholds for taking action by using a two-stage process (Kinsella and McGuire, 2005). The first stage constituted a decision tree that examined the necessity of taking action in the near term. No action was deemed necessary if (1) it was determined that a given impact was unlikely to occur before 2030, (2) the impact would not occur within the design life of the facility (for facilities with lifetimes of less than 25 years), or (3) current standards would adequately address the climate impact. If present-day action was deemed necessary, the second stage analysis determined the feasibility of taking action by comparing the costs of doing nothing, retrofitting the infrastructure, or designing all new infrastructure with future climate changes in mind.

Finally, the Climate's Long-term Impacts on Metro Boston CLIMB report develops tools for scenario analysis tools and decision support for Boston decision makers to use in understanding climate impacts. Specifically, the researchers developed a dynamic analytical modeling tool to help policy and decision makers assess changes in climate and in socioeconomic and technological developments and to understand their associated interrelated impacts on Boston's infrastructure system as a whole. The model allows users to input climate drivers in order to assess performance impacts and potential adaptation strategies for infrastructure systems, including transportation (Kirshen et al., 2004).

1.4. Conclusions Drawn from Current Literature on the State of Research

Assessing the literature on the impacts of climate change on transportation as a whole, it becomes apparent that there are a number of areas in which more research is needed on potential impacts of climate change on transportation. Many authors noted that research on the potential impacts of climate change on transportation systems is limited. Warren et al.

(2004) note that though much work has been done on adaptation to climate change in general, relatively little concerns climate impacts on transportation systems – to date, transportation research has been focused on emission-reduction strategies. Other authors noted the need for more research on specific impacts or modes. For instance, in their study of seasonal weight limits on prairie highways, Clayton et al. (2005) noted that there was essentially no transportation and climate impacts literature on their topic to draw upon.

Work in this field has so far been focused on the initial stages of risk assessment and adaptation; i.e., building a basic understanding of the issues involved. In general, the literature review shows that some work has been done on collecting data, assessing impacts, and evaluating the significance of these risks. Less work has been done to develop methodologies for assessment or to systematically evaluate adaptation strategies.

Work to develop decision support tools to facilitate these processes has received little formal attention. The state of research in each analytic area is summarized below.

Collecting data needed to assess transportation vulnerability to climate impacts. Some credible work on data collection and analysis has been done for selected modes and facilities in specific regions. Researchers have been able to make use of the good data on transportation networks and transportation engineering practice that exists for most of the developed world.

Most studies used climate projections consistent with long-term IPCC global projections as the basis for their analyses. However, few studies considered a broader range of plausible climate futures that could occur, such as scenarios, including additional feedbacks or abrupt climate change. In addition, few studies addressed the implications of changes in temperature or precipitation extremes.

In addition, there are significant gaps in data collection and analysis for several modes and for transportation infrastructure in hot or tropical climates, such as are found in the southwestern and southeastern portions of the United States. Most of the available literature addresses temperate or Arctic climates.

Developing knowledge about potential impacts. Researchers considered a wide variety of potential impacts on transportation, and significant work has been done for selected modes and facilities. However, a number of important gaps were found in the current literature, most notably the lack of quantitative assessment and dearth of literature on operations, network, performance, and secondary impacts:

- **Quantitative assessment** – Most studies to date have been qualitative. More quantitative assessments of impacts, along with the development of quantitative analytical methodologies, will provide needed information for decision makers.

- **Operations impacts** – The implications of climate change impacts on operations (both normal and emergency) are not as well explored as they are for physical infrastructure. Most of the existing literature on operations is focused on a select few issues such as waterborne freight and winter maintenance.

- **Network and performance impacts** – Relatively few studies (Kirshen et al., 2004; Suarez et al., 2005) focused on the network-level impacts of climate change. Most focused on the facility level (impacts to a type of facility, for instance, rather than system-level impacts on the whole network), and few measured performance impacts.

- **Secondary impacts –** Several secondary impacts mentioned in the literature but not discussed in-depth could provide useful avenues for further study. These include shifts in transportation demand due to climate-induced changes in economic activity and demographics; the impact of a warming climate on air quality (which influences transportation investment decisions); and other environmental impacts related to climate change that may intersect with transportation decision making in relation to ecosystem and habitat preservation, water quality and stormwater management, mitigation strategies, safety, and system and corridor planning.

Assessing the significance of these risks. Work in this area is largely qualitative. Though many researchers were able to communicate an assessment of which risks were significant enough to require further study, few produced quantitative assessments of cost or performance impacts. In particular, more work is needed regarding the economic implications of climate impacts on transportation facilities and systems. Relatively few studies addressed this quantitatively from an overall life-cycle benefits/costs framework.

Developing a methodological approach for assessment. Most studies used a similar basic approach (identify climate effects of concern, assess potential risks for specific modes/facility types, and identify potential adaptations). However, very few attempted to develop a generalized approach or consider the ramifications of translating their approach to other modes/regions.

Identifying strategies for adaptation and planning. Most studies dealt with adaptation from a facility engineering approach, rather than a strategic or systems performance level. Thus, it is largely specific design adaptations appropriate for particular types of facilities that were identified in the literature (for instance, insulating railbeds to prevent permafrost melt or raising roads to protect them against sea level rise).

Nonetheless, beginning elements of larger adaptation strategies were recognized in the literature. There is a general understanding of the differences between likely short- and long-term effects and acknowledgment that different approaches might be needed at different points in time (Meyer, forthcoming). In addition, some studies recognized that institutional change is necessary and recommended institutional processes for examining impacts and deciding on adaptations.

Significantly, almost no research has been done on how climate change can be incorporated into the long-range transportation planning process. Issues to address in future research include the mismatch between the timeframe of 20- and 30-year long-range plans and the 50- and 100-year projections of climate impacts; how to address the potential for nonlinear or abrupt changes in climate systems in a planning process; and how to make planning decisions that account for uncertainty in climate projections.

Developing decision-support tools. Very little work has been done to develop decision-support tools for transportation managers and planners. The field is sufficiently new that there has likely been little demand from transportation decision makers for such tools; rather they are only now beginning to learn about the potential impacts they might face in the future.

One of the most important gaps in this area is the lack of probabilistic approaches to address uncertainty. More sophisticated methodologies to incorporate uncertainty will need to be developed for transportation decision makers in order for them to incorporate climate change into transportation planning. Currently, uncertainty is rarely incorporated in a probabilistic sense in the literature on climate impacts on transportation (though the existence of uncertainty is acknowledged and expressed through the use of ranges in the climate factors and sometimes the use of scenarios). In addition, little attention is given to decision making practices under uncertainty, such as the development of adaptation strategies that are robust across multiple potential futures.

In summary, research on the potential impacts of climate change on transportation is an emerging field and one that has shown a remarkable upturn in interest and activity over the past few years. This has coincided with greater interest in the subject of adaptation in general, as recognition has grown that some degree of climate change is inevitable in the coming decades, even as steps are taken to reduce future emissions. Considerable work remains to be done in bringing this field to a greater level of maturity, including investigations of impacts not yet thoroughly examined and developing strategies, methodologies, and tools that decision makers at all levels can use to both assess the importance of climate impacts and identify ways to respond.

1.5. Gulf Coast Study Selection, Objectives, and Organization

1.5.1. Study Selection

To advance research on the implications of climate change for transportation, the U.S. DOT Center for Climate Change solicited and reviewed a range of project concepts. A case study approach was selected as an initial research strategy that would both generate concrete, useful information for local and regional decision makers as well as help to develop a prototype for analysis in other regions and contribute to research methodologies for broader application.

In selecting the study, U.S. DOT considered the extent to which the research would:

- Increase the knowledge base regarding the risks and sensitivities of transportation infrastructure to climate variability and change, the significance of these risks, and the range of adaptation strategies that may be considered to ensure a robust and reliable transportation network;
- Provide relevant information and assistance to transportation planners, designers, and decision makers;
- Build research approaches and tools that would be transferable to other regions or sectoral analyses;
- Produce near-term, useful results;

- Address multiple aspects of the research themes recommended by the 2002 workshop;
- Build on existing research activities and available data; and
- Strengthen U.S. DOT partnerships with other Federal agencies, State and local transportation and planning organizations, research institutions, and stakeholders.

Based on these criteria, the U.S. DOT selected a study of the Gulf Coast as the first of a series of research activities that its Center for Transportation and Climate Change will pursue to address these research priorities.

There are several intended uses for the products of this study. First, the findings of the study will help inform local and regional transportation decision makers in the central U.S. Gulf Coast region. While focused on one region of the United States, it is expected that this study will provide a prototype for analysis in other regions. The study findings will contribute to research methodologies in this new area of investigation. For example, Phase I has identified priority databases and methodologies for the integration of data for analysis in a GIS format, developed formats for mapping products, and developed criteria for assessing and ranking infrastructure sensitivities to the potential impacts of climate variability and change. Each of these outputs will offer useful information and example methodologies for use in research activities in other locations, as well as in decision making processes for transportation and planning in other areas. This research also is intended to help scientists and science agencies better understand the transportation sector's information needs, leading to improved data and better decision support.

1.5.2. Gulf Coast Study Objectives and Three Phases

The Gulf Coast Study has been organized into three phases, as depicted in figure 1.1. This report presents the findings of Phase I. The objectives of the overall study are to:

- Develop knowledge about potential transportation infrastructure sensitivities to climate changes and variability through an in-depth synthesis and analysis of existing data and trends;
- Assess the potential significance of these sensitivities to transportation decision makers in the central U.S. Gulf Coast region;
- Identify potential strategies for adaptation that will reduce risks and enhance the resilience of transportation infrastructure and services; and
- Identify or develop decision support tools or procedures that enable transportation decision makers to integrate information about climate variability and change into existing transportation planning and design processes.

The two primary objectives of Phase I of the central Gulf Coast transportation impact assessment were to: (1) collect data needed to characterize the region – its physiography and hydrology, land use and land cover, past and projected climate, current population and trends, and transportation infrastructure; and (2) demonstrate an approach for assessing risks and vulnerability of transportation at regional and local scales. The results of this analysis are presented in this report. The methodologies developed during Phase I of the study can be applied to assess transportation risk and vulnerability at a community, county, or regional level.

Phase II of the study will entail an in-depth assessment of impacts and risks to selected areas and facilities (as identified in Phase I) and will contribute to the development of risk-assessment tools and techniques that can be used by transportation decision makers to analyze the vulnerability of other areas.

The objectives of Phase III are to identify the range of potential adaptation strategies available to Federal, regional, and local transportation managers to respond to the risks identified in Phases I and II; to identify the potential strengths and weaknesses of these responses; and to develop an assessment tool that may assist transportation managers in selecting adaptation strategies appropriate to their agency, community, or facility, and to the identified sensitivity to climate change.

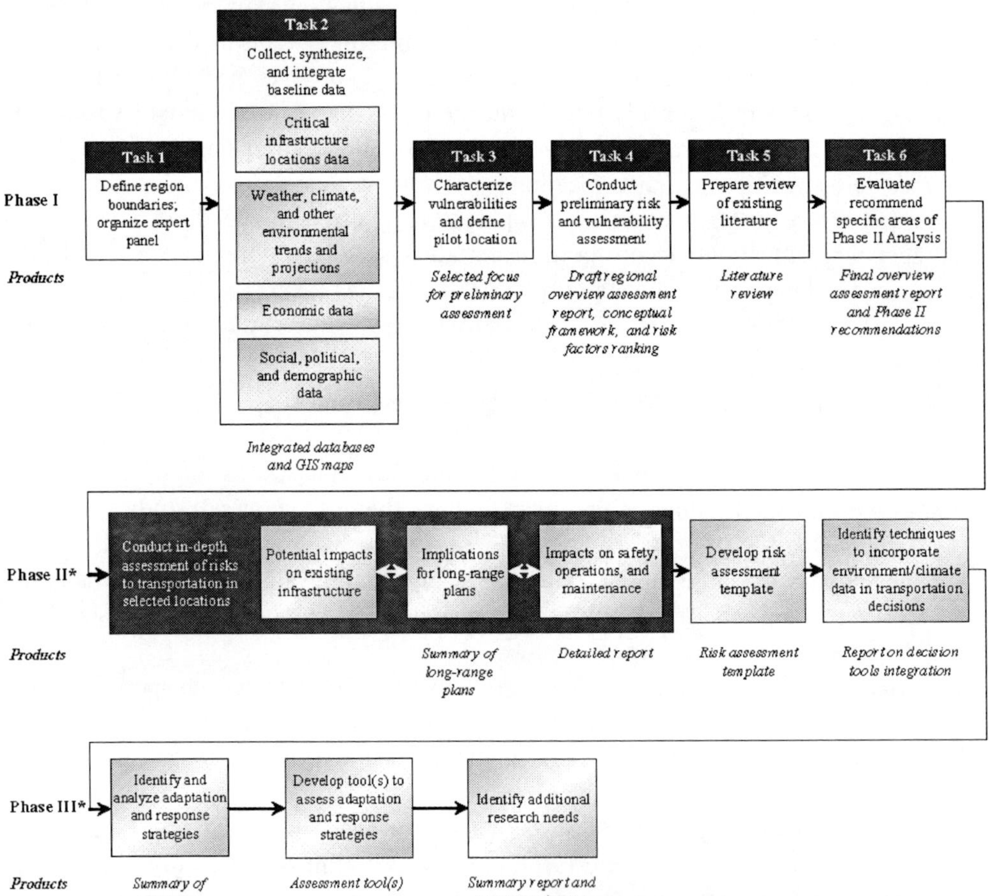

*Study design of Phases II and III will be refined based on findings of Phase I.

Figure 1.1. Gulf Coast study design

Table 1.1. Impacts of Climate Change on Transportation Identified in the Literature, 1987-2006

Climate Impact	Potential Infrastructure Impact	Potential Operations Impact	Adaptation	Source
Temperature Increase				
Increased Summer Temperatures	Highway asphalt rutting		Proper design/construction, milling out ruts, more maintenance, overlay with more rut-resistant asphalt	(1) Wooler, Sarah. 2004. (2) Andrey and Mills. 2003. (3) Hass, et al. 2006. (4) Black, William. 1990. (5) Meyers, Michael. 2006. (6) Barrett, et al. 2004. (7) Marbek Resource Consultants Ltd. 2003. (8) Kerr, Andrew, et al. 1999. (9) Warren, et al. 2004. (10) Entek U.K. Limited. 2000. (11) Entek U.K. Limited. 2004. (12) Lockwood, Steve. 2006. (13) Kinsella, Y. and McGuire, E. 2005. (14) Mills and Andrey. 2002. (15) OFCM. 2002.
	Rail buckling	Potential for derailment and malfunction of track sensors and signal sensors, increased travel time due to speed restrictions, increased risk of hazardous material spill	Speed restrictions, reducing frequency of some services, better air conditioning for signals. Improve systems to warn and update dispatch centers, crews, and stations. Inspect and repair tracks, track sensors, and signals. Distribute advisories, warnings, and updates regarding the weather situation and track conditions.	(1) Wooler, Sarah. 2004. (2) Eddowes, M.J., et al. 2003. (3) Kerr, Andrew, et al. 1999. (4) Warren, et al. 2004. (5) Entek U.K. Limited. 2000. (6) Land Use Consultants, et al. 2002. (7) Smyth, et al. 2002. (8) Kerr, Andy. 2001. (9) Entek U.K. Limited. 2004. (10) Rossetti. 2002. (11) OFCM. 2002.

Table 1.1. (Continued)

Climate Impact	Potential Infrastructure Impact	Potential Operations Impact	Adaptation	Source
	More airport runway length and fuel needed because of less dense air		New planes designed to takeoff more efficiently	(1) Wooler, Sarah. 2004. (2) Andrey and Mills. 2003. (3) Irwin and Johnson. 1990. (4) Warren, et al. 2004. (5) Entek U.K. Limited. 2000. (6) Smyth, et al. 2002. (7) Entek U.K. Limited. 2004.
	Heat/Lack of ventilation on underground urban transit systems	Personnel health/safety risk, heat exhaustion, engine/equipment heat stress. Overcrowding, failed, or delayed service will only compound the problem. Could cause passengers to avoid taking public transportation (mode shift).	Install better ventilation systems, climate monitoring systems, personnel safety and equipment monitoring systems.	(1) Wooler, Sarah. 2004. (2) Greater London Authority. 2005. (3) OFCM. 2002.
		Health and safety risks from heat stress to highway, transit, and pipeline system operators, maintenance personnel, and passengers; including increased risk of collisions/spills of hazardous cargo, control system integrity	Improve systems to advise operators, monitor personnel, and take prescribed and precautionary measures	(1) OFCM. 2002.
	Low water levels on inland waterways	Increased shipping costs; shift to other modes (rail, truck)	Changes to navigation, dredging of channels, flow augmentation	(1) Wooler, Sarah. 2004. (2) Andrey and Mills. 2003. (3) Olsen, et al. 2005. (4) Black, William. 1990. (5) Irwin and Johnson. 1990. (6) U.S. Federal Highway Administration Office of Environment and Planning. 1998. (7) U.S. Department of State. 2002.

Table 1.1. (Continued)

Climate Impact	Potential Infrastructure Impact	Potential Operations Impact	Adaptation	Source
				(8) Institute for Water Resources, U.S. Army Corps of Engineers. 2004. (9) Sousounis, Peter J. and Jeanne M. Bisanz, Eds. 2000. (10) National Assessment Synthesis Team. 2000. (11) Marbek Resource Consultants Ltd. 2003. (12) D'Arcy, Pierre. 2004. (13) Warren, et al. 2004. (14) Entek U.K. Limited. 2000. (15) Ministry of Housing, Spatial Planning, and the Environment, The Netherlands. 2001. (16) Ruth, Matthais. 2006. (17) Quinn. 2002.
	Thermal expansion of bridges	Frequent detours, traffic disruptions	Increased ongoing maintenance	(1) Cohen, Susan, Soo Hoo, Wendy K., and Sumitani, Megumi. 2005.
	Overheating of diesel engines		Adaptation of cooling systems	(1) Entek U.K. Limited. 2000.
	Increased vegetation – leaf fall	Ineffective braking of rail cars, visual obstruction	Vegetation management, plant low-maintenance vegetation as buffer	(1) Wooler, Sarah. 2004. (2) Eddowes, M.J., et al. 2003. (3) Land Use Consultants, et al. 2002. (4) Smyth, et al. 2002. (5) Kerr, Andy. 2001. (6) Entek U.K. Limited. 2004. (7) Kinsella, Y. and McGuire, E. 2005.

Table 1.1. (Continued)

Climate Impact	Potential Infrastructure Impact	Potential Operations Impact	Adaptation	Source
	Changes to landscape/biodiversity	Highway agency owns many medians. Increased pest management. Impact on wetlands commitments	Different types of vegetation may have to be considered	(1) Wooler, Sarah. 2004. (2) Kinsella, Y. and McGuire, E. 2005. (3) Mortenson and Bank. 2002.
Increased Summer Temperature and Decreased Precipitation	Less rain to dilute surface salt may cause steel reinforcing in concrete structures to corrode (Australia)		Better protect reinforcing in saline environments	(1) Norwell, Gary. 2004.
Increased Winter Temperatures	Reduction in cold weather rail maintenance	Fewer broken rails, excessive wheel wear, and frozen switches		(1) Andrey and Mills. 2002.
	Longer construction season	Drier and warmer days		(1) Andrey and Mills. 2003. (2) Kinsella, Y. and McGuire, E. 2005.
Thawing Permafrost (United States, Canada, China)	Road, rail, airport, pipeline embankments will fail and shallow pile foundations could settle	Potential for fewer construction problems in long run	Crushed rock cooling system, insulation/ground refrigeration systems, rehabilitation, relocation, mechanically stabilize embankments against ground movement, remove permafrost before construction	(1) Instanes et al.. 2005. (2) Brown, Jeff. 2005. (3) Cheng, Guadong. 2005. (4) Hass, et al. 2006. (5) Black, William. 1990. (6) Irwin and Johnson. 1990. (7) U.S. Arctic Research Commission Permafrost Task Force. 2003. (8) Weller, Gunter, et al. 1999. (9) Grondin et al. 2005. (10) Wright, Fred. 2001. (11) Warren, et al. 2004. (12) Ruth, Matthais. 2006. (13) Smith and Levasseur. 2002. (14) Caldwell et al. 2002.

Table 1.1. (Continued)

Climate Impact	Potential Infrastructure Impact	Potential Operations Impact	Adaptation	Source
Reduction of Freezing Season for Ice Roads (Arctic)	Roads unusable during certain seasons	Shorter shipping season, higher maintenance costs, higher life-cycle costs, seasonal mode shift	Reconstruction of severely damaged infrastructure with less frost-susceptible foundation (geosynthetic barrier), retrofitting road side drains	(1) Instanes et al.. 2005. (2) Lonergan, et al. 1993. (3) Andrey and Mills. 2003. (4)Hass, et al. 2006. (5) Weller, Gunter, et al. 1999. (6) Marbek Resource Consultants Ltd. 2003. (7) Clayton et al. (and Montufar). 2005. (8) Warren, et al. 2004. (9) Lockwood, Steve. 2006.
Increased Winter Precipitation – Rain/Snow	Flooding of roads/airport runways/bikeways and walkways (frequency and magnitude will increase)	Infrastructure deterioration (quicker with acid rain), impacts on water quality, travel and schedule delays, loss of life and property, increased safety risks, increased risks of hazardous cargo accidents	Seek alternative routes, improve flood protection, risk assessment for new roads, emergency contingency planning, ensure bridge openings/culverts sufficient to deal with flooding, improve drainage, improved asphalt/concrete mixtures, perform adequate maintenance, and minimize repair backlogs	(1) Wooler, Sarah. 2004. (2) Andrey and Mills. 2003. (3) Irwin and Johnson. 1990. (4) U.S. Department of State. 2002. (5) Kirshen, Paul H. and Matthais, Ruth. 2004. (6) Intergovernmental Panel on Climate Change. 2001. (7) Sousounis, Peter J. and Jeanne M. Bisanz, Eds. 2000. (8) Wilkenson, Robert. 2002. (9) Meyers, Michael. 2006. (10) Barrett, et al. 2004. (11) Kerr, Andrew, et al. 1999. (12) Warren, et al. 2004. (13) Entek U.K. Limited. 2000. (14) Land Use Consultants, et al. 2002. (15) Smyth, et al. 2002. (16) Kerr, Andy. 2001.

Table 1.1. (Continued)

Climate Impact	Potential Infrastructure Impact	Potential Operations Impact	Adaptation	Source
				(17) Entek U.K. Limited. 2004. (18) Norwell, Gary. 2004. (19) Kinsella, Y. and McGuire, E. 2005. (20) Rossiter, Lisa. 2004. (21) Smith, Orson. 2006. (22) OFCM. 2002.
	Flooding of rails	Service disruption, increased malfunctions of track or signal sensors, wash-outs and mud slides, increased risk of hazardous material spills	Engineering solutions, increase advisories, warnings and updates to dispatch centers, crews, and stations. Modify operations for current or forecast conditions.	(1) Wooler, Sarah. 2004. (2) Irwin and Johnson. 1990. (3) Eddowes, M.J., et al. 2003. (4) Entek U.K. Limited. 2000. (5) Smyth, et al. 2002. (6) OFCM. 2002.
	Bridge scour		Speed restrictions, closure to traffic, new materials, better maintenance	(1) Wooler, Sarah. 2004. (2) Hass, et al. 2006. (3) Kirshen, Paul H. and Matthais, Ruth. 2004. (4) Meyers, Michael. 2006. (5) Eddowes, M.J., et al. 2003. (6) Smith, Orson. 2006. (7) OFCM. 2002.
	Flooding of underground transit systems	Power outages (third rail blowouts), complete loss of service in affected areas, drowned passengers	Pumping systems	(1) Wooler, Sarah. 2004. (2) Zimmerman, 2002a and 2002b. (3) OFCM. 2002.
	Flooding of inland marine transportation waterways	Interruptions of river navigation and other inland waterway activities (ferries, boating, commerce, port operations, lock operations)		(1) Intergovernmental Panel on Climate Change. 2001. (2) Ning, Zhu H., et al. 2003. (3) OFCM. 2002.
	Pipeline system flooding and damage from	Disruption of fuel delivery, pipeline sensor failure, disruption of		(1) OFCM. 2002.

Table 1.1. (Continued)

Climate Impact	Potential Infrastructure Impact	Potential Operations Impact	Adaptation	Source
	scouring away pipeline roadbed or unearthing buried pipelines	construction or maintenance cycles, leaks or other pipeline failures		
Increased Precipitation and Increased Summer Temperatures	Highway, rail, and pipeline embankments at risk of subsidence/heave	Landslides	Fill cracks and carry out more maintenance	(1) Wooler, Sarah. 2004. (2) Instanes et al.. 2005. (3) Cohen, Susan, Soo Hoo, Wendy K., and Sumitani, Megumi. 2005. (4) Wilkenson, Robert. 2002. (5) Weller, Gunter, et al. 1999. (6) Eddowes, M.J., et al. 2003. (7) Konuk, Ibrahim. 2005. (8) Marbek Resource Consultants Ltd. 2003. (9) Kerr, Andrew, et al. 1999. (10) Warren, et al. 2004. (11) Entek U.K. Limited. 2000. (12) Land Use Consultants, et al. 2002. (13) Smyth, et al. 2002. (14) Entek U.K. Limited. 2004. (15) Kinsella, Y. and McGuire, E. 2005. (16) Rossiter, Lisa. 2004. (17) duVair et al. 2002. (18) OFCM. 2002.
	Concrete deterioration			(1) Wooler, Sarah. 2004. (2) U.S. Department of State. 2002. (3) OFCM. 2002.

Table 1.1. (Continued)

Climate Impact	Potential Infrastructure Impact	Potential Operations Impact	Adaptation	Source
	More frequent and larger slush-flow avalanches (Arctic)		Incorporate potential risk into planning process for new settlements, detection systems, temporary closures	(1) Instanes et al.. 2005. (2) Marbek Resource Consultants Ltd. 2003. (3) Warren, et al. 2004. (4) Stethem, Chris, et al. 2003.
	Altered runoff patterns (Arctic)	Disruption of the ice-water balance		(1) Instanes et al.. 2005.
Sea Level Rise	Erosion of coastal highways		Construction of sea walls	(1) Wooler, Sarah. 2004. (2) Black, William. 1990. (3) U.S. Federal Highway Administration Office of Environment and Planning. 1998. (4) Marbek Resource Consultants Ltd. 2003. (5) Norwell, Gary. 2004. (6) Kinsella, Y. and McGuire, E. 2005. (7) Ruth, Matthais. 2006. (8) Hyman, William, et al. 1989. (9) Titus, 2002. (10) OFCM. 2002.
	Higher tides at ports/harbor facilities	Damage to docks and terminals		(1) Wooler, Sarah. 2004. (2) Black, William. 1990. (3) U.S. Department of State. 2002. (4) Kirshen, Paul H. and Matthais, Ruth. 2004. (5) Smyth, et al. 2002. (6) Ministry of Housing, Spatial Planning, and the Environment, The Netherlands. 2001.

Table 1.1. (Continued)

Climate Impact	Potential Infrastructure Impact	Potential Operations Impact	Adaptation	Source
				(7) Caldwell et al. 2002. (8) OFCM. 2002.
	Deeper water	Permit greater ship drafts		(1) Andrey and Mills. 2003. (2) Kerr, Andrew, et al. 1999. (3) Titus. 2002.
	Low-level aviation infrastructure at risk		Relocation or protection of facilities	(1) Andrey and Mills. 2003. (2) Committee on Engineering Implications of Change in Relative Mean Sea Level. 1987. (3) Warren, et al. 2004. (4) Ruth, Matthais. 2006. (5) Hyman, William, et al. 1989.
	Less bridge clearance			(1) Cohen, Susan, Soo Hoo, Wendy K., and Sumitani, Megumi. 2005. (2) Committee on Engineering Implications of Change in Relative Mean Sea Level. 1987. (3) Norwell, Gary. 2004. (4) Hyman, William, et al. 1989. (5) OFCM. 2002.
		More search and rescue operations	Obtain more vessels with emergency towing capabilities, better weather forecasting, change seasonal classifications of waters around coast, change ship/boat design	(1) Wooler, Sarah. 2004. (2) Marbek Resource Consultants Ltd. 2003. (3) OFCM. 2002.

Table 1.1. (Continued)

Climate Impact	Potential Infrastructure Impact	Potential Operations Impact	Adaptation	Source
Storm Activity				
Storm Surges	Coastal road flooding	Increased VMT and VHT; increased number of road accidents, evacuation route delays, disruption of transit services, stranded motorists.	Seawalls, build more redundancy into system, support land use policies that discourage development on shoreline, design and material changes, pumping of underpasses, raise roads	(1) Choo, Kristin. 2005. (2) U.S. Federal Highway Administration Office of Environment and Planning. 1998. (3) Intergovernmental Panel on Climate Change. 2001. (4) Suarez, Pablo et Al. 2005. (5) Rosenzweig, Cynthia and Soleki, William. 2001. (6) Wilkenson, Robert. 2002. (7) National Assessment Synthesis Team. 2000. (8) Meyers, Michael. 2006. (9) Committee on Engineering Implications of Change in Relative Mean Sea Level. 1987. (10) Greater London Authority. 2005.
	Railway flooding	Safety risks to personnel and equipment (possible injury or death from accidents); rail and railway roadbed damage; disruption of rail traffic; rail sensor failure likely; increased risk of hazardous material spill.	Seawalls, raising rails	(1) Black, William. 1990. (2) Committee on Engineering Implications of Change in Relative Mean Sea Level. 1987. (3) Kerr, Andrew, et al. 1999. (4) Greater London Authority. 2005. (5) OFCM. 2002.
	Subway flooding		Flood barriers	(1) Choo, Kristin. 2005. (2) Black, William. 1990. (3) Greater London Authority. 2005.

Table 1.1. (Continued)

Climate Impact	Potential Infrastructure Impact	Potential Operations Impact	Adaptation	Source
				(4) Ruth, Matthais. 2006. (5) Zimmerman, 2002.
	Port flooding/damage	Damage to ports facilities from vessels tied alongside.	Reduce "cope" level at ports to reduce likelihood of water flowing across docks; construct flood defense mechanisms	(1) ABP Marine Environmental Research Ltd 2004. (2) Committee on Engineering Implications of Change in Relative Mean Sea Level. 1987. (3) Entek U.K. Limited. 2000. (4) Land Use Consultants, et al. 2002. (5) OFCM. 2002.
Increased Frequency/ Magnitude of Storms	Damage to infrastructure on roads, railways, pipelines, seaports, airports	Closures or major disruptions of roads, railways, airports, transit systems, pipelines, marine systems and ports; emergency evacuations; travel delays		(1) Instanes et al.. 2005. (2) Smyth, et al. 2002. (3) Ruth, Matthais. 2006. (4) OFCM. 2002. (5) Intergovernmental Panel on Climate Change. 2001.
Storm Activity (continued)				
Increased Wind Speeds	Bridges, signs, overhead cables, railroad signals, tall structures at risk		Design structures for more turbulent wind conditions, build with better material, use "smart" technologies to detect abnormal events	(1) Wooler, Sarah. 2004. (2) Meyers, Michael. 2006. (3) Eddowes, M.J., et al. 2003. (4) Kerr, Andrew, et al. 1999. (5) Kerr, Andy. 2001. (6) Kinsella, Y. and McGuire, E. 2005. (7) OFCM. 2002.
		Roadways: loss of visibility from drifting snow, loss of stability/maneuverability, lane obstruction (debris), treatment chemical dispersion,		(1) OFCM. 2002.

Table 1.1. (Continued)

Climate Impact	Potential Infrastructure Impact	Potential Operations Impact	Adaptation	Source
		Railways: Rail car blow over; schedule delays; increased risk of hazardous material spill		(1) OFCM. 2002.
		Ship handling becomes difficult		(1) OFCM. 2002.
		Impacts on airport ground operations: increased incidence of foreign objects present in aircraft movement areas, maintenance at high locations on large aircraft impeded/slowed, snow removal and de-icing operations affected		(1) OFCM. 2002.
Lightning/Electrical Disturbance	Disruption to transportation electronic infrastructure, signaling, etc.	Risk to personnel from lightning, maintenance activity delays, rail and aircraft refueling operations delayed, track signal sensor malfunction resulting in possible train delays and stops, threats to barge tow equipment, communications and data distribution from pipeline sensors may fail		(1) Wooler, Sarah. 2004. (2) Eddowes, M.J., et al. 2003. (3) OFCM. 2002.
Fewer Winter Storms	Less snow/ice for all modes	Improved mobility/safety, reduced maintenance costs, less pollution from salt, decrease in vehicle corrosion		(1) Andrey and Mills. 2003. (2) Black, William. 1990. (3) Irwin and Johnson. 1990. (4) Intergovernmental Panel on Climate Change. 2001. (5) Barrett, et al. 2004. (6) Marbek Resource Consultants Ltd. 2003. (7) Kerr, Andrew, et al. 1999. (8) Warren, et al. 2004. (9) Entek U.K. Limited. 2000.

Table 1.1. (Continued)

Climate Impact	Potential Infrastructure Impact	Potential Operations Impact	Adaptation	Source
				(10) Land Use Consultants, et al. 2002. (11) Entek U.K. Limited. 2004. (12) Wooler, Sarah. 2004. (13) Kinsella, Y. and McGuire, E. 2005. (14) Hyman, William, et al. 1989. (15) Pisano et al. 2002.
Ice Melting				
Reduced Ice Cover (Canada, Alaska, Great Lakes)	Reduced ice loading on structures, such as bridges or piers			(1) Instanes et al.. 2005.
	New northern shipping routes	Shorten shipping distance and delivery time, security concerns, environmental risks, law-diplomacy issues, Inuit unease	Develop a "transit management regime" for area	(1) Instanes et al.. 2005. (2) Johnston, Douglas. 2002. (3) Brigham, Lawson and Ben Ellis, Eds. 2004. (4) Office of Naval Research, Naval Ice Center, Oceanographer of the Navy. 2001. (5) National Assessment Synthesis Team. 2000. (6) Marbek Resource Consultants Ltd. 2003. (7) Warren, et al. 2004. (8) Smith and Levasseur. 2002. (9) Caldwell et al. 2002.
		Lengthened season for float planes		(1) Black, William. 1990. (2) Irwin and John

Table 1.1. (Continued)

Climate Impact	Potential Infrastructure Impact	Potential Operations Impact	Adaptation	Source
		Longer shipping season		(1) Wooler, Sarah. 2004. (2) Andrey and Mills. 2003. (3) Black, William. 1990. (4) Irwin and Johnson. 1990. (5) U.S. Federal Highway Administration Office of Environment and Planning. 1998. (6) Sousounis, Peter J. and Jeanne M. Bisanz, Eds. 2000. (7) National Assessment Synthesis Team. 2000. (8) Warren, et al. 2004. (9) Ruth, Matthais. 2006. (10) Caldwell et al. 2002.
	Multiyear ice, in low concentrations, will be hazard to ships and naval submarines		New ship/submarine design or modifications	(1) Brigham, Lawson and Ben Ellis, Eds. 2004. (2) Office of Naval Research, Naval Ice Center, Oceanographer of the Navy. 200
Earlier River Ice Breakup (United States, Canada)	Ice-jam flooding risk			(1) Instanes et al.. 2005. (2) Hass, et al. 2006. (3) Smith and Levasseur.

1.5.3. Study Organization and Oversight

The Gulf Coast Study is 1 of 21 "synthesis and assessment" products planned and sponsored by the U.S. Climate Change Science Program (CCSP). The primary objective of the CCSP is to provide scientific information needed to inform public discussions and government and private sector decision making on key climate-related issues. This project is one of seven projects organized under CCSP Goal 4, which is "to understand the sensitivity and adaptability of different natural and managed ecosystems and human systems to climate and related global changes" (CCSP, 2003, p. 20).

Led by the U.S. DOT in collaboration with the U.S. Geological Survey (USGS), this study was conducted through a groundbreaking interdisciplinary approach that integrated natural science disciplines with expertise in risk assessment, transportation, and planning. The U.S. DOT and USGS convened a research team with expertise in multiple fields based on each agency's mission and core capabilities. The USGS coordinated the provision of scientific research support, coordinating expertise in climate change science and impacts assessment; meteorology; hydrology; storm surge analysis and modeling; risk analysis; and economics. Cooperators from Louisiana State University, the University of New Orleans, and Texas A&M University assisted in the data collection aspects of Phase I and in developing a framework for assessing risk and vulnerability. (The U.S. DOT assembled expertise in transportation planning, engineering, design, and operation.) Cambridge Systematics, Inc., (CS) a transportation consulting firm, supported the coordination and design of the study, assisted in organizing the data, and provided transportation experts with expertise in ports, rail, highways and transit, pipelines, aviation, emergency management, and transportation planning and investment. The CS Transportation Analysis Team included consultant support from Wilbur Smith Associates and the Texas Transportation Institute. The U.S. DOT's Bureau of Transportation Statistics (BTS) supported geospatial and other data collection and analysis related to transportation, working in coordination with USGS geospatial experts. Collectively, this group of scientists and transportation experts has served as the research team conducting Phase I of the study.

The Secretary of Transportation, following the guidelines of the Federal Advisory Committee Act (5 U.S.C. App. 2) or "FACA," established a U.S. DOT Advisory Committee on Synthesis and Assessment Product 4.7: Impacts of Climate Variability and Change on Transportation Systems and Infrastructure – Gulf Coast Study, Phase I. The committee provides technical advice and recommendations in the development of this product for the CCSP. The committee provides balanced, consensual advice on the study design, research methodology, data sources and quality, and study findings. The committee functions as an advisory body to the two Federal agencies leading the research project.

This product adheres to Federal Information Quality Act (IQA) guidelines and Office of Management and Budget (OMB) peer review requirements. Background sources of information, included as illustrative material and to provide context, are clearly identified as such at the end of the list of sources in each chapter.

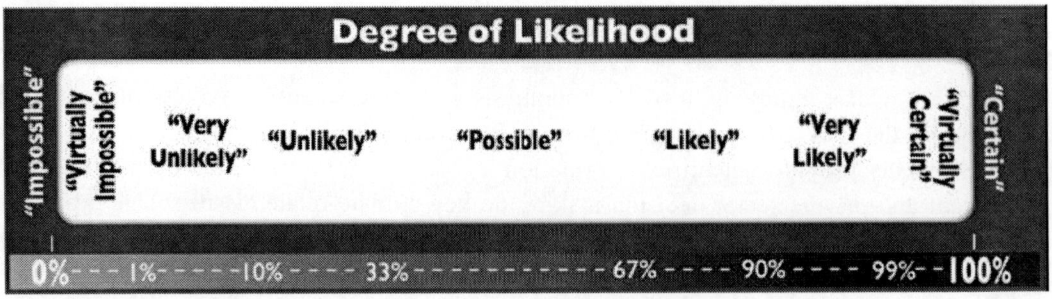

Source: Karl et al., 2006.

Figure 1.2. Lexicon of terms used to describe the likelihood of climate outcomes

1.5.4. Characterizing Uncertainty

Some degree of uncertainty is inherent in any consideration of future climate change; further, the degree of certainty in climate projections varies for different aspects of future climate. Throughout this report, the research team has adopted a consistent lexicon first developed by the IPCC to indicate the degree of certainty that can be ascribed to a particular potential climate outcome. As presented in figure 1.2, the "Degree of Likelihood" ranges from "Impossible" to "Certain," with different terminology used to describe different ranges of statistical certainty as supported by available scientific modeling and analysis. The analytic approach required to characterize uncertainty for each climate factor (e.g., temperature, precipitation, sea level rise, storm surge) is discussed in detail in the relevant section of this report.

1.6. Sources

1.6.1. References

Andrey, J. & Mills, B. N. (2003). Chapter 9 – Climate Change and the Canadian Transportation System: Vulnerabilities and Adaptations. In: *Weather and Transportation in Canada* [J. Andrey, & C. Knapper, (Eds.)], Department of Geography, University of Waterloo, Ontario, Canada.

Barrett, Byrd Associates, (2004). Weathering The Changes. *Transportation Professional* (BBA Linden House, Linden Close, Tunbridge Wells, Kent, United Kingdom), Volume 10, page 13.

Black, W. R. (1990). Global Warming: Impacts On The Transportation Infrastructure. *TR News No. 150,* Transportation Research Board, pages 2-8, 34.

Brennan, D., Akpan, U., Konuk, I. & Zebrowski, A. (2001). *Random Field Modeling of Rainfall Induced Soil Movement.* Geological Survey of Canada, February.

Board on Atmospheric Sciences and Climate (BASC), (2004). *Where the Weather Meets the Road: A Research Agenda for Improving Road Weather Service.* National Academies Press, Washington, D.C.

Brigham, L. & Ellis, B. (Eds.), (2004). *Arctic Marine Transport Workshop: September 28-30.* Held at Scott Polar Research Institute, Cambridge University, United Kingdom Hosted by: Arctic Research Committee, Arlington, Virginia. Institute of the North, Anchorage, Alaska, and International Arctic Science Committee, Oslo, Norway. Northern Printing, Anchorage Alaska.

Brown, J. L. (2005). Cold Region Design: High-Altitude Railway Designed to Survive Climate Change. *Civil Engineering – ASCE*, Volume 75, Number 4, April, page 28.

Burkett, V. R. (2002). Potential Impacts Of Climate Change And Variability On Transportation in the Gulf Coast/Mississippi Delta Region. In: *The Potential Impacts of Climate Change on Transportation Workshop, October 1-2, 2002.* Center for Climate Change and Environmental Forecasting, U.S. Department of Transportation, Washington, D.C.

Caldwell, H., Quinn, K. H., Meunier, J., Suhrbier, J. & Grenzeback, L. (2002). Potential Impacts Of Climate Change On Freight Transport. In: *The Potential Impacts of Climate Change on Transportation Workshop, October 1-2, 2002.* Center for Climate Change and Environmental Forecasting, U.S. Department of Transportation, Washington, D.C.

Cheng, G. (2005). Permafrost Studies in the Qinghai-Tibet Plateau for Road Construction. *Journal of Cold Regions Engineering. Volume 19*, Number 1, pages 19-29.

Choo, K. (2005). Heat Wave. *Planning*. Volume *71*, Number 8, August, pages 8-13.

Clayton, A., Montufar, J., Regeher, J., Isaacs, C. & McGregor, R. (2005). *Aspects of the Potential Impacts of Climate Change on Seasonal Weight Limits and Trucking in the Prairie Region.* Climate Change Impacts and Adaptation Directorate, Natural Resoures Canada, June.

Climate Risk Management and Metroeconomica, (2005). *Financial Risks of Climate Change – Summary Report.* The Association of British Insurers (ABI), June.

Committee on Engineering Implications of Change in Realitive Mean Sea Level, (1987). *Responding to Changes in Sea Level – Engineering Implications.* Marine Board Commission on Engineering and Technical Systems, National Research Council, National Academy Press, Washington, D.C.

D'Arcy, P. (2004). *Response Strategy to maintain shipping and port activities in the face of climate change – reduced water levels in the Great Lakes/St. Lawrence Seaway.* Navigation Consensus Building Committee, Transport Canada.

Dewar, J. A. & Wachs, M. (2008). Transportation Planning, Climate Change, and Decision-Making Under Uncertainty. In: *The Potential Impacts of Climate Change on U.S. Transportation.* The National Research Council, National Academy of Science, Transportation Research Board and Department of Earth and Life Sciences, Washington, D.C.

Dlugolecki, A. (2004). *A Changing Climate for Insurance – A Summary Report for Chief Executives and Policy-Makers*. The Association of British Insurers (ABI), June.

DuVair, P., Wickizer, D. & Burer, M. J. (2002). Climate Change and the Potential Implications for California's Transportation System. In: *The Potential Impacts of Climate Change on Transportation Workshop, October 1-2, 2002.* Center for Climate Change and Environmental Forecasting, U.S. Department of Transportation, Washington, D.C.

Easterling, D. R. (2002). Observed Climate Change And Transportation. In: *The Potential Impacts of Climate Change on Transportation Workshop, October 1-2, 2002.* Center for

Climate Change and Environmental Forecasting, U.S. Department of Transportation, Washington, D.C.

Eddowess, M. J., Waller, D., Taylor, P., Briggs, B., Meade, T. & Ferguson, I. (2003). *Rail Safety Implications of Weather, Climate and Climate Change: Final Report.* Rail Safety and Standards Board, United Kingdom, March.

Frey, H. C. (2006). *Incorporating Risk and Uncertainty into the Assessment of Impacts of Global Climate Change on Transportation Systems.* North Carolina State University Center for Transportation and the Environment 2nd Workshop on Impacts of Global Climate Change On Hydraulics and Hydrology and Transportation, March 27, 2006, Washington, D.C.

Greater London Authority, (2005). *Climate Change and London's Transport Systems: Summary Report.* London, United Kingdom, September.

Grenzeback, L. R. & Lukmann, A. T. (2008). Case Study of the Transportation Sector's Response to and Recovery from Hurricanes Katrina and Rita. In: *The Potential Impacts of Climate Change on U.S. Transportation.* The National Research Council, National Academy of Science, Transportation Research Board and Department of Earth and Life Sciences, Washington, D.C.

Grondin, G. (2005). Impact of Permafrost Thaw on Airfield and Road Infrastructures in Nunavik, Quebec. *Routes = Roads*, Number 326, pages 42-49.

Haas, R., Falls, L. C., Macleod, D. & Tighe, S. (2006). Climate Impacts And Adaptation On Roads in Northern Canada. In: *Transportation Research Board 85th Annual Meeting*, Washington, D.C. January 22, page 19.

Hyman, W., Miller, T. R. & Walker, J. C. (1989). Impacts of the Greenhouse Effect on Urban Transportation. *Transportation Research Record 1240.* Transportation Research Board, National Research Council, Washington, D.C., pages 45-50.

Instanes, A., Anisimov, O., Brigham, L., Goering, D., Khrustalev, L. N., Ladanyi, B., Otto Larsen, J., Smith, O. , Stevermer, A., Weatherhead, B. & Weller, G. (2005). Chapter 16 – Infrastructure: Buildings, Support Systems, and Industrial Facilities, In: *Arctic Climate Impact Assessment,* Cambridge University Press, 1,042.

Institute for Water Resources, (2005). *Climate Impacts on Inland Waterways (Final Report).* Report to U.S. Department of Transportation. Institute for Water Resources, U.S. Army Corps of Engineers, Alexandria, Virginia. July.

Intergovernmental Panel on Climate Change (IPCC), (1996). *Climate Change 1995: Impacts, Adaptations, and Mitigation of Climate Change.* IPCC. Cambridge University Press, Cambridge, United Kingdom.

Intergovernmental Panel on Climate Change (IPCC), (2001). *Climate Change 2001: Impacts, Adaptations, and Vulnerability.* IPCC. Cambridge Press. New York, New York.

Intergovernmental Panel On Climate Change (IPCC), (2007), *Climate Change 2007: The Physical Science Basis, Summary for Policy-Makers, Contribution of Working Group I to the Fourth Assessment Report of the Intergovernmental Panel on Climate Change,* IPCC. Cambridge Press. New York, New York, February

Irwin, N. A. & Johnson, W. F. (1990). Implications of Long-Term Climatic Changes for Transportation in Canada. *Transportation Research Record,* Number 1267, pages 12-25.

Johnston, D. M. (2002). The Northwest Passage Revisited. *Ocean Development & International Law.* April-June, *Volume 33*, Issue 2, pages 145-164.

Karl, T. R., Hassol, S. J., Miller, C. D. & Murray, W. L. eds. (2006). *Temperature Trends in the Lower Atmosphere: Steps for Understanding and Reconciling Differences.* U.S. Climate Change Science Program, Synthesis and Assessment Product 1.1, Washington, D.C., page 164.

Kerr, A. & McLeod, A. (2001). *Potential Adaptation Strategies for Climate Change in Scotland.* Scottish Executive Central Research Unit, Scotland, U.K.

Kerr, A., Schackley, S., Milne, R. & Allen, S. (1999). *Climate Change: Scottish Implications Scoping Study.* The Scottish Executive Central Research Unit, Scotland, U.K.

Kinsella, Y. & McGuire, F. (2005). *Climate Change Uncertainty and the State Highway Network: A moving target.* Transit New Zealand, Auckland, New Zealand.

Kirshen, P. H., Ruth, M., Anderson, W. & Lakshmanan, T. R. (2004). *Infrastructure Systems, Services, and Climate Change: Integrated Impacts and Response Strategies for the Boston Metropolitan Area (CLIMB Final Report).* U.S. Environmental Protection Agency, Washington, D.C., August 13.

Lockwood, S. (2008). Operational Responses to Climate Change. In: *The Potential Impacts of Climate Change on U.S. Transportation.* The National Research Council, National Academy of Science, Transportation Research Board and Department of Earth and Life Sciences, Washington, D.C.

Lonergan, S., Difrancesco, R. & Woo, M. K. (1993). Climate Change and Transportation in Northern Canada: An Integrated Impact Assessment. *Climate Change, 24*, 4, August, pages 33 1-352.

Marbek Resource Consultants Ltd. (2003). *Impacts of Climate Change on Transportation in Canada – Transport Canada Canmore Workshop – Final Report.* Transport Canada, March.

Marine Board, (1987). *Responding to Changes in Sea Level – Engineering Implications.* Washington, D.C, National Academy Press, page 148.

McRobert, J., Houghton, N. & Styles, E. (2003). Salinity impacts and roads – towards a dialogue between climate change and groundwater modelers. *Road and Transport Research.* Volume 12, Number 3, September, pages 45-60.

Meyer, M. D. (2008). Design Standards for U.S. Transportation Infrastructure: The Implications of Climate Change. In: *The Potential Impacts of Climate Change on U.S. Transportation.* The National Research Council, National Academy of Science, Transportation Research Board and Department of Earth and Life Sciences, Washington, D.C.

Mills, B. & Andrey, J. (2002). Climate Change and Transportation: Potential Interactions and Impacts. In: *The Potential Impacts of Climate Change on Transportation Workshop, October 1-2, 2002.* Washington, D.C.

Ministry of Housing, Spatial Planning, and the Environment, (2001). *Third Netherlands' National Communication on Climate Change Policies: Prepared for the Conference of the Parties under the Framework Convention on Climate Change.* Directorate Climate Change and Industry/IPC 650, Climate Change and Acidification Department, October, pages 79-80.

National Assessment Synthesis Team, (2000). *Climate Change Impacts on the United States: The Potential Consequences of Climate Variability and Change.* U.S. Global Change Research Program, Cambridge University Press, New York.

Ning, Z. H., Turner, R. E., Doyle, T. & Abdollahi, K. (2003). *Integrated Assessment of the Climate Change Impacts on the Gulf Coast Region: Findings of the Gulf Coast Regional Assessment.* Report for the U.S. Global Change Research Program, GCRCC, Baton Rouge, Louisiana. June

Norwell, G. (2004). *Impact of Climate Change on Road Infrastructure.* Austroads Inc., Sydney, Australia.

OFCM, (2002). *Weather Information for Surface Transportation: National Needs Assessment Report.* Office of the Federal Coordinator for Meteorological Services and Supporting Research, 302 pp., http://www.ofcm.gov/wist_report/wist-report.htm.

Office of Naval Research, Naval Ice Center, Oceanographer of the Navy, and the Arctic Research Commission, (2001). *Naval Operations in an Ice-Free Arctic Symposium, 17-18 April 2001,* Washington, D.C. Office of Naval Research. Department of the Navy, Arlington, Virginia.

Olsen, J. R., Zepp, L. J. & Dager, C. (2005). *Climate Impacts on Inland Navigation.* Conference Proceeding: Environmental and Water Resources Insitute of ASCE: Impacts of Global Climate Change.

Peterson, T., McGuirk, M., Houston, T., Horvitz, A. & Wehner, M. (2008). Climate Variability and Change with Implications for Transportation. In: *The Potential Impacts of Climate Change on U.S. Transportation.* The National Research Council, National Academy of Science, Transportation Research Board and Department of Earth and Life Sciences, Washington, D.C.

Pisano, P., Goodwin, L. & Stern, A. (2002). Surface Transportation Safety and Operations: The Impacts Of Weather Within The Context Of Climate Change. In: *The Potential Impacts of Climate Change on Transportation Workshop, October 1-2, 2002.* Center for Climate Change and Environmental Forecasting, U.S. Department of Transportation, Washington, D.C.

Potter, J. R. (2002). Workshop Summary. In: *The Potential Impacts of Climate Change on Transportation Workshop, October 1-2, 2002.* Center for Climate Change and Environmental Forecasting, U.S. Department of Transportation, Washington, D.C.

Potter, J. R. & Savonis, M. J. (2003). Transportation in an Age of Climate Change: What are the Research Priorities? *TR News,* 227, July-August, 26-31.

Quinn, F. H. (2002). The Potential Impacts Of Climate Change On Great Lakes Transportation. In: *The Potential Impacts of Climate Change on Transportation Workshop, October 1-2, 2002.* Center for Climate Change and Environmental Forecasting, U.S. Department of Transportation, Washington, D.C.

Rosenzweig, C. & Solecki, W. D. (Eds.), (2001). *Climate Change and Global City: The Potential Consequences of Climate Variability and Change – Metro East Coast.* Report for the U.S. Global Change Research Program, National Assessment of the Potential Consequences of Climate Variability and Change for the United States, Columbia Earth Institute, New York, 224 pp.

Rossetti, M. A. (2002). Potential Impacts Of Climate Change On Railroads. In: *The Potential Impacts of Climate Change on Transportation Workshop, October 1-2, 2002.* Center for Climate Change and Environmental Forecasting, U.S. Department of Transportation, Washington, D.C.

Rossiter, L. (2004). *Climate Change Impacts on the State Highway Network: Transit New Zealand's Position.* Transit New Zealand, Auckland, New Zealand, July.

Ruth, M. (Ed.), (2006). *Smart Growth and Climate Change: Regional Development, Infrastructure and Adaptation*. Edward Elgar Publishing, Cheltenham, United Kingdom.

Sato, N. & Robeson, S. M. (2006). Climate Change and Its Potential Impact on Winter- Road Maintenance: Temporal Trends in Hazardous Temperature Days in the United States And Canada. In: *TRB 85th Annual Meeting*, January 22, 2006, Washington, D.C., page 13.

Smith, O. P. & Levasseur, G. (2002). Impacts Of Climate Change On Transportation Infrastructure in Alaska. In: *The Potential Impacts of Climate Change on Transportation Workshop, October 1-2, 2002.* Center for Climate Change and Environmental Forecasting, U.S. Department of Transportation, Washington, D.C.

Smith, O. (2006). *Trends in Transportation Maintenance Related to Climate Change*. North Carolina State University Center for Transportation and the Environment 2nd Workshop on Impacts of Global Climate Change On Hydraulics and Hydrology and Transportation, March 27, 2006, Washington, D.C.

Smyth, A., Montgomery, W. I., Favis-Mortlock, D. & Allen, S. (Eds.), (2002). *Implications of Climate Change for Northern Ireland: Informing Strategy Development.* Scotland and Northern Ireland Forum for Environmental Research (SNIFFER).

Soo Hoo, W. K., Sumitani, M. & Cohen, S. (2005). *Climate Change Will Impact the Seattle Department of Transportation*. City of Seattle, Office of City Auditor, Seattle, Washington, August 9.

Sousounis, P. J., & Bisanz, J. M. (Eds.), (2000). *Preparing for a Changing Climate: The Potential Consequences of Climate Variability and Change Great Lakes Overview.* Report for the U.S. Global Change Research Program. October.

Suarez, P., Anderson, W., Mahal, V. & Lakshmanan, T. R. (2005). *Impacts of Flooding and Climate Change on Urban Transportation: A Systemwide Performance Assessment of The Boston Metro Area.* Elsevier Transportation Research Part D 10. Pages 231-244.

Three Regions Climate Change Group, (2005). *Consultation Document – Adapting to Climate Change: A Checklist for Development, Guidance on Designing Developments in a Changing Climate*. London, United Kingdom, February.

Titus, J. (2002). Does Sea Level Rise Matter To Transportation Along The Atlantic Coast? In: *The Potential Impacts of Climate Change on Transportation Workshop, October 1-2, 2002.* Washington, D.C.

Trilling, D. R. (2002). Notes on Transportation into the Year 2025. In: *The Potential Impacts of Climate Change on Transportation Workshop, October 1-2, 2002.* Center for Climate Change and Environmental Forecasting, U.S. Department of Transportation, Washington, D.C.

U.S. Arctic Research Commission Permafrost Task Force, (2003). *Climate Change, Permafrost, and Impacts on Civil Infrastructure.* Special Report 01-03, U.S. Arctic Research Commission, Arlington, Virginia.

U.S. Climate Change Science Program (CCSP), (2003). *Vision for the Program and Highlights of the Scientific Strategic Plan*. Climate Change Science Program and the Subcommittee on Global Change Research, Washington, D.C. July.

U.S. Department of State, (2002). Chapter 6: Impacts and Adaptation. In: *U.S. Climate Action Report 2002*. Washington, D.C., May.

U.S. Federal Highway Administration (FHWA), (1998). *Transportation and Global Climate Change: A Review and Analysis of the Literature.* Office of Environment and Planning,

Federal Highway Administration, U.S. Department of Transportation, Washington, D.C. June.

Wagner, F. H. (Ed. & Principal Author), (2003). *Rocky Mountain/Great Basin Regional Climate-Change Assessment.* Report for the U.S. Global Change Research Program. Utah State University, Logan, Utah: IV + 240 pages.

Warren, F., Barrow, E., Schawartz, R., Andrey, J., Mills, B. & Riedel, D. (2004). *Climate Change Impacts and Adaptation: A Canadian Perspective*, [Lemmen, D.S. & F.J. Warren, eds.]. Climate Change Impacts and Adaptation Directorate Natural Resources Canada. Ottawa, Ontario.

Weller, G., Anderson, P. & Wang, B. (1999). *Preparing for a Changing Climate – The Potential Consequences of Climate Variability and Change: Alaska.* Report for the U.S. Global Change Research Program, Center for Global Change and Arctic System Research, Fairbanks, Alaska, December.

Wilkinson, R. (2002). *Preparing for a Changing Climate – The Potential Consequences of Climate Variability and Change.* A Report of the California Regional Assessment Group. Report for the U.S. Global Change Research Program, September.

Wooler, S. (2004). *The Changing Climate: Impact on the Department for Transport.* Department for Transport, London, United Kingdom.

Wright, F., Duchesne, C., Nixon, F. M. & Cote, M. (2001). *Ground Thermal Modeling in Support of Terrain Evaluation and Route Selection in the Mackenzie River Valley – Summary Report.* Geological Survey of Canada, August.

Wu, S. Y., Yarnal, B. & Fisher, A. (2002). Vulnerability of Coastal Communities to Sea Level Rise: A Case Study of Cape May County, New Jersey, U.S.A. *Climate Research, 22*, 255-270.

Zimmerman, R. (2002a). Global Climate Change and Transportation Infrastructure: Lessons from the New York Area. In: *The Potential Impacts of Climate Change on Transportation Workshop, October 1-2, 2002.* Center for Climate Change and Environmental Forecasting, U.S. Department of Transportation, Washington, D.C.

Zimmerman, R. (2002b). Global Warming, Infrastructure, and Land Use in the Metropolitan New York Area: Prevention and Response. In: *Global Climate Change and Transportation: Coming to Terms.* Eno Transportation Foundation, pages 55-63.

1.6.2. Background Sources

ABP Marine Environmental Research Ltd. (2004). *Climate Change 2003 Port Template: Findings from the First Assessment.* Report Number R.1061, Associated British Ports – Head Office, March.

Easterling, W., III, Hurd, B. H. & Smith, J. B. (2004). *Coping with Global Climate Change: The Role of Adaptation in the United States.* Pew Center on Global Climate Change. June.

Entek U. K. Limited, (2000). *The Potential Impacts of Climate Change on the East Midlands – Summary Report.* East Midlands Sustainable Development Round Table, Warwickshire, England, July.

Entek U. K. Limited, (2004). *The Potential Impacts of Climate Change in the West Midlands.* Sustainability West Midlands, United Kingdom, January.

Land Use Consultants, CAG Consultants, and SQW Ltd. (2002). *Living with Climate Change in the East of England Summary Report.* East of England Sustainable Development Round Table, England, United Kingdom, September.

Struck, D. (2006). Melting Arctic Makes Way for Man: Researchers Aboard Icebreaker Say Shipping Could Add to Risks for Ecosystem. *Washington Post*, November 5, page A01.

In: Impacts of Climate Change on Transportation and... ISBN: 978-1-60741-424-7
Editor: Iason Pavlopoulos

Chapter 2

WHY STUDY THE GULF COAST?

Lead Authors: Virginia R. Burkett, Robert C. Hyman, Ron Hagelman, Stephen B. Hartley and Matthew Sheppard

Contributing Authors: Thomas W. Doyle, Daniel M. Beagan, Alan Meyers, David T. Hunt, Michael K. Maynard, Russell H. Henk, Edward J. Seymour, Leslie E. Olson, Joanne R. Potter and Nanda N. Srinivasan

2.1. OVERVIEW OF THE STUDY REGION

2.1.1. Regional and National Significance

The Phase I Study area includes 48 contiguous coastal counties in 4 States, running from the Galveston Bay region in Texas to the Mobile Bay region in Alabama. This region is home to almost 10 million people living in a range of urban and rural settings, contains some of the Nation's most critical transportation infrastructure, and is highly vulnerable to sea level rise and storm impacts.

This area has little topographic relief but is heavily populated. Given its low elevation and the regional climate, the area is particularly vulnerable to flooding and storm surges that accompany hurricanes and tropical storms – almost half of the Nation's repetitive flood damage claims are paid to homeowners and businesses in this region. These effects may be exacerbated by global sea level rise and local land subsidence.

In addition, the central Gulf Coast's transportation modes are both unique and economically significant. The study area contains transportation infrastructure that is vital to the movement of passengers and a variety of goods domestically and internationally. Ports and pipeline infrastructure represent perhaps the most conspicuous transport modes in the region. Some of the Nation's most important ports, such as the ports of Houston- Galveston,

South Louisiana, and New Orleans are found in the study area. The Port of South Louisiana, for example, is a critical agricultural export center. Agricultural producers in the Midwest depend on the continued operation of this port to ship their products for international sale. Likewise, disruptions in the functioning of pipelines and fuel production and shipping facilities in the study region have broad domestic and international impacts. Roughly two-thirds of all U.S. oil imports are transported through this region, and pipelines traversing the region transport over 90 percent of domestic Outer Continental Shelf oil and gas.

The importance of these marine facilities and waterways to the study area, and to the Nation as a whole, is difficult to overstate. These are vital National resources, providing essential transportation and economic services. While some of these functions could be considered "replaceable" by facilities and waterways elsewhere, many of them – by virtue of geography, connections to particular industries and markets, historic investments, or other factors – represent unique and largely irreplaceable assets.

In addition to ports and pipelines, the study region contains critical air, rail, highway, and transit infrastructure. Passenger and freight mobility depend both on the functioning of each mode and the connectivity of the modes in an integrated transport network. The efficacy of evacuation during storms is an important determinant of the safety and wellbeing of the region's population. The region sits at the center of transcontinental trucking and rail routes and contains one of only four major points in the United States where railcars are exchanged between the dominant eastern and western railroads.

The region is experiencing a population shift from rural to urban and suburban areas. Much of the population inhabiting the study area, as well as the transportation infrastructure supporting them, reside in low-lying areas vulnerable to inundation and flooding. In addition, parts of the population face challenges that may make it more difficult for them to adapt to the conditions imposed by a changing climate, such as poverty, lack of mobility, and isolation. Some of Louisiana's rural counties and the urban centers of New Orleans and Mobile County, AL, have particularly high proportions of vulnerable citizens.

2.1.2. Study Area Boundaries

This initial study focuses on the central portion of the low-lying Gulf of Mexico coastal zone. The study region extends from Mobile, AL, to Galveston, TX, as shown in figure 2.1. The study area encompasses all coastal counties and parishes along that stretch of the Gulf of Mexico as well as their adjacent inland counties (figure 2.2). In addition, the boundaries of the study area were extended so that all portions of Metropolitan Planning Organizations (MPOs) within a two-county swath of coastline would be included (figure 2.3). Table 2.1 provides the resulting list of counties and parishes included in the study area.

2.1.3. Structure of This Chapter

The following sections provide a more detailed overview of the central Gulf Coast study region, as follows:

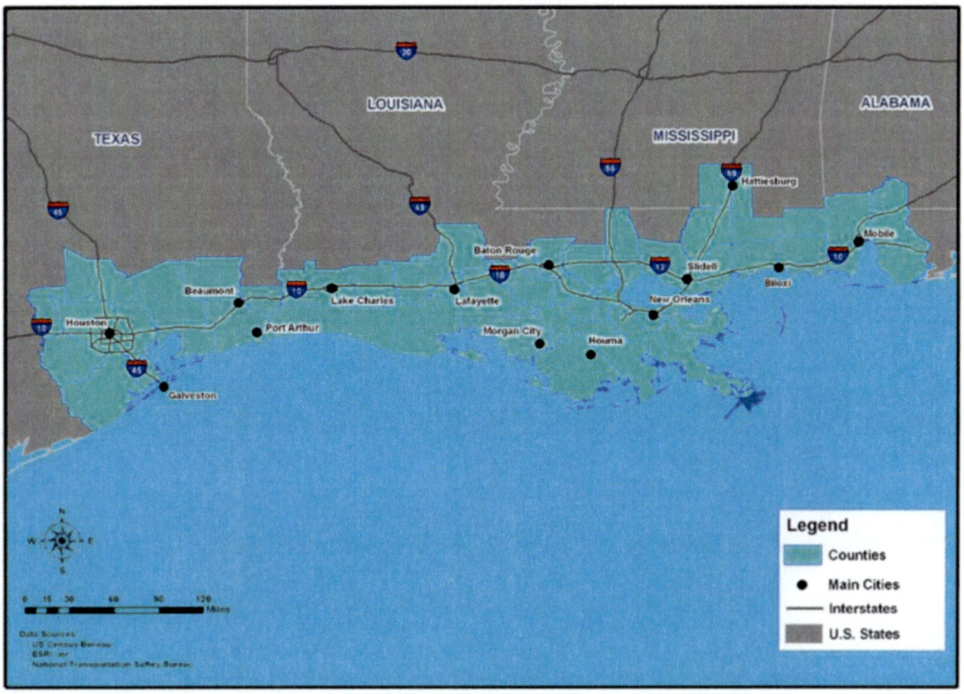

Source: U.S. Census Bureau; ESRI, Inc.; National Transportation Safety Bureau.

Figure 2.1. Map of the study area, which extends from Mobile, AL, to Houston/Galveston, TX

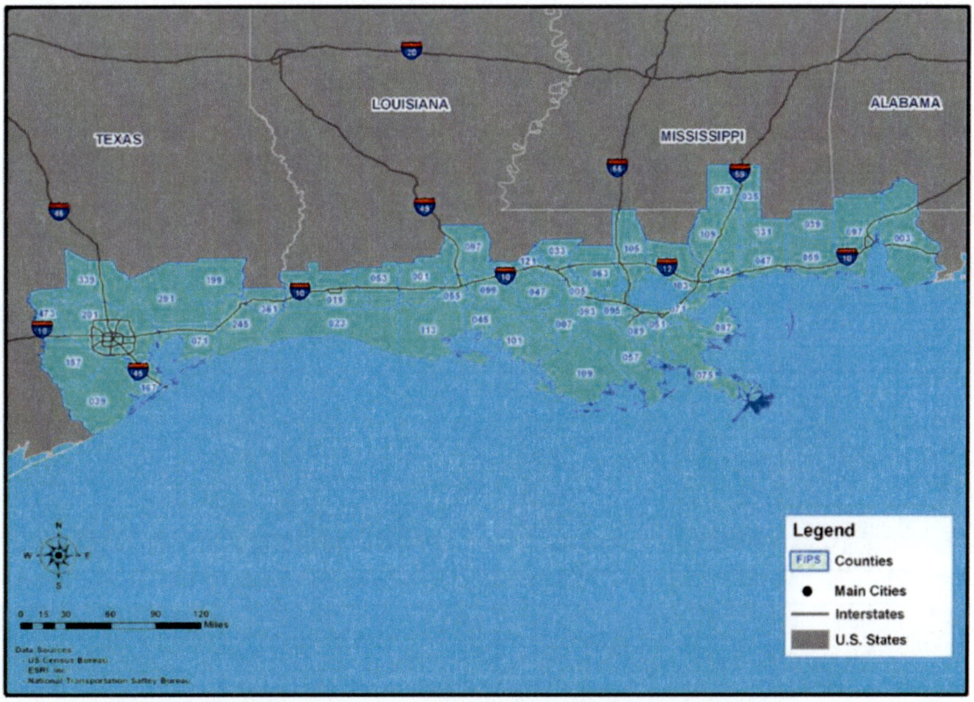

Source: U.S. Census Bureau; ESRI, Inc.; National Transportation Safety Bureau.

Figure 2.2. Study area counties and Federal Information Processing Standard (FIPS) codes

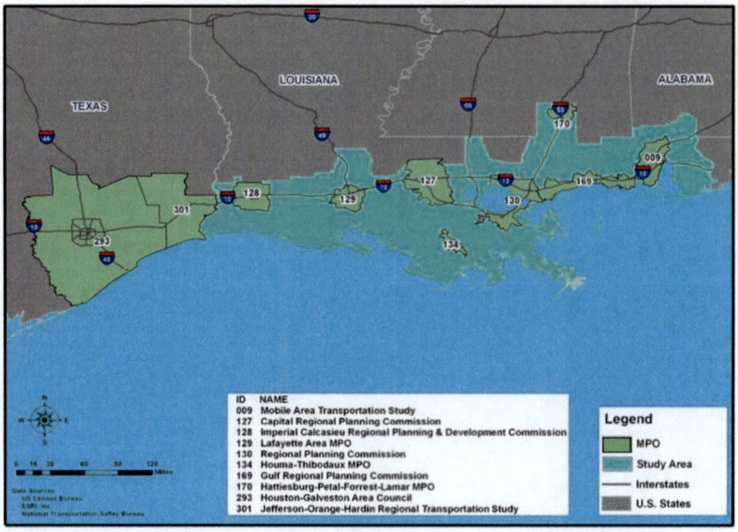

Source: U.S. Census Bureau; ESRI, Inc.; National Transportation Safety Bureau.

Figure 2.3. Metropolitan planning organizations (MPO) in the study area

Table 2.1. Study Area Counties and Federal Information Processing Standard (FIPS) Codes

County name	State	FIPS	Name	State	FIPS[1]
Baldwin	Alabama	003	St. Tammany	Louisiana	103
Mobile	Alabama	097	Tangipahoa	Louisiana	105
Acadia	Louisiana	001	Terrebonne	Louisiana	109
Ascension	Louisiana	005	Vermilion	Louisiana	113
Assumption	Louisiana	007	West Baton Rouge	Louisiana	121
Calcasieu	Louisiana	019	Forrest	Mississippi	035
Cameron	Louisiana	023	George	Mississippi	039
East Baton Rouge	Louisiana	033	Hancock	Mississippi	045
Iberia	Louisiana	045	Harrison	Mississippi	047
Iberville	Louisiana	047	Jackson	Mississippi	059
Jefferson	Louisiana	051	Lamar	Mississippi	073
Jefferson Davis	Louisiana	053	Pearl River	Mississippi	109
Lafayette	Louisiana	055	Stone	Mississippi	131
Lafourche	Louisiana	057	Brazoria	Texas	039
Livinston	Louisiana	063	Chambers	Texas	071
Orleans	Louisiana	071	Fort Bend	Texas	157
Plaquemines	Louisiana	075	Galveston	Texas	167
St. Bernard	Louisiana	087	Hardin	Texas	199
St. Charles	Louisiana	089	Harris	Texas	201
St. James	Louisiana	093	Jefferson	Texas	245
St. John the Baptist	Louisiana	095	Liberty	Texas	291
St. Landry	Louisiana	097	Montgomery	Texas	339
St. Martin	Louisiana	099	Orange	Texas	361
St. Mary	Louisiana	101	Waller	Texas	473

[1]The FIPS county code is a number that uniquely identifies each county in the United States.

- Section 2.2 describes the transportation system in the study area;
- Section 2.3 describes the physical setting and natural environment of the study area, including factors that make it more susceptible to climate change impacts; and
- Section 2.4 discusses the social and economic setting, including factors that make portions of the population more vulnerable to climate impacts.

2.2. THE TRANSPORTATION SYSTEM IN THE GULF COAST REGION

The transportation network of the Gulf Coast study area comprises a complex system of multiple modes that enables both people and goods to move throughout the region and supports national and international transport. While roadways are the backbone of the region's transportation system, the viability of the network as a whole depends on reliable service connections across all modes. Section 2.2.1 provides an introduction to passenger travel, freight transport, intermodal facilities, and emergency management in the Gulf Coast study area, while Section 2.2.2 provides an in-depth look at each of the transportation modes present in the region. Climate impacts to this transportation system are then discussed in Section 4.0. The transportation facility location information cited and shown in maps throughout the report is from the National Transportation Atlas Database (BTS, 2004).

2.2.1. Overview of the Intermodal Transportation System in the Gulf Coast Region

Passenger Travel

Passenger travel in the Gulf Coast study area is accommodated by a variety of modes, including highway, transit, rail, and aviation. Roads are the most geographically extensive system in the study area, and autos traveling on the highways serve as the principal mode for passenger travel. Some of those highways, particularly I-10/I-12, serve substantial national travel that is passing through the study area. The 27,000 km (17,000 mi) of major highways within the study area comprise about 2 percent of the Nation's major highways. These highways carry 134 billion vehicle km of travel (83.5 billion vehicle mi of travel) annually.

Public transit provides an important function – particularly in urban areas – by carrying passengers more efficiently (in densely populated areas) than they could be carried in autos and thus relieving congestion. Further, transit provides essential accessibility to those passengers who do not own or cannot rely on autos for transportation. Lower-income workers rely heavily on city and intercity bus services for basic needs: getting to and from work, transporting children to school or childcare services, and shopping. The majority of transit ridership in the study area is carried by scheduled bus services. Other transit services available include light rail, ferries, and unscheduled paratransit vans and minibuses.

Intercity passenger rail services are provided by the National Railroad Passenger Corporation (Amtrak), which operates three long-distance routes connecting the study area to other parts of the nation. Passenger rail services are not extensive, but they do supply an alternative mode of transportation and are important to certain segments of the population.

Airports are critical in connecting local, regional, and national economies, as well as the global economy. Several major airports serve the larger cities of the study area; in addition, numerous airports outside of the major metropolitan markets serve smaller municipal markets, and many provide general aviation services. Smaller regional airports are critical infrastructure elements as they are often used for the movement of emergency medical supplies and patients.

Freight Transport

The Gulf Coast Study area is a critical crossroads for the Nation's freight network, with marine, rail, pipeline, trucking, and air cargo all represented. A large portion of the Nation's oil and gas supply originates in the study area, either as domestic production or imports. New Orleans provides the ocean gateway for much of the U.S. interior's agricultural production and is a major interchange point for freight railroads. Products are shipped from the study area to points throughout the United States. Figure 2.4 depicts Federal Highway Administration (FHWA) Freight Analysis Framework data describing combined domestic truck flows originating in Louisiana (FHWA, 2004).

The pipeline network along the Gulf of Mexico coast is vital to the supply and distribution of energy for national use everywhere east of the Rocky Mountains. Approximately one- half of all the natural gas used in the United States passes through or by the Henry Hub gas distribution point in Louisiana. The pipelines originating in this region provide a low-cost, efficient way to move oil and gas long distances throughout the United States.

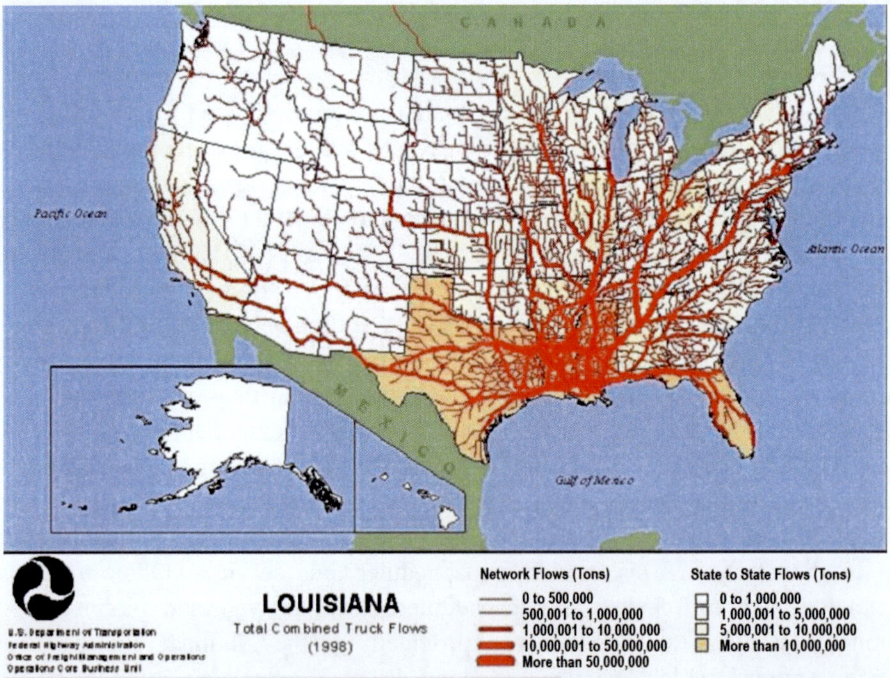

Source: U.S. Department of Transportation Federal Highway Administration, Freight Management and Operations, Office of Operations.

Figure 2.4. Combined truck flows shipped domestically from Louisiana, 1998

The study area also is home to the largest concentration of public and private freight handling ports in the United States, measured on a tonnage basis. These facilities handle a huge share – around 40 percent – of the Nation's waterborne tonnage. The study area also hosts the Nation's leading and third leading inland waterway systems (the Mississippi River and the Gulf Intracoastal Waterway) based on tonnage. The inland waterways traversing this region provide 20 States with access to the Gulf of Mexico, as shown in figure 2.5.

The rail links in the study area provide crucial connectivity to the national rail network for ports in the region and, via intermodal facilities, the major highway freight corridors. Figure 2.6 shows the network of major freight railroads nationwide, illustrating an obvious divide between the eastern railroads and the western railroads along the Mississippi River. New Orleans is one of four major gateways nationwide where the dominant eastern and western railroads interchange transcontinental shipments (Chicago, St. Louis, and Memphis are the others). At New Orleans, for example, CSX interchanges over 1,000 cars per day with the western railroads. A disruption to any of the four major gateways has implications for the entire U.S. rail network.

Intermodal Facilities

Intermodal facilities are critical infrastructure facilities that enable the transfer of goods and passengers between different transport modes. These facilities are critical to transportation logistics processes and provide a key link in industrial and public sector supply chains.

Source: U.S. Department of Transportation.

Figure 2.5. Navigable inland waterways impacting the study area, shown as named waterways

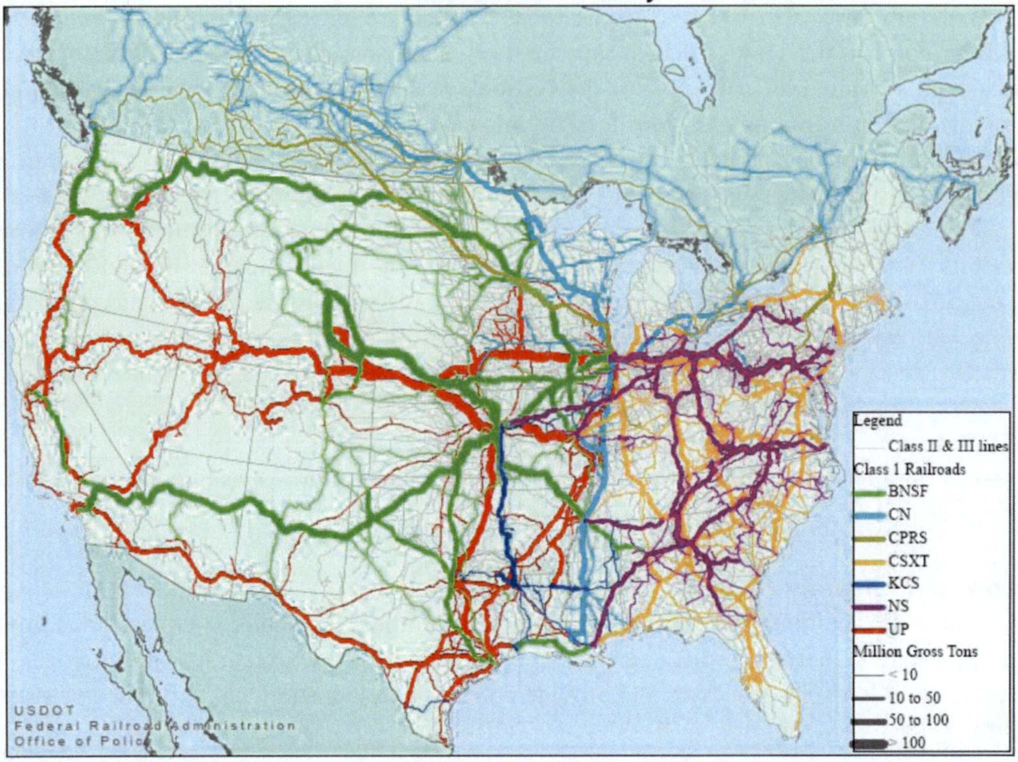

Source: Federal Railroad Administration Office of Policy, U.S. Department of Transportation.

Figure 2.6. National network of Class I railroads

There are more than 100 intermodal facilities in the study area. Figure 2.7 shows the locations of these facilities in the study area, with coded symbols for the various mode combinations handled at each. Unsurprisingly, many of these facilities are clustered in the port and rail hubs of New Orleans and Houston.

Emergency Management

Interstates and arterial roadways provide the majority of the transportation infrastructure for emergency management and evacuation along the Gulf Coast. While public transportation facilities exist, they typically rely on the highway system; there are no large scale transit systems operating on separate right-of-ways. This substantial reliance on a single mode of transportation represents a risk if the highway infrastructure is damaged or made inaccessible during an emergency.

Existing infrastructure may be able to handle local evacuations and diversions such as in the case of spilled hazardous material from a tanker truck or risk from a point source event – like a ruptured pipeline. However, network-wide roadway capacity is not designed nor built to handle large-scale evacuations or emergencies. Further, evacuation protocols require time-sensitive actions that existing roadway infrastructure cannot accommodate.

The limitations of the existing infrastructure to accommodate a major evacuation during a broad-scale emergency were dramatically illustrated during the 2005 hurricane season. As Hurricane Rita demonstrated, evacuating a substantial portion of the population from a major metropolitan area is problematic and, in many ways, difficult to accomplish in a timely and orderly fashion. The "normal" condition of the already capacity-constrained transportation infrastructure does not allow for a major ramp-up of evacuation capabilities during daylight hours in major urban areas.

Managing the transportation infrastructure and leveraging its available capacity is highly dependent upon: (1) means for gathering real-time traffic information and (2) robust and integrated communication systems that are consistent across regional jurisdictional boundaries. In this regard, the state of practice within the region varies considerably. Advanced transportation management systems such as the TranStar Traffic Management Center in Houston and a similar array of intelligent transportation system (ITS) technologies and a traffic control center in New Orleans represent relatively new and effective advancements in obtaining accurate real-time data upon which to base transportation system management decisions. On the other hand, the interagency and interjurisdictional communication systems in the Gulf Coast region are sometimes independent from one another, with multiple radio systems in use by emergency responders in each State.

2.2.2. Modal Characteristics

Highways

Highway Network and Usage

Highways provide the overwhelming majority of the public transportation infrastructure in the Gulf Coast study area. There are 28,154 centerline km (17,494 centerline mi) of highway in the study region (table 2.2, figure 2.8) (FHWA, 2005). Highway facilities in the Gulf Coast study area are primarily owned and operated by the state departments of transportation (DOT). Roads are classified as:

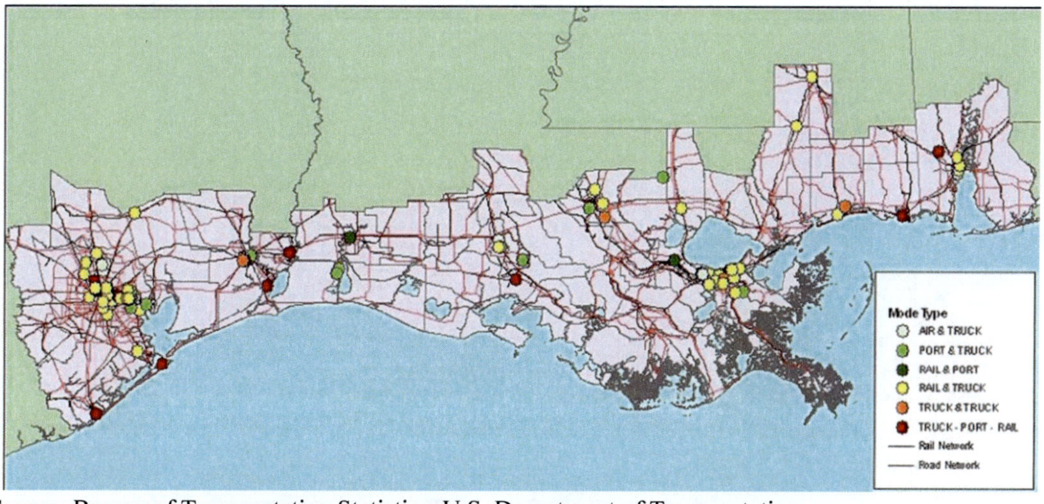

Source: Bureau of Transportation Statistics, U.S. Department of Transportation.

Figure 2.7. Intermodal facilities in the study area

- **Interstates –** Highways that are designated as part of the Dwight D. Eisenhower National System of Interstate and Defense Highways;
- **Arterials –** Highways that provide longer through travel between major trip generators (larger cities, recreational areas, etc.);
- **Collectors –** Roads that collect traffic from the local roads and also connect smaller cities and towns with each other and to the arterials; and
- **Local –** Roads that provide access to private property or low-volume public facilities.

Local roads serve mainly a land-access function, carry little of the demand for transportation compared to the Interstates and the arterial roadways, and are not included as part of the highways studied in this report.[1] State DOTs administer 100 percent of the centerline miles on interstate highways, 60 percent of the centerline miles on arterial highways, and 50 percent of the centerline miles on collector highways.

(Source: Cambridge Systematics from 2004 Highway Performance Monitoring System database for Gulf Coast study supplied by the Bureau of Transportation Statistics)

Table 2.2. Gulf Coast study area centerline miles of highway, by classification and ownership

	State	County	Municipal	Other	Total
Interstate	1,096	0	0	0	1,096
Arterials	4,484	794	2,268	105	7,651
Collector	4,390	1,776	2,016	35	8,747
Total	**9,970**	**2,570**	**4,284**	**140**	**17,494**

Source: Cambridge Systematics from 2004 Highway Performance Monitoring System database for Gulf Coast study supplied by the Bureau of Transportation Statistics)

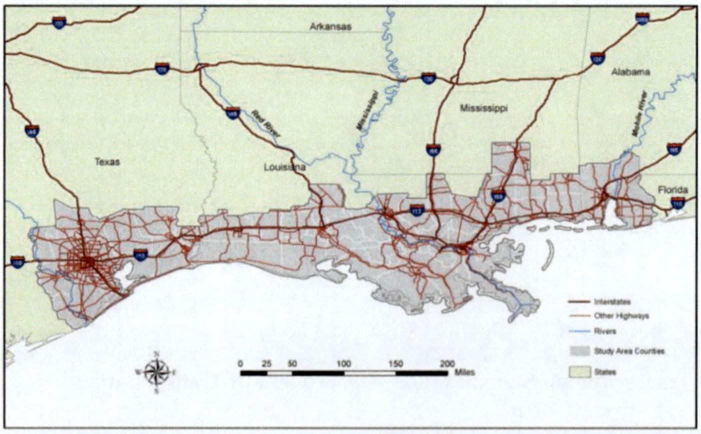

Source: Cambridge Systematics analysis of U.S. Department of Transportation data.

Figure 2.8. Highways in the study area

The volumes on the interstate, arterial, and collector roads are primarily on the State-owned highways, to an even greater extent than that of centerline miles. Of the 83.5 billion annual vehicle miles of travel (VMT) in the study area, 63.7 billion (76.3 percent) of that travel is on State-owned nonlocal roads (FHWA, 2005).

State-owned nonlocal roads carry an even larger share of truck volumes. As shown in figure 2.9, 92 percent of the truck VMT is on State roads. Additionally, while truck VMT is 7.5 percent of the total VMT, which compares closely to national truck percentages of volumes, trucks represent 9.1 percent (5.7 billion of 63.7 billion) of traffic on all State-owned roads and 10 percent of the VMT (2.4 billion of 24.4 billion) of all traffic on State-owned interstate highways (FHWA, 2005).

Intermodal Connectors

Access to intermodal facilities is most often provided by highways. Because this access function is critical to the viability of other modes, States have been given the authority to designate major intermodal passenger and freight terminals and the road connectors between these terminals and the National Highway System (NHS) as NHS intermodal connectors. The NHS intermodal connectors for the Gulf Coast study area were identified from an FHWA database (FHWA, 2006). The official listing of the NHS Intermodal Terminals and Connectors includes the following:

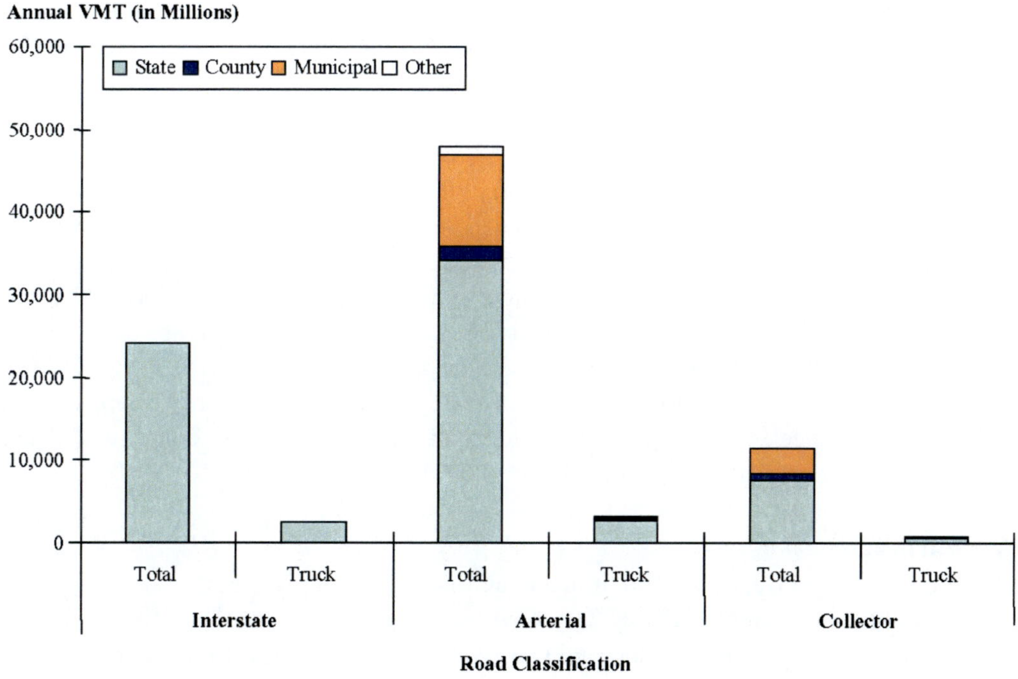

Source: Cambridge Systematics, from 2004 Highway Performance Monitoring System database for Gulf Coast study supplied by the Bureau of Transportation Statistics, U.S. Department of Transportation

Figure 2.9. Total and truck annual vehicle miles of travel (VMT) on nonlocal roads, 2003

- Ferries/Ports:
    - Five ferry terminals served by 25 intermodal connector segments totaling 478.2 km (297.1 mi); and
    - Twenty-three ports served by 54 intermodal connector segments totaling 380.9 km (236.7 mi).
- Bus/Transit:
    - One intercity bus terminal served by 12 intermodal connector segments totaling 26.7 km (16.6 mi);
    - Two multipurpose passenger terminals served by nine intermodal connector segments totaling 13.0 km (8.1 mi); and
    - Eight public transit stations served by 14 intermodal connector segments totaling 17.7 km (11.0 mi).
- Railroads:
    - Two Amtrak stations (Houston and New Orleans) served by four intermodal connector segments totaling 3.9 km (2.4 mi); and
    - Thirteen rail freight terminals served by 23 intermodal connector segments totaling 49.4 km (30.7 mi).
- Pipelines:
    - Four pipeline terminals served by seven intermodal connector segments totaling 30.7 km (19.1 mi).
- Airports:
    - Six airports served by 24 intermodal connector segments totaling 44.7 km (27.8 mi).

Bridges

Highway bridges are structures that carry the highway over a depression or an obstruction, such as water, a highway, or railway. As shown in figure 2.10 there are almost 8,200 bridges that serve nonlocal roads in the study area. The overwhelming majority, 80 percent, of those bridges are owned by the States. Of those State bridges, almost 80 percent serve interstate or arterial highways. Seventy-five percent of the bridges in the study area pass over water, making them susceptible to scour of their piers by water runoff (FHWA, 2001).

Eighty-one percent of the bridge structures are concrete compared to 15 percent of the bridges which are steel, and 80 percent of the road surface on bridge decks are concrete compared to 16 percent that are asphalt (FHWA, 2001).

Other Facilities

While roads and bridges are the primary facilities that comprise the highway system in the Gulf Coast Study area, highway agencies own and operate many ancillary facilities necessary to operate and maintain the highway system. These facilities include maintenance buildings and facilities, truck weight and inspection stations, rest areas, toll booths, traffic controls/signs, luminaries, fences, guardrails, traffic monitoring equipment, etc.

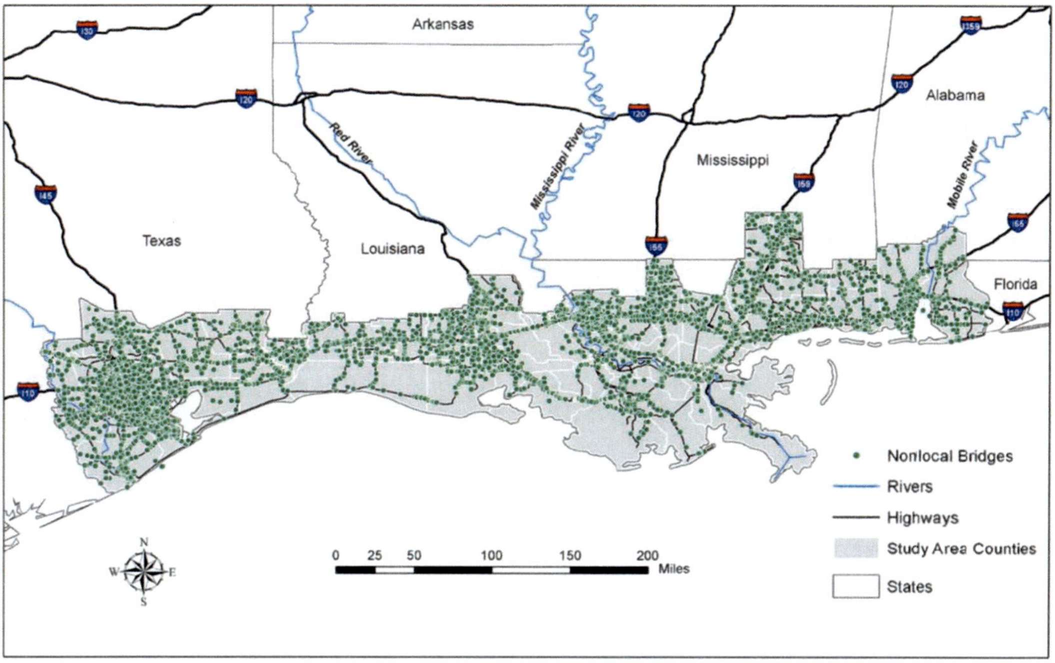

Source: Cambridge Systematics analysis of U.S. Department of Transportation data.

Figure 2.10. Nonlocal bridges in the study area (National Bridge Inventory [NBI] latitude and longitude location)

Transit

The American Public Transportation Association (APTA) lists over 136 public transit providers that serve the Gulf Coast study area (APTA, 2005). Most of those providers offer transportation as a social service to elderly, disabled, or low-income passengers. These transit providers include 13 major transit agencies that receive funding from the Federal Transit Administration (FTA) and are included in the National Transit Database (NTD) (FTA, 2005). Statistical information on transit services in the study region have been drawn from this database.

By far the largest transit networks in the study area are found in Houston and New Orleans. As an illustration, in 2003 the NTD listed Houston as having almost $88 million in citywide transit revenues and New Orleans with almost $35 million –while no other city in the study area topped $4 million.

Fixed Guideway (Light Rail)

There are three transit agencies that operate fixed guideway rail service in the Gulf study Area. Fixed guideway rail service carries passengers in vehicles moving on fixed light rails. The service operated by the Regional Transit Authority (RTA) in New Orleans and Metro in Houston consists of street cars operated by overhead power lines, over 47 km (29 mi) and 27 km (17 mi) of routes, respectively. The service operated by Island Transit in Galveston consists of heritage streetcars powered by diesel and operated on rails, on 29 km (18 mi) of route. These light-rail services account for a relatively small portion of total transit passengers in the study area: the New Orleans light rail service carried 8.9 million passengers in 2004,

Houston's carried 5.4 million, and Galveston's carried 40,000. By comparison, fixed-route bus services in the study area carried 10 times as many passengers in 2004 (FTA, 2005).

Fixed-Route Buses

Not including the ridership for HART/Hub City Transit (Hattiesburg, MS), Lake Charles Transit System, LA, and Saint Bernard Urban Rapid Transit which was not reported, fixed-route bus service in the Gulf Coast study area in 2004 carried 139 million passengers traveling 650 million passenger mi for an average trip length of 7.6 km (4.7 mi).

Table 2.3 shows data on equipment, service levels, and ridership for fixed-route bus service of the 13 major transit agencies in the Gulf Coast study area. Houston's Metro, New Orleans' RTA, and Jefferson Transit provide a small portion of this service as Bus Rapid Transit (BRT).[2] A total of 586 route km (364 route mi) of BRT are provided in the study area, of which 558 km (347 mi) are in the Houston area (FTA, 2005).

Paratransit

Transit agencies also provide special services to elderly, disabled, and other disadvantaged passengers. These paratransit services are offered in addition to accessible service on the fixed routes and are typically offered in smaller buses or vans with door-to-door service for passengers on a demand-responsive, flexible schedule. Twelve agencies in the study area offer paratransit service annually carrying 2.3 million passengers over 24 million passenger mi for an average trip of 17.1 km (10.6 mi) per trip. By far the largest paratransit provider in the study area is Houston's Metro, which accounts for 80 percent of the paratransit vehicles in the region, 64 percent of the passengers, and 69 percent of the passenger miles.

Other Facilities

In addition to transit vehicles and guideways, transit agencies may own other facilities to serve vehicles or riders. According to the 2004 NTD, within the Gulf Coast Study area 10 transit agencies own 86 terminals and transfer stations. Those terminals are most numerous in the light-rail systems operated by the New Orleans RTA and the Houston Metro. Also included are the terminals associated with passenger ferries within the study area.

Other facilities include vehicle maintenance facilities, of which the NTD lists six major facilities owned by six transit agencies. In addition, transit agencies also own numerous small passenger shelters and signs and other controls that are neither inventoried nor located in the NTD.

Rail

The Gulf Coast region has an extensive rail network, with east-west lines linking the southern United States, north-south lines paralleling the Mississippi River, and diagonal lines connecting the region to the northeastern and northwestern U.S. Six of the seven class I railroads in the United States serve the study region, along with several short lines.[3,4] These railroads support important regional industries, such as chemicals, paper, lumber, and international trade. The Gulf Coast region also serves as a critical junction for national freight movements, with New Orleans serving as a major gateway between the eastern and western railroads (most rail freight using New Orleans infrastructure is interchanging rather than originating or terminating in New Orleans).

Intercity passenger rail services are provided by the National Railroad Passenger Corporation (Amtrak). Amtrak operates nationwide routes through the region over track owned by the class I railroads. Passenger rail services are not extensive, but they do supply an alternative mode of transportation and are important to certain segments of the population.

Table 2.3. Equipment, Annual Service, and Passengers for Fixed-Route Bus Operations in the Study Area, 2004

Agency	Urban Area	Vehicles	Type of Vehicle[1]	Passengers		Revenue	
				(000)	(000) Miles	Miles (000)	Hours (000)
Metropolitan Transit Authority of Harris County, MTAHC (Metro)	Houston, Texas	1,434	210 articulated diesel buses, 1224 diesel buses	87,940	504,902	44,097	3,051
New Orleans Regional Transit Authority (RTA)	New Orleans, Louisiana	367	367 diesel buses	38,202	92,252	10,655	748
Capital Area Transit System (CATS)	Baton Rouge, Louisiana	74	5 CNG buses, 51 diesel buses, 18 diesel vans	4,805	15,749	3,172	159
Jefferson Transit (JeT)	New Orleans, Louisiana	63	59 diesel buses, 4 diesel vans	4,192	19,581	2,276	149
Lafayette Transit System (LTS)	Lafayette, Louisiana	22	22 diesel buses	1,156	4,856	536	41
Island Transit (IS)	Galveston, Texas	20	11 diesel buses, 9 diesel vans	940	1,454	555	45
The Wave Transit (The Wave)	Mobile, Alabama	31	26 diesel buses, 5 diesel vans	860	5,233	1,371	97
Beaumont Municipal Transit System	Beaumont, Texas	19	1 CNG bus, 18 diesel buses	662	2,858	729	52
Coast Transit Authority	Gulfport-Biloxi, Pascagoula, Mississippi	18	16 diesel buses, 2 LPG buses	534	2,672	770	61
Port Arthur Transit (PAT)	Port Arthur, Texas	11	10 diesel buses, 1 diesel van	125	935	235	14
Hattiesburg Area Readi Transit, Hub City Transit (HART)	Hattiesburg, Mississippi	5	3 gasoline buses, 2 diesel vans	N/A	N/A	N/A	N/A
Lake Charles Transit System (LCTS)	Lake Charles, Louisiana	8	8 diesel buses	N/A	N/A	N/A	N/A
Saint Bernard Urban Rapid Transit(SBURT)	New Orleans, Louisiana	9	8 diesel buses, 1 diesel van	N/A	N/A	N/A	N/A
Total		**2,081**		**139,416**	**650,492**	**64,396**	**4,417**

[1] CNG – Compressed Natural Gas

LPG – Liquified Propane Gase

Source: Cambridge Systematics from 2004 National Transit Database

Freight Rail

Six class I railroads operate in the study region: Burlington Northern Santa Fe (BNSF); Canadian National Railway (CN); CSX; Kansas City Southern Railroad (KCS); Norfolk Southern (NS); and Union Pacific (UP).

Figure 2.11 shows the annual density of traffic on the rail lines in the Gulf Coast study region (BTS, 2004). The most densely used lines (60 million to 99.9 million gross ton- miles per mile per year [mgtm/mi]) are short segments in Houston, TX, and New Orleans, LA. In the 40 to 59.9 mgtm/mi category is part of the UP line between Houston and New Orleans, some segments around Houston, and the CSX line east of Mobile. The 20 to 39.9 mgtm/mi range includes the remainder of the UP line into New Orleans, the CSX line between Mobile and New Orleans, the NS line into New Orleans, and several lines around Houston.

In addition to track infrastructure, there are 94 major freight facilities (owned and served by rail lines) in the study region, including rail yards, intermodal terminals, and transloading facilities.[5] These facilities originate and terminate rail traffic, reclassify inbound railcars to outbound trains for through traffic, and interchange railcars between railroads. They include facilities owned by the railroads and nonrailroad-owned facilities that depend on rail service, such as ports. Although these facilities can be found throughout the region, there are clearly two major hubs: New Orleans and Houston.

Table 2.4 provides a more complete description of the railroads operating in the Gulf Coast study area, showing the geographical service area and primary commodities hauled by each. A complete list of freight rail facilities in the study area is provided in appendix C.

Table 2.4. Freight railroads in the Gulf Coast study area

Railroad	Class	Service Area	Primary Commodities
Acadiana Railway	III	Crowley, LA, through Eunice and Opelousas, to Bunkie, LA.	Agricultural products, edible oils, and general freight.
Alabama and Gulf Coast Railway	III	Pensacola, FL, to Columbus, AL. Extensions to Mobile, AL, via Norfolk Southern trackage.	Paper industry: logs, wood-chips, chlorine, sodium chl-orate, hydrogen peroxide, rolled and boxed paper, and kaolin clay.
Burlington Northern Santa Fe Railway	I	Over 32,000 route miles in western U.S. Operate between Houston and New Orleans.	Coal, grains, intermodal, lumber, and chemicals.
Canadian National Railway (formerly Illinois Central Gulf)	I	Over 19,000 route miles in U.S. and Canada. Serves Mobile and New Orleans via north-south route.	Petroleum, chemicals, grain, fertilizers, coal, metals, minerals, forest products, intermodal, and automotive.
CSX Transportation	I	Over 22,000 route miles in eastern U.S. Operate between Florida and New Orleans along I-10 corridor.	Coal, chemicals, autos, min-erals, agricultural products, food, consumer goods, met-als, forest and paper produ-cts, and phosphates and fertilizer.

Table 2.4. (Continued)

Railroad	Class	Service Area	Primary Commodities
Kansas City Southern	I	Operates approximately 3,100 route miles in central and southeastern U.S. Serves New Orleans and Lake Charles, LA, Port Arthur and Galveston, TX, and Mexico.	Agriculture, minerals, general merchandise, intermodal, autos, and coal.
Lake Charles Port and HarborDistrict	Switc-hing	Owned by the Port. Switches traffic for Union Pacific.	Port traffic.
Louisiana and Delta Railroad	III	Multiple branches connected by trackage rights on Union Pacific between Lake Charles and Raceland, LA.	Carbon black, sugar, mola-sses, pipe, rice, and paper products.
Mississippi Export Railroad		Escatawpa River at Evanston, MS, to port at Pascagoula, MS.	Transloading services for intercoastal and river barges or vessels.
New Orleans and Gulf Coast Railway	III	Westwego, LA, to Myrtle Grove, LA.	Food products, oils, grains, petroleum products, chemi-cals, and steel products.
New Orleans Public Belt Railroad	Switc-hing	Serves Port of New Orleans along the Mississippi River and Industrial Canal.	Exports: lumber, wood pro-ducts, and paper. Imports: metal products, rubber, pla-stics, and copper. Domestic: clay, cement, and steel plate.
Norfolk Southern Corporation	I	Over 21,000 route miles in eastern U.S. Operate from Birmingham to Mobile, AL, and New Orleans, LA.	Agriculture, autos, chemi-cals, coal, machinery, inter-modal, metals, construc-tion material, paper, clay, forest products.
Pearl River Valley Railroad	III	Goodyear, MS, to Nicholson, MS.	Lumber and forest products.
Port Bienville Railroad	Switc-hing	Port Bienville Industrial Park, Hancock County, MS.	Plastic resins and other goods for industrial park tenets.
Sabine River and Northern Railroad	III	Between Buna and Orange, TX.	Wood chips, chemicals, and other raw materials for the paper industry. Finished paper and lumber products.
Terminal Railway Alabama State Docks	Switc-hing	Operates over 75 miles in the Mobile, AL, area, serving the port and local industries.	Port cargo.
Timber Rock Railroad Company	III	De Ridder, LA, west through Merryville to Kirbyville, TX.	Forest products and rock.
Union Pacific Railroad	I	Over 32,000 route miles in western U.S. Operate between Houston and New Orleans.	Chemicals, coal, food, for-est products, grains, interm-odal, metals, minerals, and autos.

Source: Bureau of Transportation Statistics, 2004.

Passenger Rail

The National Railroad Passenger Corporation (Amtrak) offers three intercity passenger rail services in the Gulf Coast study Region: City of New Orleans, Crescent, and Sunset

Limited. The City of New Orleans provides north-south passenger service between New Orleans, LA, and Jackson, MS, Memphis, TN, and Chicago, IL, over track owned by CN. The Crescent provides service between New Orleans, Atlanta, GA, Washington D.C., Philadelphia, PA, and New York City, NY. Both the City of New Orleans and the Crescent services travel north from New Orleans and have relatively little track mileage in the study area.

The Sunset Limited, however, traverses a distance of 4,448 km (2,764 mi) between Orlando, FL and Los Angeles, CA, and makes stops throughout the Gulf Coast study region, as shown in figure 2.12. East of New Orleans, the service runs along the coast and has been indefinitely suspended since Hurricane Katrina occurred in 2005. However, even before Katrina, the Sunset Limited was one of the lowest ridership long-distance trains operated by Amtrak, with fewer than 100,000 passengers per year according to Amtrak ridership reports. A complete list of Amtrak stations in the study area is provided in appendix C.

Marine Facilities and Waterways

Freight Handling Ports and Waterways

Ports can be comprised of a single facility or terminal, but most are actually made up of a mix of public and private marine terminals within a given geographic region along a common body of water. The U.S. Army Corps of Engineers identifies almost 1,000 public and private freight handling facilities throughout the study area, including different terminals within various defined port areas. These are mapped in figure 2.13. Major port complexes include, from west to east:

- Port of Freeport, TX;
- Ports of Houston, Texas City, and Galveston, TX;
- Ports of Port Arthur and Beaumont, TX;
- Port of Lake Charles, LA;
- Mississippi River ports of Baton Rouge, South Louisiana, New Orleans, St. Bernard (included in the New Orleans district by the U.S. Army Corps of Engineers), and Plaquemines, LA;
- Ports of Bienville, Gulfport, Biloxi, and Pascagoula, MS; and
- Port of Mobile, AL.

Waterborne Freight Types and Volumes

Table 2.5 shows that four of the top five ports in the United States, as measured by annual tonnage of goods handled by the port, are located in the study area. South Louisiana – at almost 199 million tons – is the Nation's leading tonnage port, while Houston – at over 190 million tons – ranks second. Collectively, study area ports handle almost 40 percent of all tonnage moved through all U.S. ports.

The study area also includes 4 of the Nation's top 30 container ports[6], including Houston, TX (number 11), New Orleans, LA (number 19), Gulfport, MS (number 21), and Freeport, TX (number 30) (AAPA, 2004).

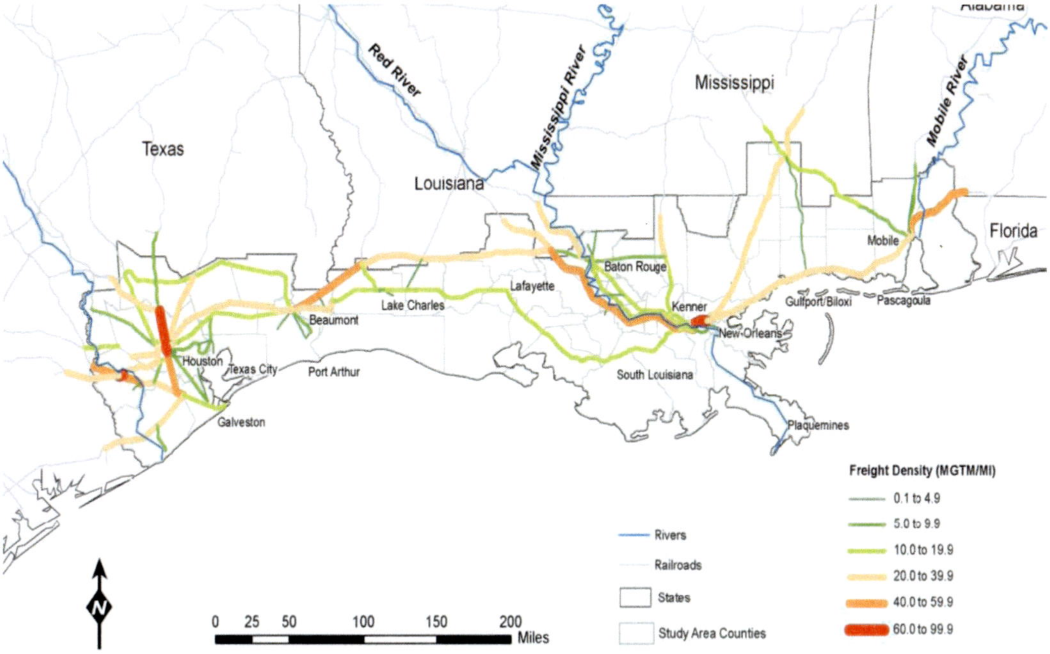

Source: Bureau of Transportation Statistics, U.S. Department of Transportation.

Figure 2.11. Freight railroad traffic density (annual millions of gross ton-miles per mile) in the study area

Source: Amtrak.

Figure 2.12. Sunset Limited route map, Houston, TX, to Mobile, AL, segment

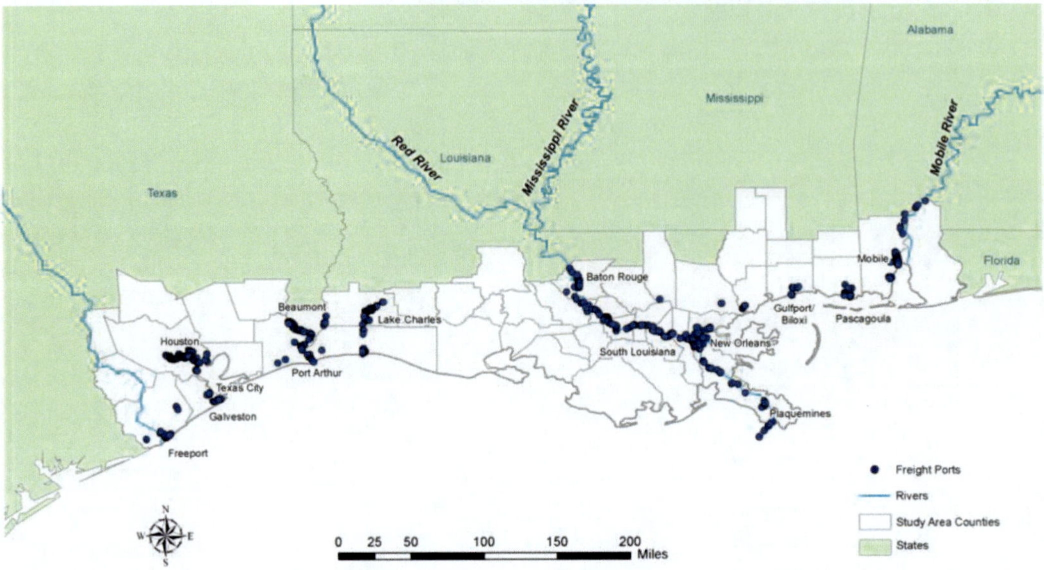

Source: Cambridge Systematics analysis of U.S. Army Corps of Engineers data.

Figure 2.13. Freight handling ports and waterways in the study area

Along with these fixed marine facilities, the study area hosts critically important navigable marine transportation networks. Among the most significant are the Gulf Intracoastal Waterway, a protected coastal route running from the Texas-Mexico border to Appalachee Bay in Florida; the Mississippi River and its tributaries; and the Tombigbee, Tennessee, and Black Warrior rivers, feeding the Mobile River in Alabama. These inland waterways and their associated lock structures (numbering in the hundreds) provide 20 States with access to the Gulf of Mexico, mostly through the Mississippi River and the Tennessee-Tombigbee River systems. Tonnage data (table 2.6) shows that largest volumes are on the Mississippi River (almost 213 million tons between Baton Rouge and New Orleans and 116 million tons between New Orleans and the Gulf of Mexico) and the Gulf Intracoastal Waterway (almost 118 million tons) (Institute for Water Resources, 2003). In fact, these two systems comprise the Nation's leading and third leading inland waterway systems by tonnage. Agriculture and other industries depend on efficient, reliable inland water transportation to move goods downriver to ports in Louisiana and Alabama, where goods are transloaded from domestic barges to international vessels. Petroleum, chemicals, and bulk products utilize the Gulf Intracoastal Waterway as an alternative to congested highway and rail corridors within the region.

Key Commodities and Industries

Overall, more than half of the tonnage (54 percent) moving through study area ports is petroleum and petroleum products – gasoline, fuel oil, natural gas, etc. This is not surprising, as the Gulf is a major petroleum producing and processing region, and an estimated 60 percent of U.S. petroleum imports passes through Gulf gateways. Of the rest, the majority – around 18 percent – is made up of food and farm products such as grains and oilseeds. Around 12 percent is chemicals, and the remaining commodities – around 4 percent to 6 percent each – are crude materials, manufactured goods, and coal (Institute for Water Resources, 2003).

Table 2.5. Domestic and International Waterborne Tonnage of Study Area Ports, 2003

National Rank	Port	2003 Short Tons
1	South Louisiana, Louisiana	198,825,125
2	Houston, Texas	190,923,145
4	Beaumont, Texas	87,540,979
5	New Orleans, Louisiana	83,846,626
9	Texas City, Texas	61,337,525
10	Baton Rouge, Louisiana	61,264,412
11	Plaquemines, Louisiana	55,916,880
12	Lake Charles, Louisiana	53,363,966
14	Mobile, Alabama	50,214,435
23	Pascagoula, Mississippi	31,291,735
24	Freeport, Texas	30,536,657
27	Port Arthur, Texas	27,169,763
	Gulf Coast Study Area Total	**932,231,248**
	National Total	**2,394,251,814**

Source: U.S. Army Corps of Engineers, Navigation Data Center.

Table 2.6. Tonnage on Study Area Inland and Coastal Waterways, 2003

Waterways Segments Within Study Area	2003 Short Tons (Millions)
Mississippi River, Baton Rouge to New Orleans, LA	212.9
Mississippi River, Mouth of Ohio to Baton Rouge, LA	185.5
Gulf Intracoastal Waterway, TX-FL	117.8
Mississippi River, New Orleans, LA, to Gulf of Mexico	115.8
Gulf Intracoastal Waterway, Port Allen Route, LA	24.3
Black Warrior and Tombigbee Rivers, AL	21.0
Atchafalaya River, LA	9.8
Tennessee-Tombigbee Waterway, AL and MS	5.2
Red River, LA	4.2
Chocolate Bayou, TX	3.3
Petit Anse, Tigre, Carlin bayous, LA	2.5
Ouachita and Black Rivers, AR and LA	2.2
Bayou Teche, LA	1.4
Subtotal for Waterway Segments Within Study Area	**705.9**
Subtotal for Full Gulf Coast and Mississippi River Systems, including Waterway Segments Within or Connecting to Study Area	**1,650.5**
National Total of All Major Inland and Coastal Waterway Segments	**1,717.0**

Source: U.S. Army Corps of Engineers, Waterborne Commerce of the United States, 2003.

Source: U.S. Army Corps of Engineers.

Figure 2.14. Barge tow on the Mississippi River

There are important differences between ports in different parts of the study area. The Alabama and Mississippi ports specialize in coal, petroleum, manufactured (containerized) goods, and crude materials. In contrast, around 38 percent of tonnage through the Mississippi River ports consists of food and farm products, much of it related to the transloading of barge traffic from the Nation's interior, with petroleum accounting for another 30 percent of tonnage. The western Louisiana and Texas ports are dominated by petroleum, which represents 75 percent of their tonnage.

Nonfreight Marine Facilities

The study area also hosts a large array of nonfreight maritime uses. The U.S. Army Corps of Engineers database lists around 800 nonfreight facilities (including unused berths) in the study area. These serve a variety of functions, including commercial fishing; vessel fueling, construction, repair, and outfitting (including shipyards); marine construction services (channel dredging and maintenance, construction of berths and other facilities); government and research facility docks; recreational and commercial vessel berthing; passenger ferry and cruise docks; and support for offshore oil facilities.

Aviation

The system of airports analyzed in the Gulf Coast Study includes 61 publicly owned, public-use airports. Private facilities are excluded from the sample as are the 387 heliports located in the study area.[7] Twenty-eight of these airports (more than 45 percent) are in Louisiana, 16 are in Texas, 9 are in Mississippi, and 8 are in Alabama.

There are over 3,800 aircraft based at publicly owned, public-use airports in the study area. Over 3.4 million aircraft takeoffs and landings take place at these airports annually, with the majority of operations taking place at commercial service airports.

Of these 61 airports, 44 are general aviation airports, 11 are commercial service, 4 are industrial, and 2 are military, as described below:

- **Commercial service airport (CS)** – Commercial service airports primarily accommodate scheduled passenger airline service. Two Houston airports led the region in passenger enplanements in 2005 (George Bush Intercontinental Airport [IAH] and William P. Hobby [HOU]), followed by Louis Armstrong New Orleans International [MSY].

- **Military airfield (MIL)** – Military Airfields accommodate strictly military aircraft and are off limits to civilian aircraft. The two active military airfields in the study area are Keesler Air Force Base [AFB] in Mississippi and the New Orleans Naval Air Station/Joint Reserve Base. Keesler AFB is notable for being the home of the 53rd Weather Reconnaissance Squadron, the "Hurricane Hunters," who fly aircraft into tropical storms and hurricanes to gather weather data.

- **Industrial airport (IND)** – Industrial airports are airports that can accommodate both commercial and privately owned aircraft. Typically, an industrial airport is used by aircraft service centers, manufacturers, and cargo companies, as well as general aviation aircraft. The four industrial airports in the study area are former military airfields, designed to accommodate the largest aircraft. None of them have scheduled passenger service.

- **General aviation airport (GA)** – General aviation airports accommodate aircraft owned by private individuals and businesses.

In addition to leading the region in passenger enplanements, George Bush Intercontinental IAH in Houston also is the leading airport in the study area for cargo tonnage, processing 75 percent of all cargo enplaned in the study area. It ranks 17th nationally for cargo, with 387,790 annual tons (ACI, 2005). Louis Armstrong New Orleans International ranked second for cargo, followed by Mobile Downtown, an industrial airport.

Table 2.7 details the passenger enplanements and cargo tonnage for the major study area airports. Figure 2.15 identifies the location of airports in the study area.

Pipelines

The pipeline system in and around the Gulf Coast is a major transporter of gas, petroleum, and chemical commodities. It links many segments of the country with energy sources located on the Gulf Coast. Unlike other transportation systems, pipelines are singularly a transportation system for bulk commodities that have little or no time sensitivity for product delivery. The entire pipeline network is privately funded and held. The onshore portion is principally regulated by the Office of Pipeline Safety (OPS), within the United States. Department of Transportation, Pipeline and Hazardous Materials Safety Administration (PHMSA). Regulation focuses on safe operations to protect people, the environment, and the national energy supply. Off-shore pipelines are regulated by the U.S. Department of the Interior, Minerals Management Service.

Table 2.7. Passenger Enplanements and Cargo Tonnage for Select Commercial Service and Industrial Airports in the Study Area, 2005

Associated City	FAA Code	State	County	Airport Name	Airport Type	2005	
						Passenger Enplanements	Cargo Tonnage
Mobile	MOB	Alabama	Mobile	Mobile Regional	CS	638,953	582
Mobile	BFM	Alabama	Mobile	Mobile Downtown	IND	0	44,000[a]
Lake Charles	LCH	Louisiana	Calcasieu	Lake Charles Regional	CS	43,250[a]	2[a]
Lake Charles	CWF	Louisiana	Calcasieu	Chennault International	IND	0	75
Baton Rouge	BTR	Louisiana	East Baton Rouge	Baton Rouge Metropolitan, Ryan Field	CS	973,625	5,663
New Orleans	MSY	Louisiana	Jefferson	Louis Armstrong New Orleans International	CS	7,775,147	66,123
Lafayette	LFT	Louisiana	Lafayette	Lafayette Regional	CS	343,301	6,774
Hattiesburg	HBG	Mississippi	Forrest	Bobby L Chain Muni	CS	8,000[a]	
Gulfport	GPT	Mississippi	Harrison	Gulfport-Biloxi International	CS	769,669	
Houston	EFD	Texas	Harris	Ellington Field	CS	53,947	15
Houston	HOU	Texas	Harris	William P. Hobby	CS	8,252,532	7,000
Houston	IAH	Texas	Harris	George Bush Intercontinental/ Houston	CS	39,684,640	387,790
Beaumont/ Port Arthur	BPT	Texas	Jefferson	Southeast Texas Regional	CS	43,038[1]	
Study Area Total						**58,586,102**	**517,418**
National Total						**738,629,000**	**30,125,644**

Source: Alabama airports from http://www.brookleycomplex.com/cargo/statistics.asp. Louisiana airports from the Airports Council International and U.S. DOT BTS T100 data. Texas airports from http://www.citydata.com/us-cities/The-South/Houston-Economy.html. Wilbur Smith Associates. National totals from Bureau of Transportation Statistics (http://www.bts.gov/programs/airline_information/air_carrier_traffic_statistics/airtraffic/annual/1981-2001.html) and Airports Council International.

Note: CS: Commercial Service Airport

IND: Industrial Airport

[1] Estimated.

There is a total of 42,520 km (26,427 mi) of onshore liquid (oil and petroleum product) transmission and natural gas transmission pipelines in the Gulf Coast area of study, with some extended sections beyond its boundaries. This includes 22,913 km (14,241 mi) of onshore natural gas transmission pipelines and 19,607 km (12,186 mi) of onshore hazardous liquid pipelines (PHMSA, 2007). The liquid pipelines are concentrated in Texas while the natural gas pipelines are concentrated in Louisiana.

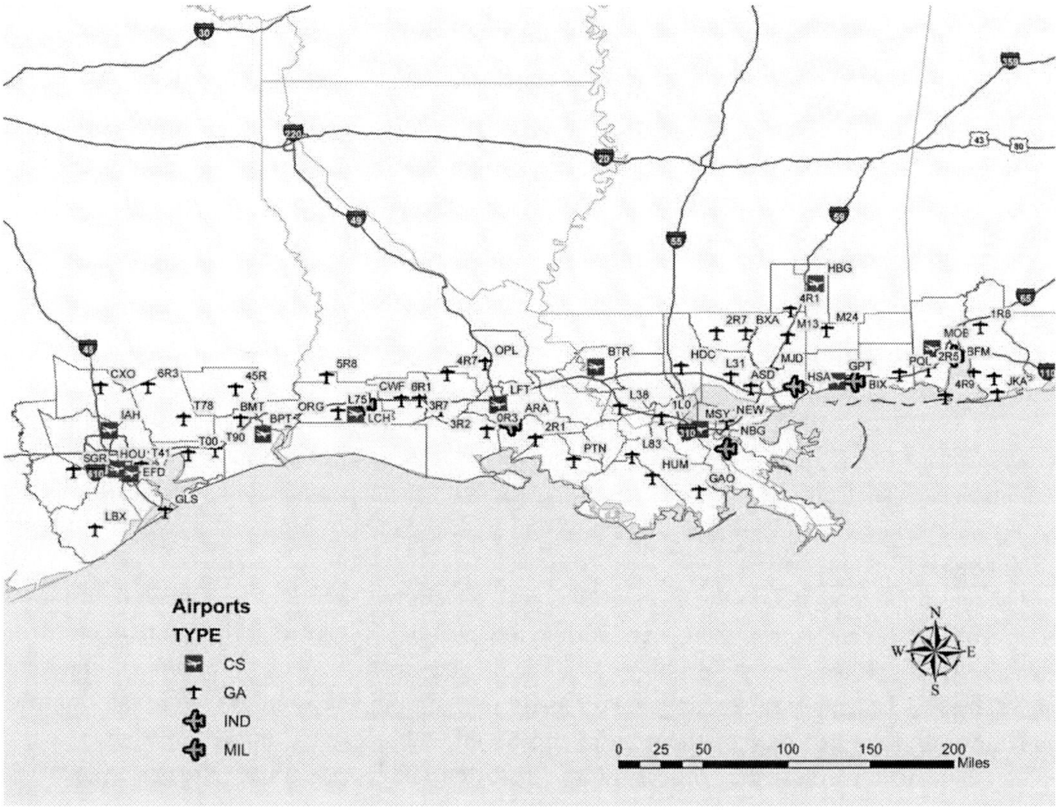

Note: CS: Commercial Service, GA: General Aviation, IND: Industrial, MIL: Military
Source: Bureau of Transportation Statistics, U.S. Department of Transportation.

Figure 2.15. Study area airports.

Approximately 49 percent of U.S. wellhead natural gas production either occurs near the Henry Hub, which is the centralized point for natural gas futures trading in the United States, or passes close to the Henry Hub as it moves to downstream consumption markets. The Henry Hub is located near the town of Erath in Vermilion Parish, LA. The Henry Hub interconnects nine interstate and four intrastate pipelines, including: Acadian, Columbia Gulf, Dow, Equitable (Jefferson Island), Koch Gateway, LRC, Natural Gas Pipe Line, Sea Robin, Southern Natural, Texas Gas, Transco, Trunkline, and Sabine's mainline.

2.3. Gulf Coast Physical Setting and Natural Environment

The unique natural environment and geology of the Gulf Coast study region brings its own set of considerations and challenges in designing the built environment. Some of these physical characteristics, such as low topography, high rates of subsidence, and predilection for coastal erosion, significantly increase the vulnerability of the area to climate change impacts. A robust transportation system must accommodate the natural features of this landscape.

A variety of physical datasets were compiled for phase I of the Gulf Coast study and posted on a Web site for review and use by the project research team (appendix A). Most of the spatial data is organized in GIS-type formats or "layers" that can be integrated for the purposes of assessing the vulnerability and risks of the transportation infrastructure in the study area and informing the development of adaptation strategies in phases II and III of the study, respectively. Examples of the spatial data products developed for the study are presented in the following sections.

2.3.1. Geomorphology

The Gulf Coast region of the United States is in the physiographic province called the southeastern Coastal Plain, which is a broad band of territory paralleling the Gulf and South Atlantic seacoast from North Carolina to Texas, with a deep extension up the Mississippi River valley. The Coastal Plain is relatively flat, with broad, slow-moving streams and sandy or alluvial soils (figure 2.16).

Much of the land area in the Coastal Plain is overlain with sediments deposited during the Holocene or Recent Age epoch, i.e., during the past 10,000 years. The remainder of the Coastal Plain surface consists primarily of late Cretaceous deposits (65 to 100 million years old). These sedimentary rocks, deposited mostly in a marine environment, were later uplifted and now tilt seaward; part of them form the broad, submerged Continental Shelf. Coastal Plain deposits overlap the older, more distorted, Paleozoic and Precambrian rocks immediately to the north and west (more than 250 million years old) (U.S. Geological Survey [USGS], 2000a).

The center of the study area is dominated by the Mississippi Embayment, a geologic structural trough in which the underlying crust of the Earth forms a deep valley that extends from the Gulf Coast inland to the confluence of the Ohio and Upper Mississippi Rivers. The Lower Mississippi Valley occupies the center of the inland part of the embayment and ranges from 30– to 180–km (20–to 110–mi) wide. Large rivers, such as the Mississippi, Arkansas, and Ohio Rivers, have flowed through this region, carved the surface, and deposited clay, silt, sand, and gravel, collectively called alluvium.

Nearly annually, the Mississippi River and its tributaries flood vast areas of the lower alluvial valley. Traditionally, these floods have lasted for several months and a few for even longer periods. For example, the great flood of 1927 occurred from April to June when the lower Mississippi River system stored the equivalent of 60 days of discharge in its 22–million-acre alluvial valley. The river flows through the Lower Mississippi River Valley in a 15- to 30-km (10- to 20-mi) wide meander belt, and historical and prehistoric records indicate the river is continually creating new channels and abandoning old ones. The alluvium provides the rich soils for massive agricultural development.

Where the Mississippi River empties into the Gulf of Mexico, old deltas are abandoned and new ones formed. This Mississippi River deltaic plain lies at the center of the Gulf Coast study area. During the formation of the deltaic plain, millions upon millions of tons of sediment were deposited in a series of overlapping delta lobes that are presently in various phases of abandonment and deterioration. The barrier island chains off the coast of Louisiana are remnant features of old deltas that are naturally eroding and retreating landward as sea level rises. Erosional forces dominate this part of the central Gulf Coast landscape.

Due largely to its sedimentary history, land along the central Gulf Coast tends to be low and flat and is dissected by numerous slow-moving streams or bayous that drain runoff from the Coastal Plain and the adjacent uplands. The central Gulf coastal zone includes many barrier islands and peninsulas, such as Galveston Island, TX, Grand Isle, LA, and the land between Gulfport and Biloxi, MS. These landforms protect numerous bays and inlets. The low-lying areas of the central Gulf Coast region are (or were) primarily marshland and wetland forests.

Erosion, sediment transport and deposition, and changes in elevation relative to mean sea level (i.e., subsidence, discussed in greater detail below) are the main land surface processes that interact with climate change and variability in a manner that could adversely affect transportation in the study area. Erosion is exacerbated by increased water depth, increased frequency or duration of storms, and increased wave energy – and all of these changes could potentially accompany an increase in the temperature of the atmosphere.

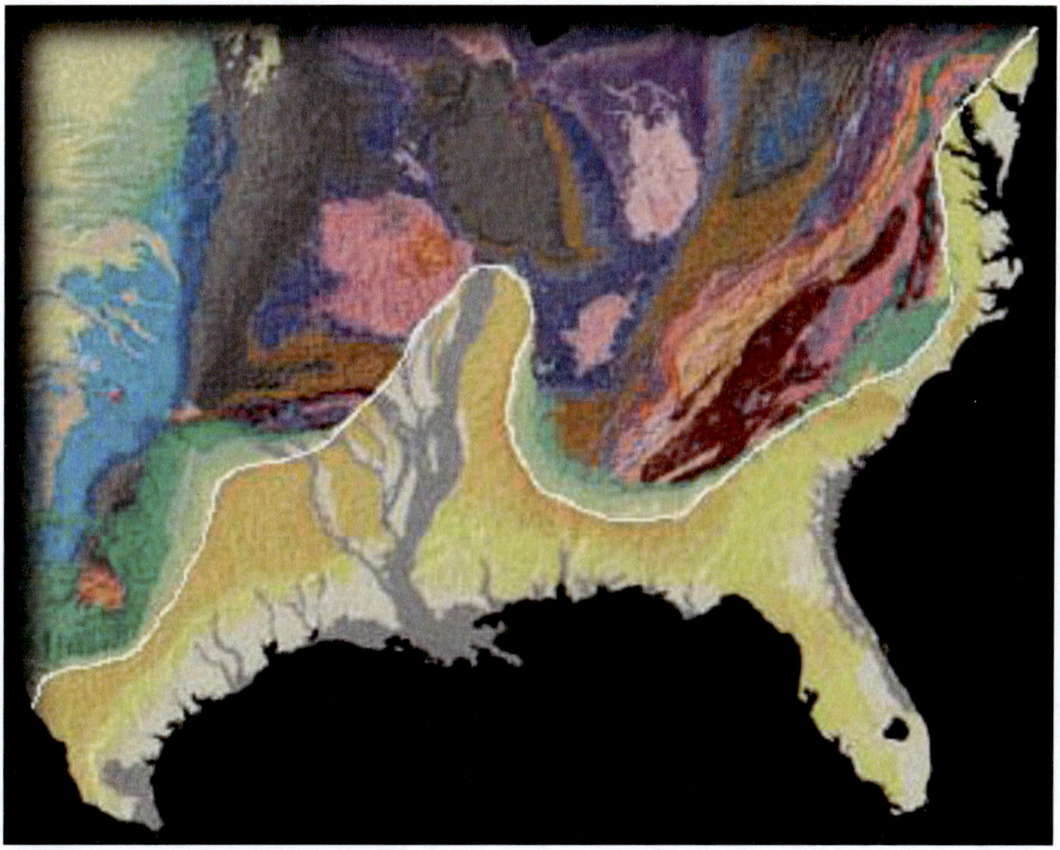

Source: U.S. Geological Survey, 2000a.

Figure 2.16. Surface geology of the southeastern United States. White line denotes inland extent of the Gulf Coastal Plain, and grey area is Holocene alluvium.

2.3.2. Current Elevation and Subsidence

The great majority of the study area lies below 30 meters in elevation (figure 2.17) (USGS, 2004). Due to its low relief, much of the central Gulf Coast region is prone to flooding during heavy rainfall events, hurricanes, and lesser tropical storms. The propensity for flooding is higher in areas that are experiencing subsidence (i.e., the gradual lowering of the land surface relative to a fixed elevation). Near the coastline, the net result of land subsidence is an apparent increase in sea level.

Land subsidence is a major factor in the study region. The rate of subsidence varies across the region and is influenced by both the geomorphology of specific areas as well as by human activities. Parts of Alabama, Texas, and Louisiana are experiencing subsidence rates that are much higher than the 20th–century rate of global sea level rise of 1-2 mm/year (IPCC, 2001). For example, in the New Orleans area the average rate of subsidence between 1950 and 1995 was about 5 mm/year (Burkett et al., 2003), with some levees, roads, and artificial-fill areas sinking at rates that exceed 25 mm/year (Dixon et al., 2006). As a result of subsidence, which was accelerated by the forced drainage of highly organic soils and other human development activity, most of the city of New Orleans is below sea level.

Subsidence in the Houston-Galveston-Baytown region is associated primarily with groundwater withdrawals, which peaked in the 1 960s. By the mid 1970s, industrial groundwater withdrawals had caused roughly two meters of subsidence in the vicinity of the Houston Ship Channel, and almost 8,300 km^2 (3,200 mi^2) of land in this region had subsided more than one foot. The growing awareness of subsidence-related flooding in southeastern Texas prompted the 1975 Texas Legislature to create the Harris-Galveston Coastal Subsidence District, which was authorized to regulate ground water withdrawals and promote water conservation programs (Coplin and Galloway, 1999). Shallow oil and gas withdrawals also have contributed to subsidence in southeast Texas (Coplin and Galloway, 1999) and coastal Louisiana (Morton et al., 2005). Recent geological and geophysical investigations suggest that subsidence across the Central Gulf Coast is occurring more rapidly than previously thought (Shinkle and Dokka, 2005; Dixon et al., 2006).

Recognizing the increasing trend in flooding in the region, the Federal EmergencyManagement Agency (FEMA) currently is updating its Base Flood Elevations Maps of the region. However, even new elevation maps can be outdated within just a few years due to the high rates of subsidence in some parts of the study area (American Geophysical Union [AGU], 2006).

While the Gulf Coast is considered at very low risk for earthquakes, it does have hundreds of subsurface faults that can be expressed at the surface by differences in elevation, by the zonation of plant communities, or by patterns of wetland loss (Morton et al., 2005). Generally, these faults run parallel to the shoreline and are displaced "down to the coast" due to the slow sliding of thick sediments towards the Gulf of Mexico. Subsidence and subsurface fluid withdrawals can activate shallow faults and cause ground failure along highways and beneath buildings. Since the late 1930s, 86 active faults in the Houston- Galveston area have offset the land surface by slow seismic creep at rates of up to 2.5 cm per year (Holzer and Gabrysch, 1987; Coplin and Galloway, 1999).

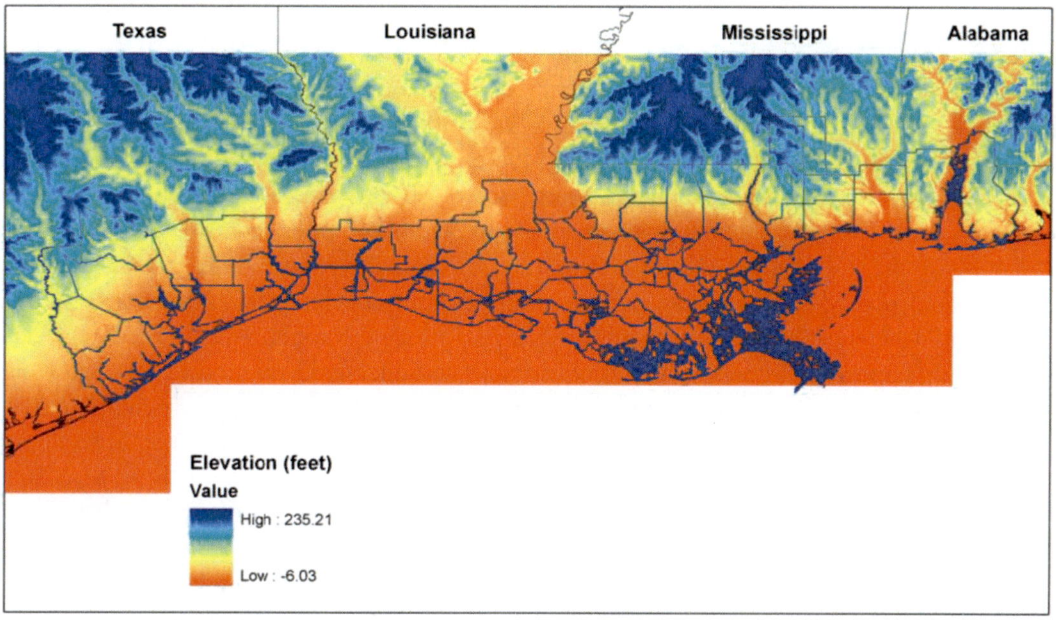

Source: U.S. Geological Survey.

Figure 2.17. Relative elevation of counties in the study area (delineated in blue). All areas shown in bright orange are below 30–m elevation.

2.3.3. Sediment Erosion, Accretion, and Transport

The northern Gulf of Mexico coastal zone is highly dynamic due to a unique combination of geomorphic, tectonic, marine, and atmospheric forcings that shape both the shoreline and interior land forms. Most of coastline of the study area is classified as "highly vulnerable" to erosion (Theiler and Hammar-Close, 1999). The retreat of shoreline of the reticulated marshes that dominate much of the coastal zone is often translated to "wetland loss," which occurs via submergence of land or erosion of the land/water interface. Highest erosion and wetland loss rates are associated with tropical storms and frontal passages. It is estimated that 56,000 ha (217 mi^2) of land were lost in Louisiana alone during Hurricane Katrina (Barras, 2006).

The barrier islands of the central Gulf Coast region are shaped continually by wind and wave action and changes in sea level, including the short-term increase in sea level associated with storm surge. The Chandeleur Islands, LA, which serve as a first line of defense for the New Orleans region, are extremely vulnerable to intense tropical storms, having lost 85 percent of their surface area during Hurricane Katrina (USGS, 2007). As barrier islands and mainland shorelines erode and submerge, onshore facilities in low-lying coastal areas become more susceptible to inundation and destruction. Many Gulf Coast barrier islands are retreating and diminishing in size, with the most significant breaching and retreat occurring during storms and frontal passages. The combined effects of beach erosion and storms can lead to the erosion or inundation of other natural coastal systems. For example, an increase in wave heights in coastal bays is a secondary effect of sandy barrier island erosion in Louisiana where increased wave heights have enhanced erosion rates of bay shorelines, tidal creeks, and adjacent wetlands (Stone and McBride, 1998; Stone et al., 2003).

Theiler and Hammar-Close (1999) assessed the relative importance of six variables that influence coastal erosion rates and developed a coastal vulnerability index (CVI) for the Gulf Coast region. Their analyses indicated that geomorphology and tide range are the most important variables in determining the CVI for the Gulf of Mexico coast, since both variables reflect very high vulnerabilities along nearly the entire shoreline. Wave height, relative sea level rise, and coastal slope explain the large-scale (50-200 km alongshore) variability of erosion rates. They concluded that erosion and accretion rates contribute the greatest variability to the CVI at short spatial scales. Rates of shoreline change, however, are the most complex and poorly documented variable in this dataset developed by the USGS. To best understand where physical changes may occur, large-scale variables must be clearly and accurately mapped, and small-scale variables must be understood on a scale that takes into account their geologic and environmental influences. Marshes that receive sufficient inputs of mineral or organic sediments, for example, can offset the potential for submergence due to subsidence and sea level rise (Rybczyk and Cahoon, 2002).

Sediments eroded by winds, tides, and waves are transported generally towards shore and continually reworked into a mosaic of wetlands, shallow bays, and barrier islands. Some sediments, however, are lost to the Gulf or deposited along the shoreline to the east or west of the study area. Nearshore currents east of the mouth of the Mississippi River carry sediments eastward. To the west of the Mississippi River delta, the predominant direction of this nearshore drift is westward.

At the geographic center of the study area, the Mississippi River alluvial or deltaic plain has been built on the continental shelf during the past 6,000 years, during a period of relatively slow sea level rise when most of the world's present deltas were formed (Woodruffe, 2003). In recent times, sediments that would be delivered to the Mississippi River delta marshes via seasonal overbank flooding have been cut off by levees and deep channel dredging of the Mississippi River for navigation (Reed, 2002). Thousands of miles of smaller navigation channels, access canals to oil and gas fields, and other development activities have contributed to the vulnerability of the Mississippi River deltaic plain to sediment deprivation and land loss (Minerals Management Service [MMS], 1994).

2.3.4. Land Use and Land Cover

Land use of the Gulf Coast study area was defined by using the National Land Cover Dataset (NLCD). The NLCD consists of 21 classifications, of which 19 were found in this study area. The data were collected from the Landsat Thematic Mapper satellite in the early to middle 1990s and are of 30–meter resolution. Table 2.8 summarizes this data for the study area.

The central Gulf Coast study area covers an area of approximately 1 million ha (23.4 million acres or 36,485 mi^2). Land cover is dominated by wetlands (32.4 percent), agriculture (19/1 percent), and upland forests (17.7 percent). The study area can be broadly divided into six ecological units based on Bailey's classification of U.S. ecoregions (Bailey, 1976) (figure 2.18). Land cover within the study area has strong similarities from east to west across the study area and appears to be influenced more by soils, topography, and human activity than by climatic differences. Natural plant community distributions are generally oriented along north/south gradients, reflecting salinity, water level, and disturbance regimes.

Nonurbanized land use in the region is devoted mainly to Federal/State protected lands, large-scale commercial agriculture, and relatively undeveloped wetlands associated with the Mobile River in Alabama; the Pearl River in Mississippi and Louisiana; the Mississippi, Atchafalaya, and Calcasieu Rivers in Louisiana; and the Neches, Sabine, and Trinity Rivers in Texas. In addition to contributing to the formation of wetlands running inland from the coast, each of these rivers intersects or connects with the Gulf Intracoastal Waterway, and each forms the basis for urbanized port areas, of varying sizes, adjacent to the coast.

2.4. Social and Economic Setting

Transportation networks exist to facilitate the movement of people and goods and are an **integral part of a region's social and economic fabric**. The need for these networks, or transportation demand, therefore, is defined by demographic and economic considerations – connecting population centers, providing access to economic resources, etc. It is important, therefore, to understand the people and the economy that exist in the Gulf Coast study region in order to assess the significance of climate impacts on its transportation systems.

Table 2.8. Land use of the Central Gulf Coast Study Area as Defined by the 1992 National Land Cover Dataset

Land Use Category	Area (Hectares)	Percent of Total
Water	508,735	5.38%
Low-Intensity Residential	250,032	3.00%
High-Intensity Residential	106,637	1.13%
Commercial, Industrial, Transportation	152,744	1.62%
Bare Rock, Sand, Clay	14,126	0.15%
Quarries, Strip Mines, Gravel Pits	3,921	0.04%
Transitional from Barren	92,835	0.98%
Deciduous Forest	492,245	5.21%
Evergreen Forest	1,175,278	12.44%
Mixed Forest	861,726	9.12%
Shrubland	23,096	0.24%
Orchard, Vineyard	5	Negligible
Grasslands, Herbaceous	123,576	1.31%
Pasture, Hay	1,213,343	12.84%
Row Crops	591,105	6.26%
Small Grains	694,855	7.35%
Urban, Recreation Grasses	83,476	0.88%
Woody Wetlands	1,087,093	11.50%
Emergent Herbaceous Wetlands	1,974,788	20. 90%
Total	**9,449,615**	

Source: National Land Cover Dataset, U.S. Geological Survey.

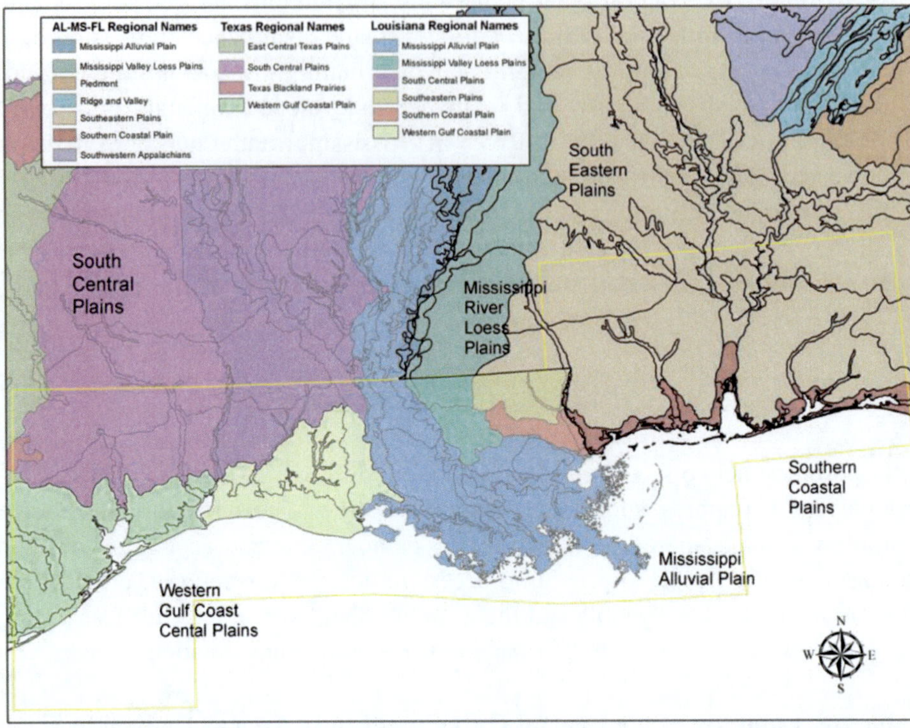

Figure 2.18. Map of terrestrial ecoregions within and adjacent to the study area. (Modified from Bailey, 1975)

The Gulf Coast study region, like many parts of the country, has been growing in population and economic activity and has become increasingly urbanized in recent decades. These trends were seriously disrupted by the 2005 hurricanes, which caused massive property damage and wide-scale relocation of residents in affected areas. It is too early to know what long-term impacts Hurricanes Katrina and Rita will have on the region's population distribution.

According to the U.S. Census Bureau estimates for 2004, the 48 counties of the designated study area are home to about 9.7 million people. Within the region are 419 cities, towns, and villages (defined as "places" by the U.S. Census Bureau), ranging in population from less than 50 residents to nearly 2 million. A quick perusal of the interstate and highway map illustrates, to some degree, the interconnectedness of the region. The majority of these places are served by a vast land- and water-based transportation grid designed to move people and goods eastward and westward along the coast, as well as into and out of the United States via Gulf of Mexico port facilities.

Figure 2.19 illustrates the degree to which urbanized zones have spread throughout the study area. Population growth and industrialization in the region are continuing to urbanize the central coast of the Gulf of Mexico. Nonetheless, major contrasts remain among urban, suburban, and rural settings within the region.

Mean household income for the study area population was lower than for the nation ($53,600 per household compared to $56,500 in the Nation). The study region also experiences higher poverty rates (15.6 percent of all persons compared to 12.4 percent in the

Nation), and higher rates of children below 5 years living in poverty (17.4 percent compared to 12.5 percent nationally). The demographic distribution showed a slightly younger population when compared to the Nation (52.8 percent of the population was less than 35 years, compared to 49.3 percent nationwide).

2.4.1. Population and Development Trends

Before the impacts of the hurricanes in 2005 were fully realized, the region had experienced an average population growth rate from 1990 to 2000 of 16 percent, with an additional 5 percent growth estimated for the period from 2000 to 2004 (figures 2.21 and 2.22). Measured in terms of building permits issued, the region has experienced an overall housing growth rate of 12 percent during the period of 1997 to 2002. However, a wide variation in growth rates exists among counties in the study area, including 17 counties (primarily rural) that have experienced declines in building permit issuance over this period.

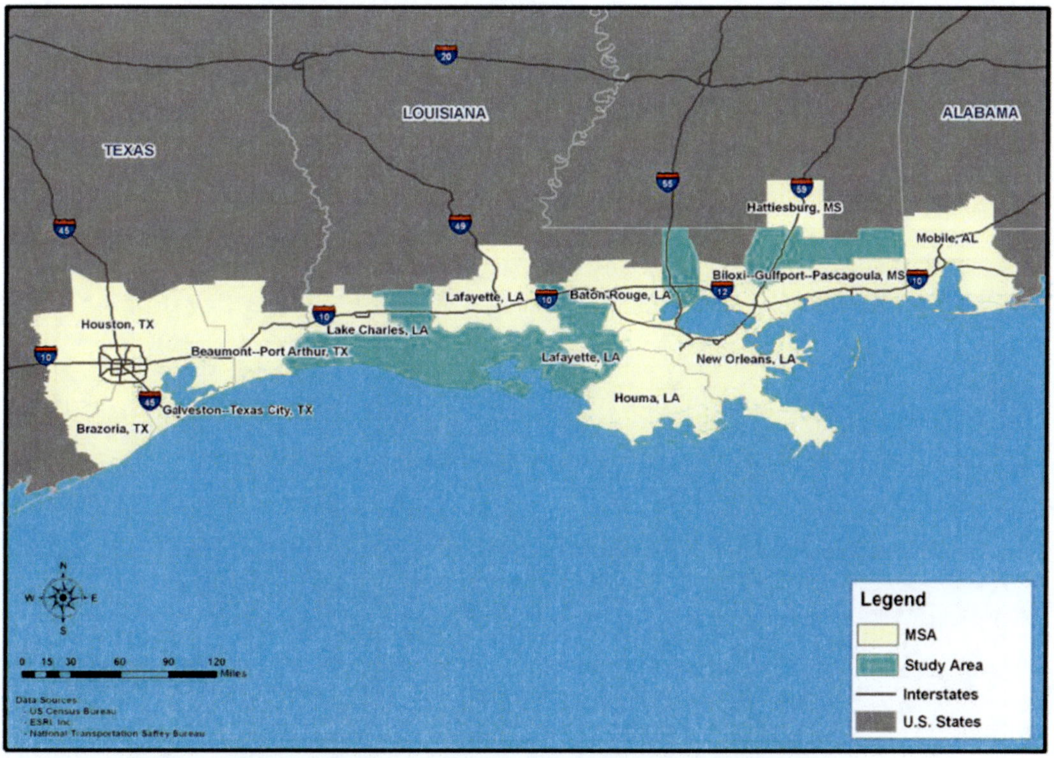

Source: U.S. Census Bureau; ESRI, Inc.; National Transportation Safety Bureau.

Figure 2.19. U.S. Census Bureau Metropolitan Statistical Areas within the study area

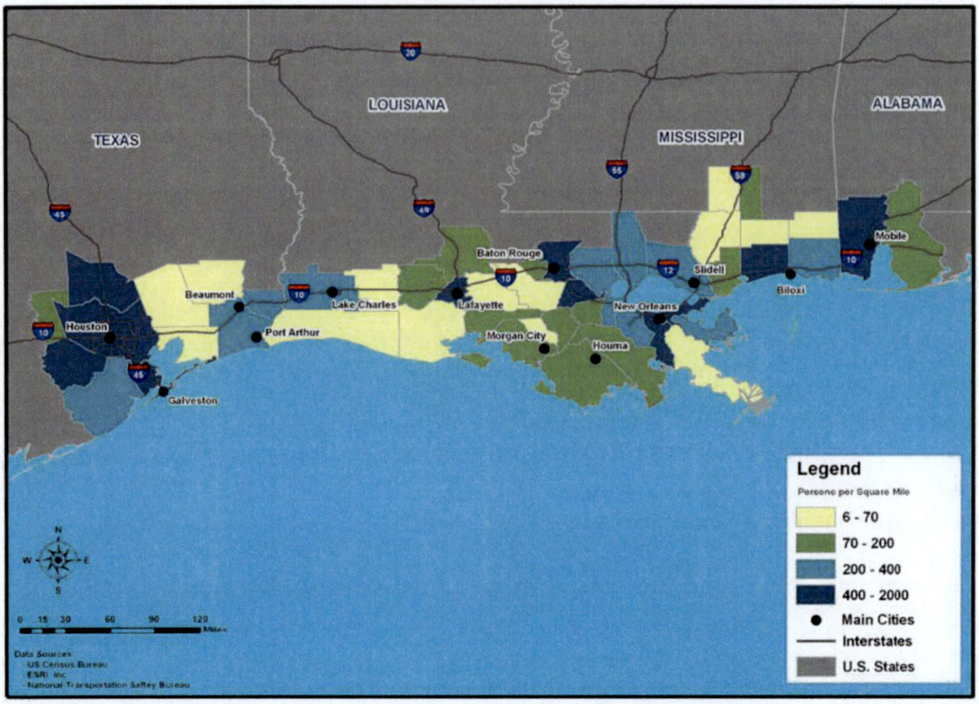

Source: U.S. Census Bureau; ESRI, Inc.; National Transportation Safety Bureau.

Figure 2.20. Population density in the study area, 2004

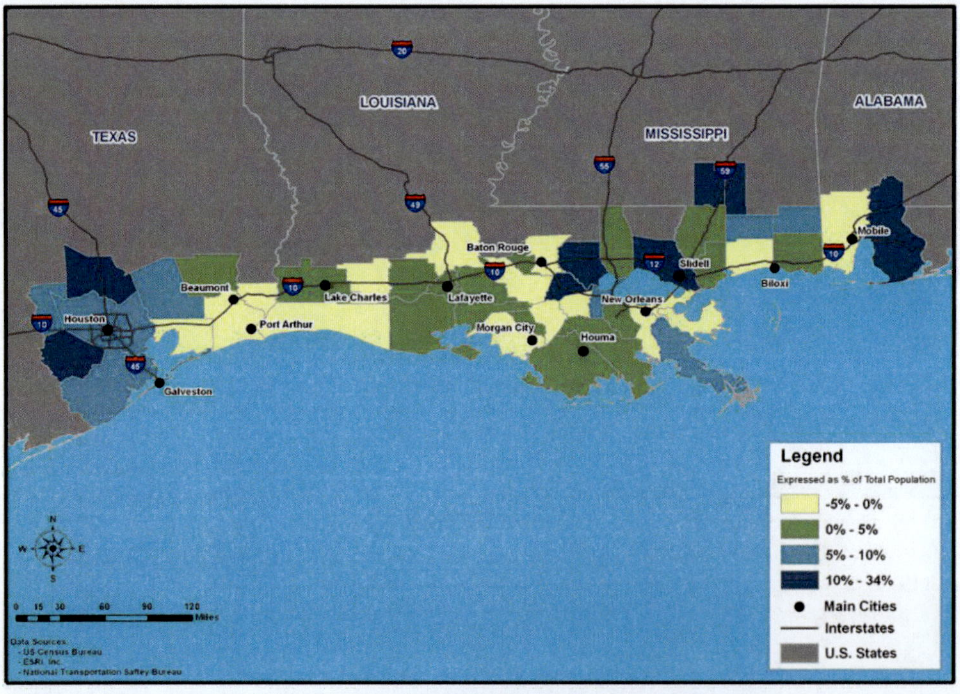

Source: U.S. Census Bureau; ESRI, Inc.: National Transportation Safety Bureau.

Figure 2.21. Estimated population change in the study area, 2000 to 2005

Population and housing growth patterns for the region are dominated by urban-rural migration and the increasing suburbanization of the larger urban areas of Houston/Galveston, TX, Baton Rouge/New Orleans, LA, Hattiesburg, MS, and Mobile, AL. Rural counties along the western and central portions of the Louisiana coast, which tend to be dominated by wetland landscapes of the Atchafalaya and Mississippi Rivers, have experienced low and/or declining population growth over this period. These counties primarily host agricultural economies, and, like many similar rural counties in the United States, they have been experiencing slowly declining population growth rates for many decades.

Urban growth has been primarily characterized by spatial expansion around existing urbanized areas. In the case of Houston/Galveston, growth has been focused on those counties surrounding the core county of Harris, especially due to the residential and commercial expansion along I-10 to the west and I-45 to the south and east. The Baton Rouge/New Orleans area is experiencing a similar suburbanization process focused on the "Northshore" of Lake Pontchartrain. This growth in "bedroom" communities on the Northshore is supported by commuter pathways along I-12 and I-10 and the Lake Pontchartrain Causeway. Baton Rouge continues to grow eastward toward these Northshore counties, and the New Orleans metro area has been undergoing the same cross-lake residential migration for many years. One of the numerous impacts of Hurricane Katrina appears to be an acceleration of this trend among residents of Orleans and St. Bernard Counties,[8] as many residents are finding the Northshore communities more affordable or attractive despite the greater commute into New Orleans. Mobile, AL, appears to be experiencing a similar pattern of suburbanization as the greatest growth is taking place in the less densely populated county of Baldwin east of Mobile Bay. Figure 2.22, "Mean Travel Time to Work," illustrates this trend toward suburbanization in the region.

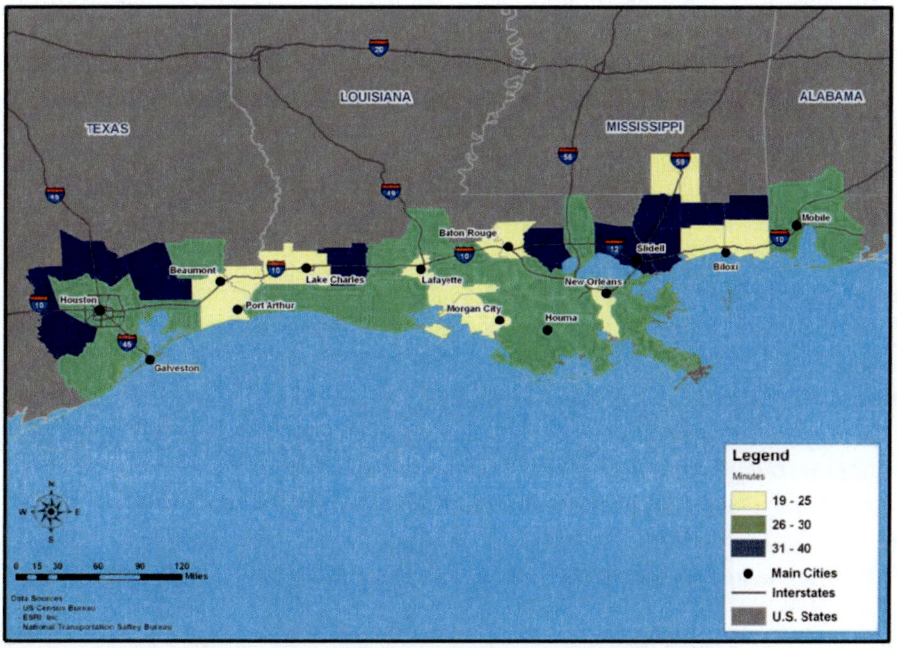

Source: U.S. Census Bureau; ESRI, Inc.; National Transportation Safety Bureau.

Figure 2.22. Mean travel time to work in the study area

Table 2.9. Top 10 Industries in the Study Area by Employment Percentage, 2000

Industry	Percent of Study Area Employment
Retail Trade	11.6
Manufacturing	11.6
Health Care and Social Assistance	10.2
Educational Services	8.9
Construction	8.6
Accommodation and Food Services	6.4
Professional, Scientific, and Technical Services	6.2
Other Services (except Public Administration)	5.2
Transportation and Warehousing	4.8
Public Administration	4.3

Source: United States Census 2000, U.S. Census Bureau, 2007.

It is still too early to know what the long-term impacts of Hurricane Katrina will be on regional demographics. Some locations, particularly New Orleans, experienced major shifts. According to the 2005 American Community Survey Special Product for the Gulf Coast Area (U.S. Census Bureau, 2005), in the months following the storm, the New Orleans Metropolitan Statistical Area [MSA] showed a 30 percent drop in population, accompanied by a nearly 4-year increase in median age (from 37.7 years to 41.6 years).

The civilian labor force dropped from nearly 600,000 to about 340,000. The survey measured higher median incomes for those remaining, indicating that more higher-income workers in relatively stable professions have tended to stay in place, while lower-income, low-skilled workers have been more likely to relocate. Many people have moved to other locations within the study area, such as the Houston-Galveston and Baton Rouge areas, while others left the study area entirely.

2.4.2. Employment, Businesses, and Economic Drivers

Energy production, chemical manufacturing, and commercial fishing dominate the economy of the study region. While the economy in the overall area has grown, certain parts of the region have not shared in this development. Table 2.9 shows the top 10 industries in the study area by employment, according to the 2000 Census (U.S. Census Bureau, 2007). On the whole, these mirror national-level census results. Differences include a smaller share of workers employed in manufacturing (11.6 percent in the study region, compared to 14 percent in the Nation) and a larger share in construction (8.6 percent in the Gulf Coast area compared to 6.8 percent in the Nation). In addition, a much larger share of study area workers are employed in extraction industries (2.2 percent in the study area, versus 0.3 percent nationally).

The study region is host to nationally significant concentrations of several industries:

- **Oil and natural gas production and refining** – Much of the U.S. domestic oil production is supported by facilities in the Gulf of Mexico region – fixed oil platforms and mobile rigs, transportation systems, refineries, storage facilities, and distribution systems. An estimated 60 percent of all U.S. energy imports come through port facilities in the Gulf of Mexico region.

- **Chemical and petrochemical manufacturing** – Due to the presence of petroleum and natural gas supplies and infrastructure, the Gulf is a leading center for the U.S. chemical industry, which generally relies on expensive investments in fixed infrastructure.

- **Commercial fishing** – This is a multibillion dollar industry that is critical to the economies of many Gulf States.

As of 2003, the study area hosted approximately 214,768 private business establishments employing approximately 3,691,883 employees. The region experienced a 4 percent growth both in the number of establishments during the period from 1998 to 2003, and in the total number of employees. Despite this overall growth, certain counties have experienced decline and/or stagnation in businesses development. The growth versus decline patterns very closely match the same patterns as the population and housing discussed earlier, with suburbanizing counties on the periphery of the larger urban areas realizing most of the growth. Most notable again are the counties currently expanding westward and southward around Houston/Galveston, TX, west of Baton Rouge, LA, the counties of Louisiana's Northshore area, and Baldwin County west of Mobile Bay, AL. Orleans and Jefferson Counties, LA, (constituting the bulk of metro New Orleans) again stand out as having a relatively high rate of business decline in recent years, while the counties to the east and north have flourished.

Most rural counties have experienced decline or stagnation in terms of total businesses and total employees. These patterns again reflect the overall development and growth that is characterized by suburbanization in the region. In some areas, this trend may be more related to technological change in agriculture or petroleum extraction methods than a true decline in the general economy.

Counties with port facilities or Mississippi River access dominate the manufacturing shipments measured in dollar amounts (figure 2.23). Retail sales patterns, on the other hand, exhibit a less rational spatial pattern and seem to be tied to idiosyncratic changes in a small sample of counties. For instance the county of Waller, TX, in the farthest northwestern corner of the Houston area, registers a top value in terms of retail sales but a low value in terms of manufacturer's shipments. Much of this can be explained by the establishment of the Katy Mills Mall, which has caused the county to develop from one dominated by agriculture and industry to one based on a growing retail economy in recent years. Small-scale changes in the economic structure or productivity of specific sectors may be behind other local trends.

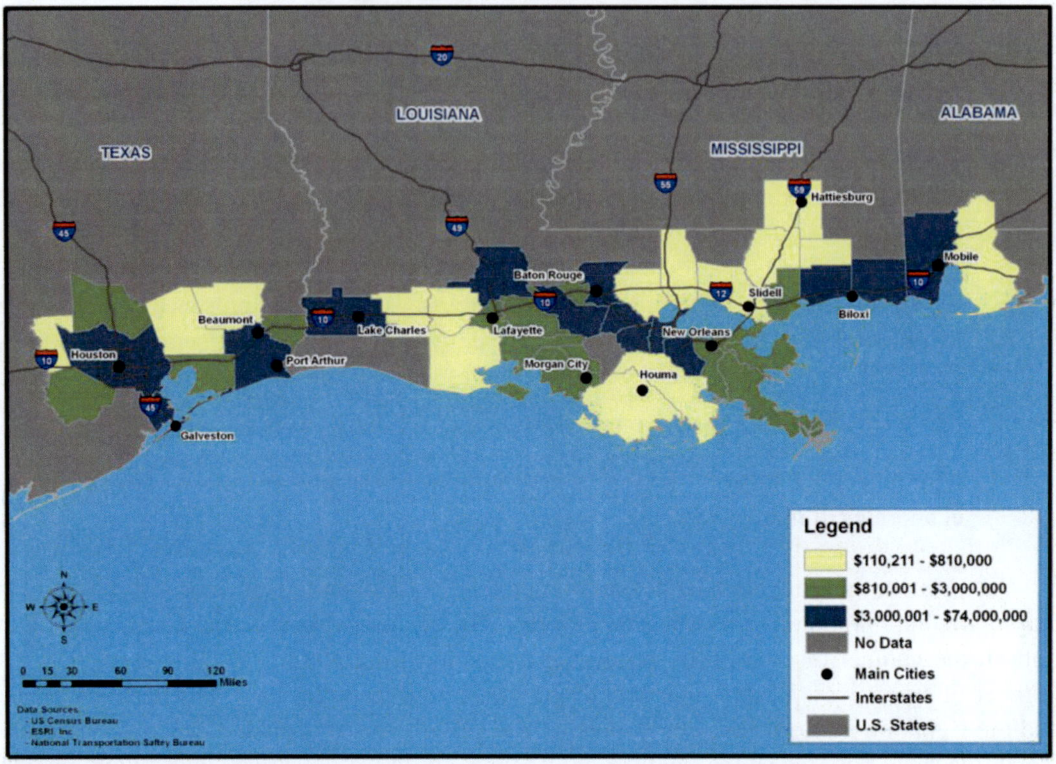

Source: U.S. Census Bureau; ESRI, Inc.; National Transportation Safety Bureau.

Figure 2.23. Manufacturers' shipments in thousands of dollars, 1997

2.4.3. Societal Vulnerability

Social vulnerability measures are important both as general background to the regional demographics but also to understand implications for future infrastructure needs and for emergency management. In this case, vulnerability refers to the inability of a social group to respond to, adapt to, or avoid negative impacts resulting from extreme or significant long-term deviations from average environmental conditions.

Generally, vulnerability assessments are conducted in respect to a single risk or hazard (flooding, radioactive release, drought, hurricane evacuation, etc.). For this study, the "hazards" are the anticipated impacts of climate change and variability, specifically as it relates to transportation interests. Since this encompasses multiple changes over a protracted time period, it is difficult, at this spatial and temporal scale, to comprehensively measure those features of the current social landscape that will be most vulnerable to future changes as they occur. Therefore, numerous social measures were included in this analysis in an effort to describe the most general patterns of vulnerability. The attributes included in this social vulnerability index are:

1. Percent persons reporting disabilities for civilian noninstitutionalized population five years and over;

2. Percent total population: Age 14 and below;
3. Percent total population: 65 years and over;
4. Percent households: Two-or-more-person household; family households; maritally single; with own children under 18 years;
5. Percent households: All languages; linguistically isolated;
6. Percent population 25 years and over: No high school graduate (includes equivalency);
7. Percent below study area median household income in 1999;
8. Percent households: With public assistance income;
9. Percent population for whom poverty status is determined: Income in 1999 below poverty level;
10. Percent housing units: Mobile home;
11. Percent housing units: Built 1969 or earlier;
12. Percent occupied housing units: No vehicle available;
13. Percent occupied housing units: Renter occupied;
14. Specified owner-occupied housing units: Percent below study area median value; and
15. Specified owner-occupied housing units: Percent housing units with a mortgage; contract to purchase; or similar debt; with either a second mortgage or home equity loan, but not both.

To illustrate how these multiple attributes can be agglomerated, these 15 measures were subjected to an indexing process to create a continuum of vulnerability at both the county- and block-group scale (most vulnerable, more vulnerable, less vulnerable, and least vulnerable). In future phases of this research, particularly for in-depth analysis of one site, the attributes included in this index can be changed or statistically weighted in response to particular transportation management or other concerns at that site. Figure 2.24 maps this vulnerability index for the study region. Maps depicting conditions within the region for each of the 15 societal attributes are contained in appendix B.

A number of patterns emerge from these measures of vulnerability. The first is the obvious pattern of counties with high degrees of social vulnerability expressed in the central portion of the Louisiana section of the study area. These counties correspond with the physical feature of the Atchafalaya River valley, the western portions of the Mississippi River valley, and the wetland landscapes produced by both. One can interpret from this analysis that these populations, if faced with extreme changes in their physical environments, will find coping with those changes extremely difficult. Many of these counties are traditionally rural, impoverished areas (figure 2.25). Also included is the urban-core county of Orleans, which ranks extremely high on many of the vulnerability measures included here.

However, poverty alone does not explain the higher rankings. These counties also tend to rank high in presence of disabled populations, persons over 65 (figure 2.26), absence of a vehicle per household, presence of single parents, linguistic isolation, and a number of other attributes. It can be argued that these are all dimensions of impoverishment. However, it is not the simple fact that a person is poor that makes them vulnerable; rather it is the context that widespread poverty can create in terms of public services, durability of infrastructure, access to egress, etc., acting together that make a community vulnerable to extreme environmental change.

To a lesser degree, this pattern of vulnerability extends southeastward into the delta region of central Louisiana. Other counties with similar characteristics outside central Louisiana tend to be rural and tertiary to urban-suburban growth. Exceptions to this statement are the heavily industrialized counties around Beaumont and Port Arthur, TX, Lake Charles, LA, and St. Bernard County, LA. The rapidly urbanizing county of Mobile, AL, also falls into this category of vulnerability.

Counties that tend to have fewer vulnerability characteristics are those on the periphery of large urban areas that were described earlier as undergoing the fastest rates of suburbanization. Again, this trend is tied heavily to overall income patterns but is not fully explained by that single attribute. For instance, these counties also tend to have higher rates of children per capita and more manufactured housing. It can be assumed that, at least for the time being, the populations of these counties will be better prepared to cope with the negative impacts of extreme environmental change.

From a transportation perspective, it also might be assumed that these areas will have special needs for transportation infrastructure in coming years. Vulnerable areas may need more services and infrastructure in the future to help them reduce their vulnerability – and to cope with destructive natural events – such as severe storms – as they occur.

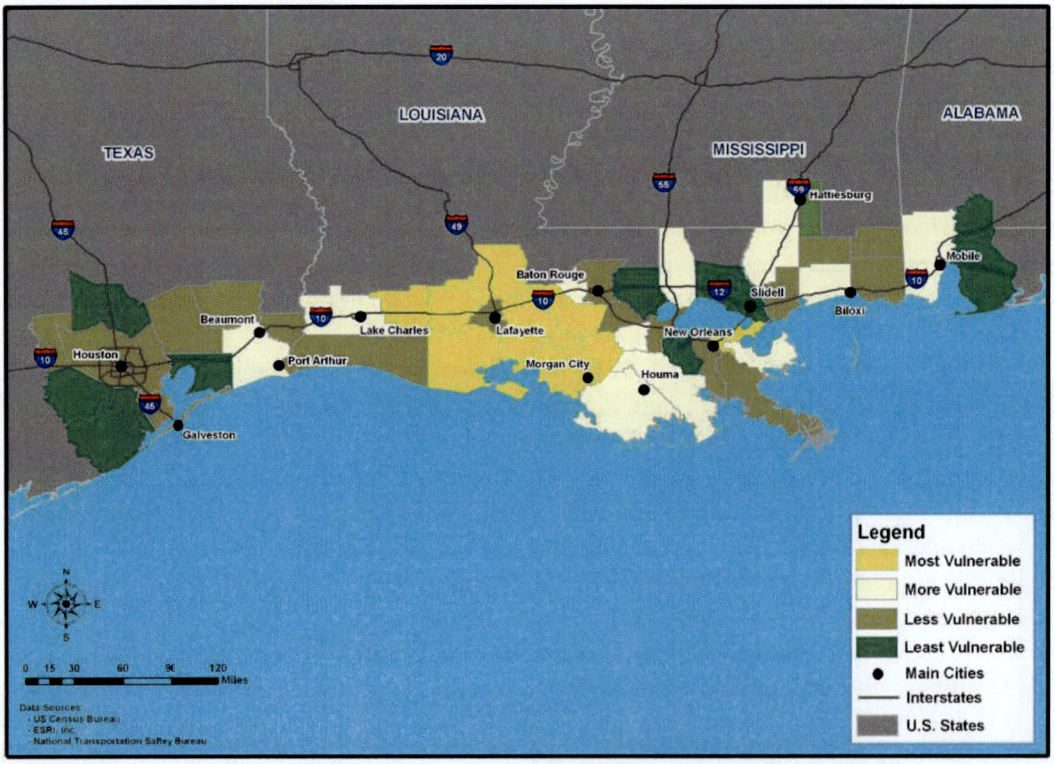

Source: U.S. Census Bureau; ESRI, Inc.; National Transportation Safety Bureau.

Figure 2.24. Social vulnerability index for the study area

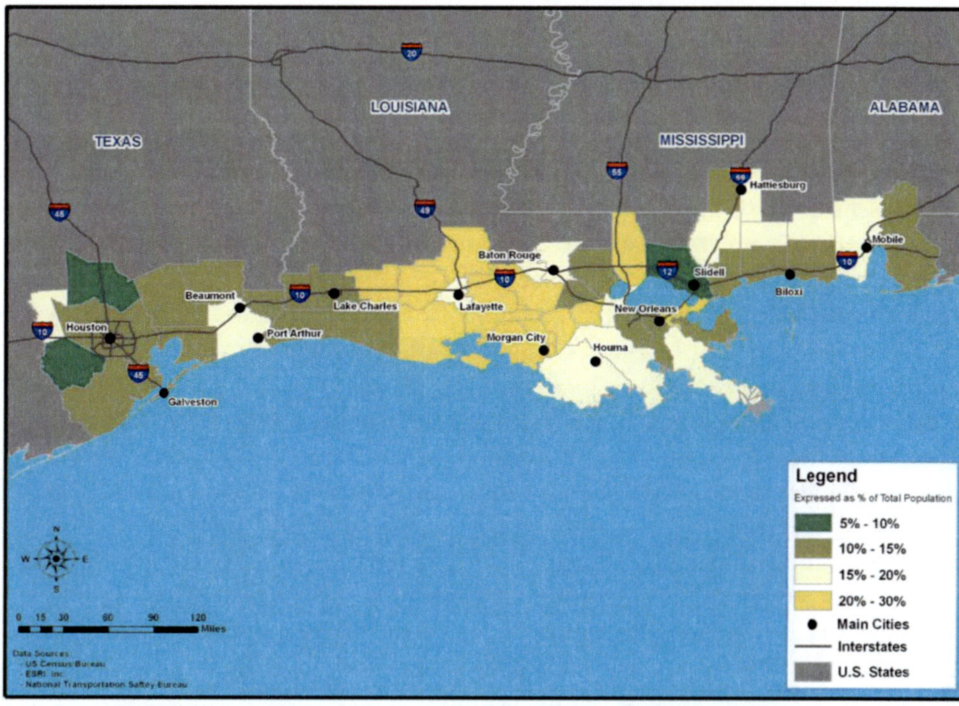

Source: U.S. Census Bureau; ESRI, Inc.; National Transportation Safety Bureau.

Figure 2.25. Persons in poverty in the study area

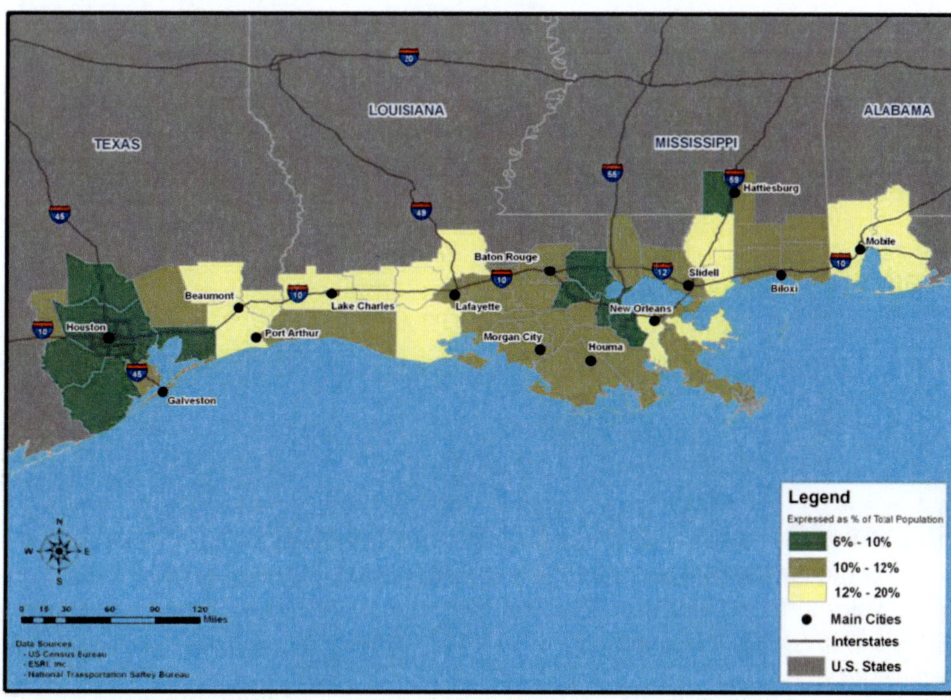

Source: U.S. Census Bureau; ESRI, Inc.; National Transportation Safety Bureau.

Figure 2.26. Persons aged 65 and older in the study area

2.5. Conclusions

The central Gulf Coast study area contains transportation infrastructure that is vital not just to the movement of passengers and goods within the study area but also to the national transportation network and economy. However, the geomorphology of the region makes it particularly sensitive to certain climate impacts. Due largely to its sedimentary history, the region is low-lying – much of it below 5 m – with little topographical relief. Much of the region experiences high rates of subsidence as these sediments naturally compact over time, while high rates of erosion mean that sections of coastline are literally washed away after tropical storms and hurricanes. As a result, the region is particularly vulnerable to the effects of sea level rise and storm activity.

In keeping with national trends, the region is experiencing a shift in population from rural to urban areas and increasing suburbanization of the larger urban areas. Much of the infrastructure supporting this population is located in vulnerable, low-lying areas. Parts of the population face vulnerabilities that may make it more difficult for them to adapt to the conditions imposed by a changing climate. This pattern of vulnerability is most focused in the rural counties of central coastal Louisiana, the urban core of New Orleans, and to a lesser extent southeastward into the delta region of Louisiana, and also into the rapidly urbanizing Mobile County, AL. On average, the population of the study area shows lower-income levels and higher poverty rates than the rest of the nation.

The following section will present the climate changes projected for the study area, while section 4.0 will discuss the resulting impacts to transportation systems in the central Gulf Coast region.

2.6. Sources

2.6.1. References

Airports Council International (ACI) (2005). *2005 North America Final Traffic Report.* Airports Council International – North America, Washington, D.C.

American Public Transportation Association (APTA) (2005). *Public Transportation Fact Book.* Washington, D.C., page 86. (Available on-line at http://www.apta.com/research/stats.)

Bailey, R. G. (1976). *Ecoregions of the United States*. USDA Forest Service, Intermountain Region, Ogden, Utah, map, scale 1:7,500,000.

Barras, J. A. cited (2006). Land area change in coastal Louisiana after the 2005 hurricanes – a series of three maps. U.S. Geological Survey Open-File Report 2006-1274. (Available on-line at http://pubs.usgs.gov/of/2006/1274/. Accessed October 18, 2006.)

Bureau of Transportation Statistics (BTS) (2004). *National Transportation Atlas Database (NTAD) 2004.* U.S. Department of Transportation, Washington, D.C., CD-ROM. (Available on-line at http://www.bts.gov/publications/national_transportation_atlas_database/.)

Burkett, V. B., Zilkoski, D. B. & Hart, D. A. (2003). Sea level rise and subsidence: Implications for flooding in New Orleans, Louisiana. In: *USGS Subsidence Interest*

Group Conference (Prince, K.R., and D.L. Galloway (Eds.)). Proceedings of the Technical Meeting, Galveston, Texas, 27–29 November 2001, Open-File Report Series 03-308, Water Resources Division, U.S. Geological Survey, Austin, Texas, pages 63-70.

Coplin, L. S. & Galloway, D. (1999). Houston-Galveston, Texas. Managing coastal subsidence. In: *Land Subsidence in the United States* (Galloway, D., D.R. Jones, and S.E. Ingebritsen (Eds.)). U.S. Geological Survey Circular 1182, U.S. Geological Survey, pages 35-48.

Dixon, T. H., Amelung, F., Ferretti, A., Novali, F., Rocca, F., Dokka, R., Sella, G., Kim, S. W., Wdowinski, S. & Whitman, D. (2006). Subsidence and flooding in New Orleans. *Nature*, *441*, 587-588.

Federal Highway Administration (FHWA) (2001). *2001 National Bridge Inventory Database.* Office of Infrastructure, Washington, D.C.

Federal Highway Administration (FHWA) (2004). *Freight Analysis Framework.* Freight Management and Operations, Office of Operations, Washington, D.C. (Dataset)

Federal Highway Administration (FHWA), (2005). *Highway Statistics 2004.* Office of Highway Policy Information, Washington, D.C.

Federal Highway Administration (FHWA), cited (2006). *Official NHS Intermodal Connector Listing*. Office of Planning, Environment, and Realty, Washington, D.C. (Available on-line at http://www.fhwa.dot.gov/hep10/nhs/intermodalconnectors/.)

Federal Transit Administration (FTA) (2005). *2004 National Transit Database.* Washington, D.C. (Available on-line at http://www.ntdprogram.com/NTD/ntdhome.nsf.)

Holzer, T. L. & Gabrysch, R. K. (1987). Effect of ground-water level recoveries on fault creep, Houston, Texas. *Ground Water*, *25*, 392–397.

Institute for Water Resources. (2003). *Waterborne Commerce of the United States.* Waterborne Commerce Statistics Center, Navigation Data Center, U.S. Army Corps of Engineers, Arlington, Virginia.

Minerals Management Service (MMS) (1994). *Backfilling Canals as a Wetland Restoration Technique in Coastal Louisiana.* MMS Publication 94-0026, Minerals Management Service, Washington, D.C.

Morton, R. A., Bernier, J. C., Barras, J. A. & Ferina, N. F. (2005). *Rapid Subsidence and Historical Wetland Loss in the Mississippi Delta Plain: Likely Causes and Future Implications*. U.S. Geological Survey Open-File Report 2005-12 16, 115 pages.

Pipeline and Hazardous Materials Safety Administration (PHMSA), (2007). *Annual Reports: National Pipeline Mapping System.* PHMSA, U.S. Department of Transportation, Washington, D.C.

Reed, D. J. (2002). Sea level rise and coastal marsh sustainability: geological and ecological factors in the Mississippi delta plain. *Geomorphology, 48*, 23 3–243.

Rybczyk, J. M. & Cahoon, D.R. (2002). Estimating the potential for submergence for two subsiding wetlands in the Mississippi River delta. *Estuaries*, *25*, 985–998.

Shingle, K. D. & Dokka, R. K. (2004). *Rates of Vertical Displacement at Benchmarks in the Lower Mississippi Valley and the Northern Gulf Coast.* NOAA Technical Report NOS/NGS 50, National Oceanic and Atmospheric Administration, U.S. Department of Commerce, Washington, D.C., 135 pages.

Stone, G. W., Morgan, J. P., Sheremet, A. & Zhang, X. (2003). *Coastal land loss and wave-surge predictions during hurricanes in Coastal Louisiana: implications for the oil and gas industry.* Coastal Studies Institute, Louisiana State University, Baton Rouge, page 67.

Stone, L., Huppert, A., Rajagopalan, B., Bhasin, H. & Loya, Y. (1999). Mass coral bleaching: a recent outcome of increased El Niño activity? *Ecology Letters, 2,* 325- 330.

Thieler, E.R. and E.S. Hammar-Klose. (2000). *National Assessment of Coastal* Vulnerability to Future Sea Level Rise: Preliminary Results for the U.S. Gulf of *Mexico Coast.* U.S. Geological Survey Open-File Report 00–1 79. Available on-line at <http://pubs.usgs.gov/dds/dds68/>, viewed June 28, 2005.

U.S. Census Bureau (2005). *Special Product for the Gulf Coast Area.* American Community Survey Office, Washington, D.C. (Available on-line at http:// www.census.gov/acs/www/Products/Profiles/gulf_coast/index.htm.)

U.S. Census Bureau, cited 2007: *United States Census 2000.* Washington, D.C.

U.S. Geological Survey (USGS) (2000). The coastal plain. In: *A Tapestry of Time and Terrain: The Union of Two Maps – Geology and Topography.* U.S. Geological Survey, Reston, Virginia. Available on-line at http://tapestry.usgs.gov/features/ 13coastalplain.html, viewed June 20, 2006.

U.S. Geological Survey (USGS) (2004). *National Elevation Dataset (NED).* U.S. Geological Survey, EROS Data Center, Sioux Falls, South Dakota. (Available on-line at http://ned.usgs.gov/.)

U.S. Geological Survey (USGS) (2007). Predicting flooding and coastal hazards. *Sound Waves Monthly Newsletter,* U.S. Geological Survey, Coastal and Marine Geology Program, June, page 1.

Woodroffe, C. D. (2003): *Coasts: Form, Process and Evolution.* Cambridge University Press, Cambridge, United Kingdom, page 623.

2.6.2. Background Sources

American Association of Port Authorities (AAPA), (2004). *North America Port Container Traffic 2004.* American Association of Port Authorities, Alexandria, Virginia. (Available on-line at http://www.aapa-ports.org/Industry/content.cfm?ItemNumber =900&navItemNumber=55 1).

American Geophysical Union. (2006). *Hurricanes and the U.S. Gulf Coast: Science and Sustainable Rebuilding.* American Geophysical Union, Washington, D.C., page 30. (Available on-line at http://www.agu.org/report/hurricanes/, viewed June 29, 2006.)

End Notes

[1] According to FHWA's Highway Statistics, while local roads represent 75 percent of the miles of the Nation's highways (Table HM-18), they carry less than 0.2 percent of the Nation's vehicle miles of travel (VMT) (Table HM-44) (FHWA, 2005).

[2] i.e., scheduled bus service on fixed guideways or HOV lanes.

[3] Railroad classification is determined by the Surface Transportation Board. In 2004, a class I railroad was defined as having $289.4 million or more in operating revenues. A class II railroad, often referred to as a regional railroad, was defined as a non-class I line-haul railroad operating 563 km (350 mi) or more with operating revenues of at least $40 million. Class III railroads, or short lines, are the remaining non-class I or II line-haul railroads. A switching or terminal railroad is a railroad engaged primarily in switching and/or terminal services for other railroads.

[4] Canadian Pacific Railway is the only North American class I railroad not serving the study region.

[5] A transloading facility handles "nonflowing" commodities transferred between railcars and trucks for customers without direct rail service. Examples include steel, lumber, and paper. A transflow facility handles "flowing" commodities transferred between railcars and trucks, such as corn syrup, petroleum products, and plastic pellets.

[6] Ports with the ability to load and unload container ships, and transfer the shipping containers to or from other modes of travel, usually rail or truck.

[7] Heliports primarily serve hospitals, office buildings, and oil and gas industry facilities.

[8] The U.S. Census Bureau term "County" is used here for consistency in Louisiana, rather than the more common term "Parish." Both indicate the same political unit.

In: Impacts of Climate Change on Transportation and… ISBN: 978-1-60741-424-7
Editor: Iason Pavlopoulos

Chapter 3

How Is the Gulf Coast Climate Changing?

Lead Authors: Barry D. Keim, Thomas W. Doyle and Virginia R. Burkett

Contributing Authors: Claudia Tebaldi, Ivor Van Heerden, S. Ahmet Binselam, Michael F. Wehner, Tamara G. Houston and Daniel M. Beagan

The central Gulf Coast is one of warmest, wettest regions in the United States, where annual rainfall averages over 150 cm (60 inches) per year (Christopherson, 2000). Since there is very little topographic relief, changes in precipitation and runoff could have a dramatic impact on fragile Gulf Coast ecosystems and coastal communities by changing the hydroclimatology of the region. Changes in runoff are important to virtually all transportation modes in the Gulf Coast region. Interstate highways in Houston and New Orleans, for example, are occasionally flooded by locally intense rainfall, and several State and local highways are closed due to high rainfall at least once every five years. Even ports can be affected by high rainfall and runoff to shallow coastal waterways. Changes in temperature and moisture regime also are relevant to many aspects of transportation planning, construction, and maintenance. Airport runway length requirements, for example, are determined by mean maximum temperature for the hottest month of the year. As the climate and sea surface warm, we can anticipate an increase in the intensity of hurricanes making landfall along the Gulf of Mexico coastline. As the ocean warms and ice sheets decline, sea level rise is likely to accelerate, which has serious implications for the Gulf Coast region where much of the land is sinking (subsiding) due to local geological processes and human development activity.

This chapter summarizes the direct and indirect effects of climate change that are most likely to affect transportation in the Gulf Coast region. The key climate "drivers" examined in the study region are:

- Temperature;
- Precipitation;

- Sea level rise; and
- Hurricanes and less intense tropical storms.

The interactive effects of these drivers, coupled with ongoing environmental processes in the region, are discussed in the following sections. This chapter presents scenarios of future climate change in addition to analysis of historical trends. While the environmental trend information for the study area is compiled from region-specific data sources, a regional model of future climate was not available for the Gulf Coast study area. One approach that is widely used to identify plausible changes in climate at a regional scale is to extract output from general circulation models run at a global scale. This approach was used in this study and is described in the following sections of the report. Specific implications of the scenarios of future climate for each mode of transportation are discussed in the subsequent chapter of this report.

Intended Use of Climate and Emissions Scenarios in the Context of This Report

A "scenario" is a plausible description of possible future conditions and is generally developed to inform decision making under uncertainty. Building and using scenarios can help people explore what the future might look like and the likely challenges of living in it (Shell International, 2003). Scenarios are distinct from assessments, models, and similar decision-support activities, although they can provide important inputs to these activities. Scenarios also can be distinguished from precise statements about future conditions, which may be referred to as "forecasts" or "predictions." Compared to these, scenarios tend to presume lower predictive confidence, because they pertain to processes for which weaker causal understanding or longer time horizons increase uncertainties (Parson et al., 2007).

Climate scenarios describe potential future climate conditions and are used to inform decision making relative to adaptation and mitigation. Scenarios can be constructed for higher order aspects of climate change and its impacts, such as future changes in sea level, drought and storm intensity, or vegetation distribution. Scenarios of relative sea level rise, for example, in a subsequent section of this report were constructed by combining climate-change scenarios with information about coastal subsidence and other specific regional characteristics. The climate and sea level rise scenarios discussed in this report identify plausible potential future conditions for the Gulf Coast region. They are intended to frame the analysis of potential risks and vulnerability within the transportation sector.

The Earth's climate is determined, in part, by the concentration of atmospheric greenhouse gases and particulates that absorb infrared radiation (heat) reflected from the Earth's surface. Human activity is increasing greenhouse gas and particulate emissions, which has resulted in an increase in the Earth's temperature (Intergovernmental Panel on Climate Change [IPCC], 2001, 2007). In order to assess how the climate may change in the future, future emissions must be specified. The Intergovernmental Panel on Climate Change (IPCC) has conducted three exercises to generate scenarios of 21st century greenhouse-gas emissions, the most recent being the IPCC Special Report on Emissions Scenarios (SRES) (Nakicenovic and Swart, 2000). To explore the potential effects on transportation, we selected a range of emissions futures from the SRES report, including the low-emissions B1 scenario, the mid-range A1B scenario, and the high-emissions A2 scenario. The AIFI scenario, which assumes the highest reliance on fossil fuels during this century, also was added to the SRES scenarios used to assess the effects of sea level rise.

The SRES A1B scenario assumes a balance across all energy sources, meaning it does not rely too heavily on any one particular source, including fossil fuels. It is, therefore, based on the assumption that improvement rates apply to all energy sources and end-use technologies. The A2 scenario assumes that economic development is primarily regionally oriented and that per capita economic growth and technological change are more fragmented and slower than for the other emission scenarios. The B1 scenario assumes a high level of social and environmental awareness with an eye toward sustainability. It includes an increase in resource efficiency and diffusion of cleaner technologies (IPCC, 2001). These three emission scenarios are among the six "marker/illustrative scenarios" selected for climate model simulations in the IPCC's Third and Fourth Assessment Reports (IPCC, 2001, 2007) (figure 3.1). The B1 scenario lies at the lower extreme end of the potential changes in atmospheric CO_2 concentrations during this century, while the A1B emission scenario is considered a middle-of-the-range scenario in terms of the hypothesized rate of greenhouse gas emissions. The A2 scenario is among the higher end of the SRES scenarios in terms of both CO_2 and SO_2 emissions. The influence of SRES emission scenarios on global temperature simulations is presented in table 3.1.

3.1. Temperature, Precipitation, and Runoff

The climate of the study area is influenced by remote global factors, including the El Niño Southern Oscillation, and regional factors such as solar insolation. Due to the influence of the nearby Gulf of Mexico, the region is warmer and moister than most other continental regions at this latitude. Rainfall across the study area has little seasonality, with slightly higher rainfall values in spring and summer relative to fall and winter. The region enjoys mild winters, which are occasionally interrupted by cold air masses extending far south from the northern pacific or the Arctic, which brings low temperatures and freezing conditions. Rainfall in the region is dependent upon a variety of processes, including frontal passages in the winter and spring (Twilley et al., 2001). Short-lived, unorganized thunderstorms fueled by afternoon heating and moisture are common in the study area and associated, in part, with a prominent sea/land breeze (Ahijevych et al., 2003).

The Gulf Coast, like much of the world, has experienced significant changes in climate over the past century. With continued increases in atmospheric greenhouse gases and their radiative forcing, the Earth's climate is expected to change even more rapidly during the 21st century (IPCC, 2007). Computer-based climate simulation models are used to study the present climate and its responses to past perturbations like variation in the sun's output or major volcanic eruptions. They also are used to assess how the future climate would change under any specified scenario of greenhouse-gas emissions or other human activity (Parson et al., 2007).

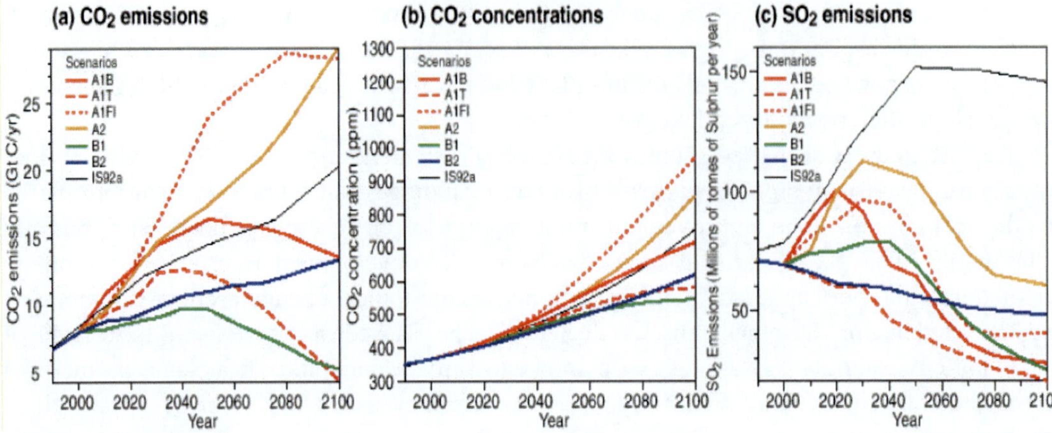

Source: IPCC, 2001.

Figure 3.1. CO_2 emissions, SO_2 emissions, and atmospheric CO_2 concentration through 2100 for the six "marker/illustrative" SRES scenarios and the IS92a scenario (a "business as usual" scenario, IPCC [1992])

Table 3.1. Projected Global Average Surface Warming and Sea Level Rise at the End of the 21st Century (IPCC, 2007). These Estimates are Assessed from a Hierarchy of Models that Encompass a Simple Climate Model, Several Earth Models of Intermediate Complexity (EMIC), and a Large Number of Atmosphere-Ocean Global Circulation Models (AOGCM). Sea Level Projections do not Include Uncertainties in Carbon-Cycle Feedbacks Because a Basis in Published Literature is Lacking (IPCC, 2007)

Case	Temperature Change (°C from 2090-2099 Relative to 1980-1999)		Sea Level Rise (m) from 2090-2099 Relative to 1980-1999)
	Best Estimate	Likely Range	Model-Based Range, Excluding Future Rapid Dynamical Changes in Ice Flow
Constant Year 2000 Concentrations	0.6	0.3-0.9	NA
B1 Scenario	1.8	1.1-2.9	0.18-0.38
A1T Scenario	2.4	1.4-3.8	0.20-0.45
B2 Scenario	2.4	1.4-3.8	0.20-0.43
A1B Scenario	2.8	1.7-4.4	0.21-0.48
A2 Scenario	3.4	2.0-5.4	0.23-0.51
A1F1 Scenario	4.0	2.4-6.4	0.26-0.59

3.1.1. Historical Data Sources

Changes in the historical climatology of the study area were investigated from an empirical perspective relying on instrumental records. The assessment of the present climate and 20th-century trends was built around climatic data from the United States Climate

Division Datasets (CDD) (Guttman and Quayle, 1996) and the United States Historical Climate Network (USHCN) (Karl et al., 1990; Easterling et al., 1996). Since CDD were used in a portion of this analysis, caution needs to be taken with data from 1905 to 1930, which are synthesized from statewide data as described by Guttman and Quayle (1996) and therefore are not true averages of data from within a climate division.

Empirical trends and variability were analyzed for temperature and precipitation at the CDD level for the climate divisions along the Gulf Coast from Galveston, TX, to Mobile, AL, including Texas Climate Division 8, Louisiana Divisions 6-9, Mississippi Division 10, and Alabama Division 8 (figure 3.2).

Table 3.2. United States Historical Climatology Network (USHCN) Stations Within the Seven Climate Divisions of the Central Gulf Coast Region

Climate Division (CD)	USHCN Stations
Texas CD 8	Danevang, Liberty
Louisiana CD 7	Jennings[1]
Louisiana CD 8	Franklin, Lafayette
Louisiana CD 9	Donaldsonville, Houma, New Orleans, Thibodaux
Louisiana CD 6	Amite, Baton Rouge, Covington
Mississippi 10	Pascagoula, Poplarville, Waveland
Alabama CD 8	Fairhope

[1] The Jennings climate record only dates back to the late 1960s. As a result, CD 7 is made up of an average of Liberty, TX, to the west and Lafayette, LA, to the east.

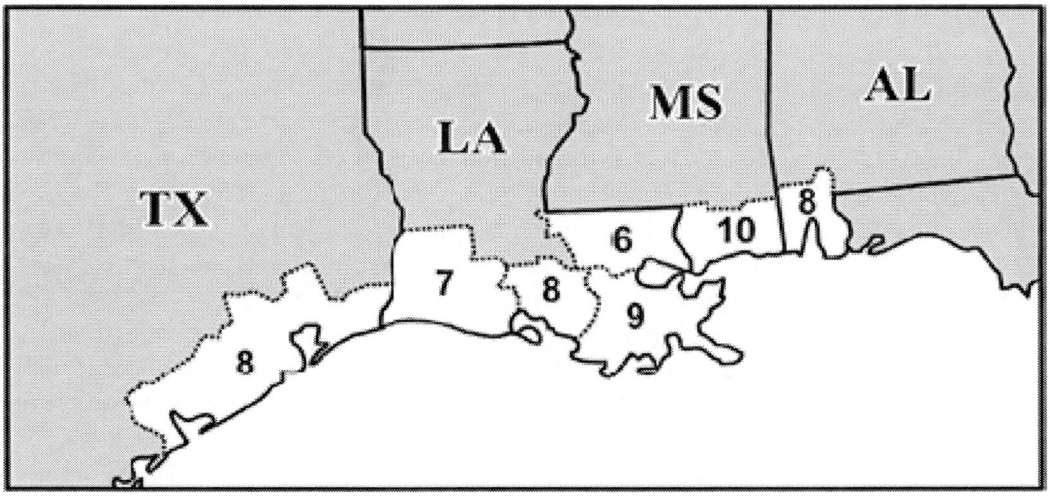

Figure 3.2. United States climate divisions of the central Gulf Coast study area. Empirical trends and variability were analyzed for temperature and precipitation at the Climate Division Dataset-(CDD) level for the climate divisions along the Gulf Coast from Galveston, TX, to Mobile, AL, including Texas Climate Division 8, Louisiana Divisions 6-9, Mississippi Division 10, and Alabama Division 8. These climatic divisions cover the entire central Gulf Coast study area

Keim and others (2003) showed that CDD can have spurious temperature trends. Our analysis synthesized CDD consisting of averages of stations within each division from the USHCN (table 3.2). The Filnet data have undergone numerous quality assurances and adjustments to best characterize the actual variability in climate. These adjustments take into consideration the validity of extreme outliers, time of observation bias (Karl et al., 1986), changes in instrumentation (Quayle et al., 1991), random relocations of stations (Karl and Williams, 1987), and urban warming biases (Karl et al., 1988). Furthermore, missing data were estimated from surrounding stations to produce a nearly continuous dataset for each station.

Monthly averages from the USHCN stations from 1905 to 2003 within each climate division were then averaged annually, thereby constructing an alternative "divisional data" annual time series. The year 1905 was selected as a starting point because it represents a common period of record for all but one of the USHCN stations utilized in the study – the exception is Fairhope, AL, beginning in 1919. Fairhope was maintained because it is the only USHCN station available in Alabama Climate Division 8. Only USHCN FILNET stations with a continuous monthly record of temperature from January 1905 through December 2003 were included in the analysis, with the exception of Fairhope. USHCN precipitation data were not as serially complete as temperature, and there were fewer stations available. As a result, this study incorporated the original CDD for precipitation, which seems reasonable given results of Keim and others (2005).

3.1.2. General Circulation Model Applications for the Study Area

The scenarios of future climate referenced in this report were extracted from an ensemble of up to 21 different atmosphere-ocean general circulation model (GCM) efforts, which contributed the results of their simulations in support of the IPCC's Fourth Assessment Report, and are labeled "Coupled Model Intercomparison Project 3" (CMIP3). Gridded output limited to the study area was extracted from each GCM. Figure 3.3 shows the study region and the boundaries used to subset the global grid of a typical GCM output. Results are presented as spatial averages across the entire area. The GCMs were run under three forcings, the low-emissions B1, the high-emissions A2, and the mid-range A1B scenarios from the IPCC's SRES (Nakicenovic and Swart, 2000).

Scenarios of future temperature and precipitation change for the middle of the 21st century were derived from the regional GCM runs. Scatter diagrams were produced to convey the range of output of the models with respect to present conditions following the procedures of Ruosteenoja et al. (2003) (figure 3.4). Probability density (or distribution) functions (PDF) were developed by applying the method of Tebaldi and others (2004, 2005). Data forming the basis of the PDF estimation is an ensemble of historical and future climate simulations (from which temperature and precipitation are extracted). Output of temperature and precipitation (averaged for area l and seasonal fluctuations) from up to 21 different GCMs under the three different scenarios was considered for two 20-year periods, one representative of recent climatology (1980-1999) and one representative of the future mid-century time slice (2040-205 9). Thus scenarios of "climate change" are to be interpreted with respect to these two time periods and conditional on the SRES A1B, A2, and B1 scenarios (Nakicenovic and

Swart, 2000). While the results from the GCM runs are indeed plausible, they should be interpreted as mid-, high-, and low-range results, respectively, among the SRES scenarios of the potential changes in temperature and precipitation.

The statistical procedure synthesizes the multimodel ensemble of projections into a continuous PDF by applying a Bayesian method of estimation. At the core of the method is the idea that both observed and modeled temperature and precipitation contribute information to the estimate, so that different models will be differently "weighted" in the final probabilistic projections on the basis of their differential skill in reproducing observed climate. The method used also considered the convergence of different models when producing future trajectories, rewarding models that agree with one another and downweighting outliers. In the version of the statistical procedure applied here, the latter criterion is discounted, ensuring that even model projections that disagree with the consensus inform the shape of the final PDFs. This choice is made as a result of two considerations: the ensemble of GCMs at our disposal is not made of independent models (there are components and algorithms in common, for example), so rewarding agreement is somewhat questionable when one can argue that the agreement is not independently created. The second consideration has to do with the width of the PDFs produced, since enforcing the convergence criterion has the effect of narrowing the width of the PDFs to a range even smaller than the original ensemble range. It is well understood that the range of uncertainty addressed by this particular ensemble of models is limited when compared to the whole range of sources of uncertainty that can be listed, when examining climate change projections. Thus we preferred to produce conservative estimates of the uncertainty (i.e., larger rather than smaller). The result of applying the statistical analysis to the GCM output are PDFs of temperature and precipitation change (the latter as absolute values or percent change with respect to historical precipitation averages) from which any percentile can be derived.

3.1.3. Water-Balance Model

The primary tool used to investigate the hydroclimatology of the study area was a modified Thornthwaite Water Balance Model as described by Dingman (2002). The Thornthwaite model is simply an accounting of hydroclimatological inputs and outputs. Monthly values of temperature, precipitation, and potential evapotranspiration – called reference evapotranspiration – were entered into the budget, and parameters such as rain/snow ratios, soil moisture, soil moisture deficits, and runoff were calculated. The water balance was modified slightly by using an alternative reference evapotranspiration (ETo) term than that originally used by Thornthwaite to provide a better estimate of ETo in the central Gulf Coast region. As with any monthly water balance, atmospheric and terrestrial variables (such as ET_0, soil moisture, runoff, etc.) were parameterized by using bulk terms. A description of the procedures used to estimate evapotranspiration, soil moisture, and other components of the water balance model are presented in appendix D.

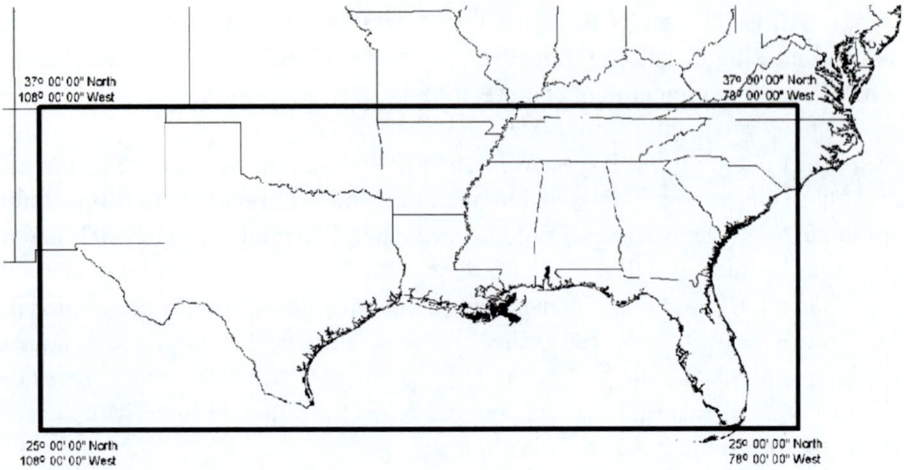

Figure 3.3. Grid area for the GCM temperature and precipitation results presented in section 3.15 of this report, which is a subset of the global grid of a typical GCM output

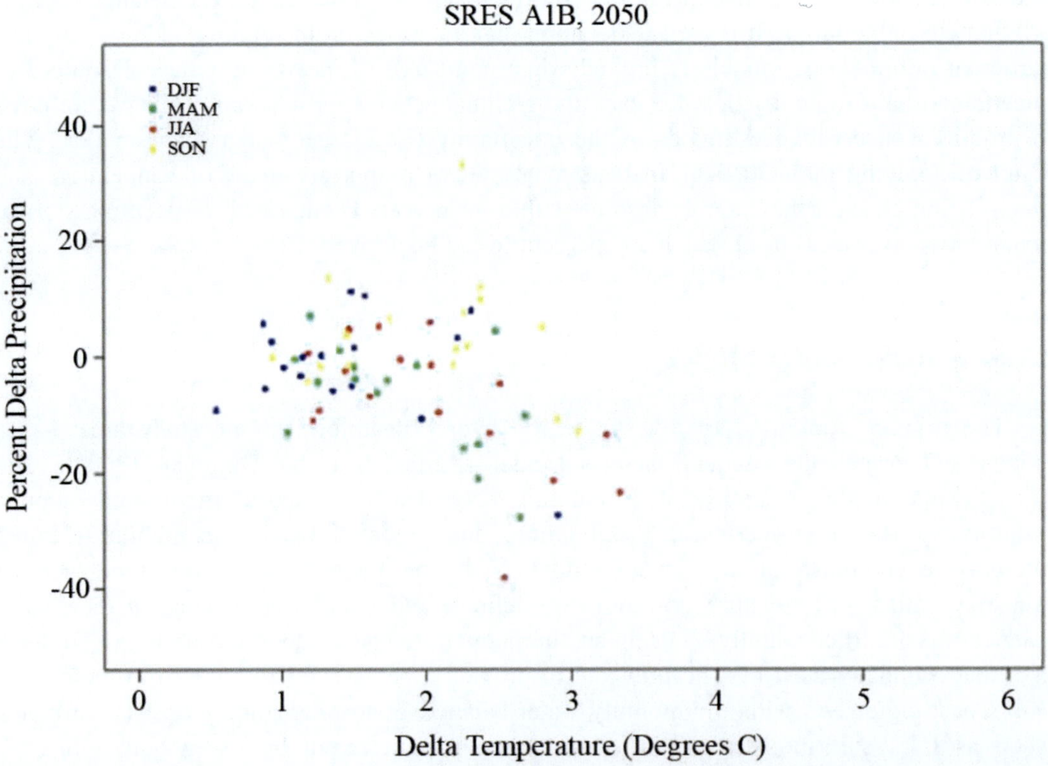

Figure 3.4. Scatterplot of seasonal temperature and precipitation predictions by an ensemble of GCMs for the Gulf Coast region in 2050 created by using the SRES A1B emissions scenario

3.1.4. Temperature and Runoff Trends

Results from our analysis of temperature variability during 1905 to 2003 indicate that the 1920s or 1930s was generally the warmest decade for the various Gulf Coast climate divisions (figure 3.5). After a step down in the temperature in the late 1950s, the coolest period occurs in the 1 960s, while a warming trend is evident for all seven climate divisions beginning in the 1970s and extending through 2003. Of the seven climate divisions, LA6, LA8, and MS10 have slight but significant cooling trends at an α ≤.05 over the 98-year period of record. Precipitation variability shows that the 1 940s and 1 990s were the wettest decades, while the 1 950s was generally the driest (figure 3.6). Although all of the climate divisions at least suggest long-term patterns of increasing rainfall, only MS10 and AL8 have trends that are significant at an α ≤.05.

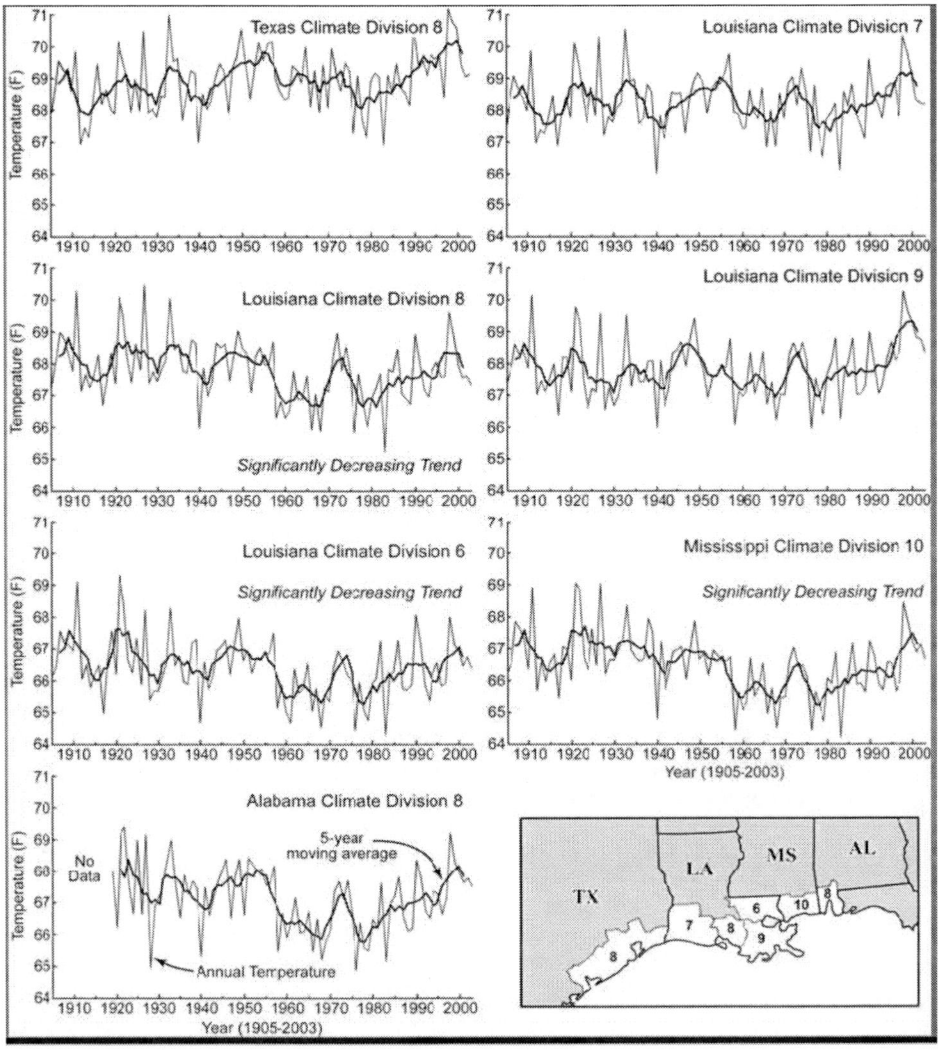

Figure 3.5. Temperature variability from 1905 to 2003 for the seven climate divisions making up the Gulf Coast study area. The level of significance in long-term temperature trend within each division was determined at a≤0.05

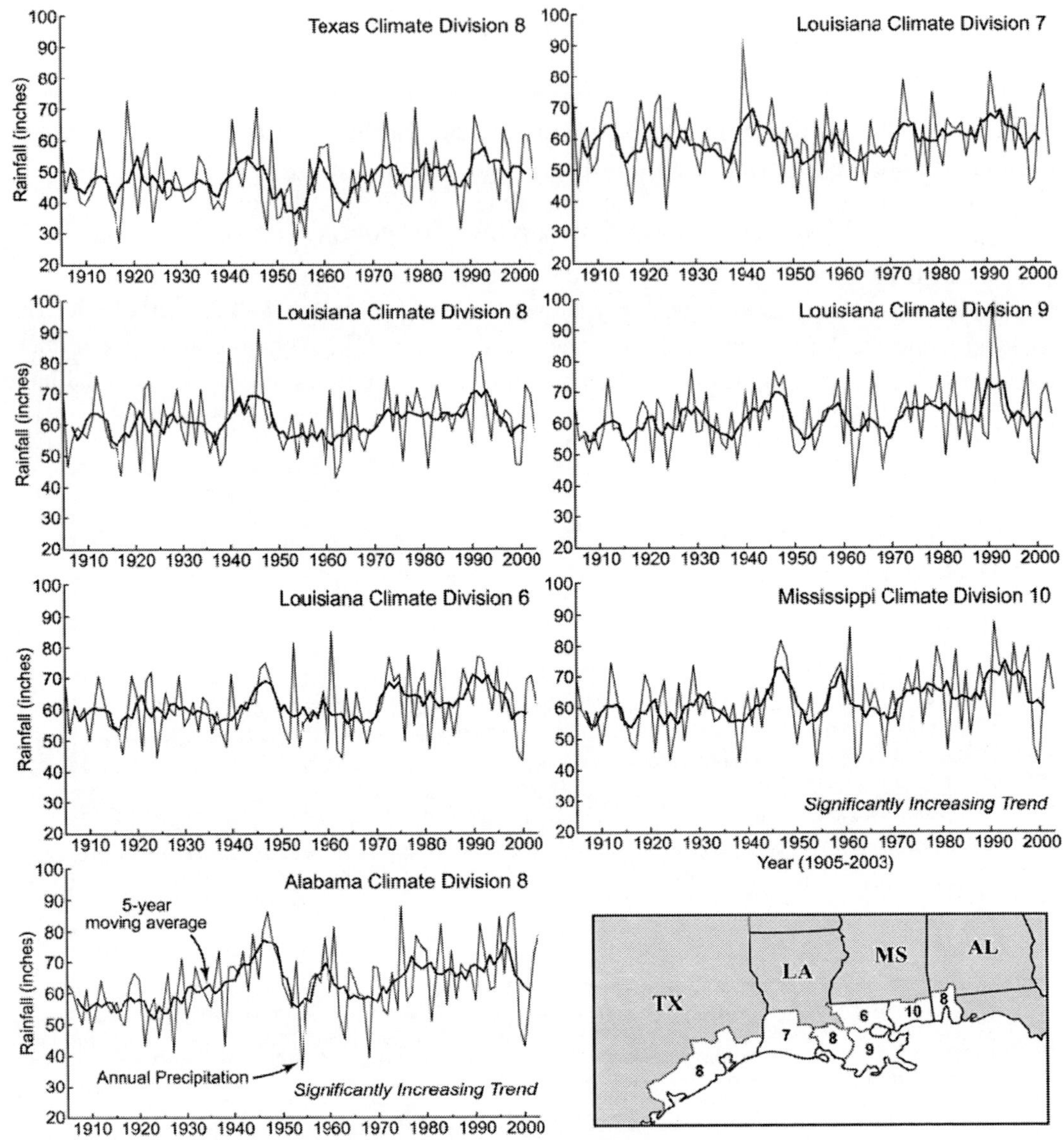

Figure 3.6. Precipitation variability from 1905 to 2003 for the seven climate divisions making up the Gulf Coast study area. The level of significance in long-term precipitation trend within each division was determined at a≤0.05

Data for each of the seven climate divisions were amalgamated into a regional dataset, by month, and the continuous monthly water balance model was run. In a typical year, ET_0 is low in winter and early spring, and most rainfall is converted to runoff because soil moisture storage remains at, or near, capacity. As temperatures rise in late spring and early summer and the number of hours of daylight increases, ET_0 also increases. Evapotranspiration will often exceed rainfall in July, August, and September, which leads to soil moisture utilization, on the average. Then in late fall, precipitation often exceeds ET_0 leading to recharge of soil moisture. Regional trends in model-derived runoff shows large inter-annual variability with the high values in the 1 940s and from 1975 to 2003 (figure 3.7). Despite the variability, a long-term trend was detected in the data at an $\alpha \leq .05$, and the trend line indicates a 36 percent increase in runoff over the time period. Moisture deficits show high values from the mid-1940s through the mid-1960s, with 1998 to 2000 also high (figure 3.7) but without any long-term trends during that period.

Historical monthly extremes of precipitation, runoff, and deficit in the Gulf Coast Region were analyzed to provide a focus for this portion of the analysis. In the empirical record, there is some evidence of an increase in precipitation extremes in the United States and in the Gulf south. Karl et al. (1995) shows that one-day extreme rainfall events have increased in portions of the United States, and Keim (1997) shows heavy rainfall events have increased in the south-central United States. These heavy rainfall events have very likely contributed the increases in runoff found in this study.

The period 1971 to 2000 serves as the baseline climatology for this analysis. Using water balance output for this 30-year period, partial duration series (PDS) are generated for the three variables. A PDS includes the number of events (monthly extremes) equal to the number of years under examination, which is 30 in this case. As such, the 30 largest monthly totals of precipitation, runoff, and deficit were extracted and then fit to the beta-p distribution, as recommended by Wilks (1993), and the 2-, 5-, 10-, 25-, 50-, and 100-year quantile estimates are determined for each. These data serve as a baseline for assessing potential future changes in extremes of precipitation, runoff, and deficit.

3.1.5. General Circulation Model Results and Future Climate Scenarios

To explore how the regional climate may change over the next 50 years, output from an ensemble of GCM runs used by IPCC for the Fourth Assessment Report (2007) was analyzed. Scatterplots and probability density functions of average temperature and precipitation change were derived from the GCM ensemble output for the IPCC SRES greenhouse gas emissions scenarios labeled A1B, A2, and B1. The results presented in the following discussion are based on GCMs (table 3.3) that contributed runs to the IPCC archive used in the IPCC's Fourth Assessment Report and are consistent with the temperature and precipitation projections reflected in IPCC Fourth Assessment Report (2007).

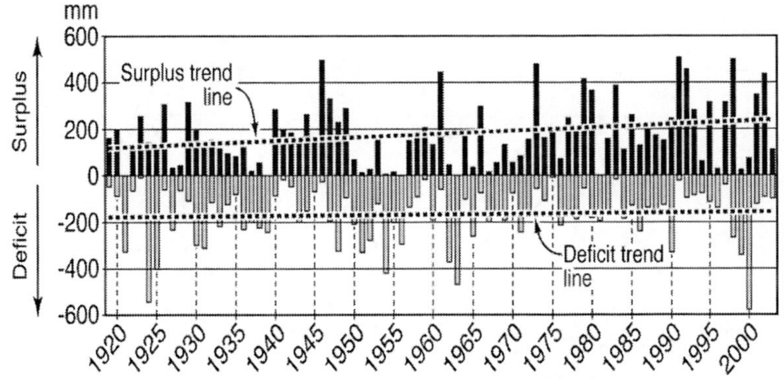

Figure 3.7 .Variability and trends in model-derived surplus (runoff) and deficit from 1919 to 2003 for the Gulf Coast study area

The GCM results run with the A1B, A2, and B1 emissions scenarios suggest a warmer Gulf Coast region, with the greatest increase in temperature occurring in summer and lowest increases in winter (tables 3.4, 3.6, and 3.8). This is consistent with another analysis of historical data that shows a significant increase in summer minimum temperature across the Gulf Coast study area between 1950 and 2002 (Groisman et al., 2004).

Although the climate model output for the A1B, A2, and B1 emissions scenarios demonstrate a large degree of similarity, the A1B scenario was retained for more detailed analysis since it is considered "mid-range" of the IPCC emissions scenarios. Also, we note that the major differences in CO_2 concentrations under the IPCC SRES scenarios occur after 2040 (figure 3.1), which helps explain why temperature and precipitation do not vary widely among the GCM experiments with the high-, low-, and mid-range emissions scenarios (tables 3.4 to 3.9). Stated another way, the climate scenarios presented in these tables are not likely to change significantly during the next three to four decades by mitigation measures that would reduce emissions, although mitigation measures could substantially affect the climate in the latter half of this century. Probability density functions for seasonal temperature and precipitation change through 2050 are presented in figures 3.8 and 3.9, respectively.

Hourly or daily precipitation extremes cannot be reliably simulated by current GCM experiments. The percentiles (i.e., the 5th, 50th , and 95th percentiles) from the A1B PDFs were used as a proxy for assessing potential changes to hydrological extremes across the region. These percentiles stretch the range of output from all 21 GCMs, while also providing the middle of the PDF, or region under the curve where there is most agreement between the models (i.e., 50th percentile). The 1971-2000 temperature and precipitation data therefore were modified seasonally according to the predicted changes presented in tables 3.5, 3.7, and 3.9 for each of the three quartiles. The water-balance model was then rerun using the three quartile datasets to simulate the hydrology under these altered climate conditions. These datasets provided the means necessary to produce new PDS of precipitation, runoff, and deficit for additional extreme value statistical testing.

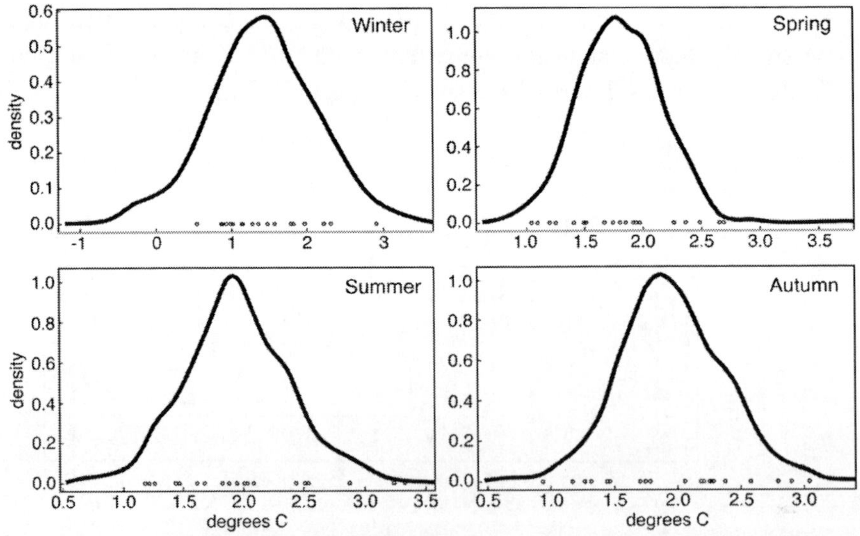

Figure 3.8. Probability density functions for seasonal temperature change in the Gulf Coast study area for 2050 created by using the A1B emissions scenario

Table 3.3. List of GCMs Run with the Three SRES Emissions Scenarios (A1B, A2, and B1) for this Study. Not All Model Runs Were Available from the IPCC Data Centre for Each SRES Scenario

A1B		A2		B1	
Temperature	**Precipitation**	**Temperature**	**Precipitation**	**Temperature**	**Precipitation**
CCCMA	CCCMA.T63	BCCR	BCCR	BCCR	BCCR
CCCMA.T63	CNRM	CCCMA	CNRM	CCCMA	CCCMA.T63
CNRM	CSIRO	CNRM	CSIRO	CCCMA.T63	CNRM
CSIRO	GFDL0	CSIRO	GFDL0	CNRM	CSIRO
GFDL0	GFDL1	GFDL0	GFDL1	CSIRO	GFDL0
GFDL1	GISS.AOM	GFDL1	GISS.ER	GFDL0	GFDL1
GISS.AOM	GISS.EH	GISS.ER	INMCM3	GFDL1	GISS.AOM
GISS.EH	GISS.ER	INMCM3	IPSL	GISS.AOM	GISS.ER
GISS.ER	IAP	IPSL	MIROC.MEDRES	GISS.ER	IAP
IAP	INMCM3	MIROC.MEDRES	ECHAM	IAP	INMCM3
INMCM3	IPSL	ECHO	MRI	INMCM3	IPSL
IPSL	MIROC.HIRES	ECHAM	CCSM3	IPSL	MIROC.HIRES
MIROC.HIRES	MIROC.MEDRES	MRI	PCM	MIROC.HIRES	MIROC.MEDRES
MIROC.MEDRES	ECHAM	CCSM	HADCM3	MIROC.MEDRES	ECHAM
ECHO	MRI	PCM	HADGEM1	ECHO	MRI
ECHAM	CCSM3	HADCM3		ECHAM	CCSM3
MRI	PCM	HADGEM1		MRI	PCM
CCSM	HADCM3			CCSM	HADCM3
PCM				PCM	
HADCM3				HADCM3	
HADGEM1					

Table 3.4. Scenarios of Temperature Change (°C) from an Ensemble of GCMs for the 5th, 25th, 50th, 75th, and 95th Percentiles for the A1B Scenario for 2050 Relative to 1971-2000 Means

	5th	25th	50th	75th	95th
Winter	0.18	0.95	1.42	1.89	2.56
Spring	1.22	1.55	1.80	2.04	2.38
Summer	1.24	1.66	1.94	2.23	2.70
Autumn	1.31	1.69	1.93	2.22	2.62

Table 3.5. Scenarios of Precipitation Change (Percent) from an Ensemble of GCMs for the 5th, 25th, 50th, 75th, and 95th Percentiles for the A1B Scenario for 2050 Relative to 1971-2000 Means

	5th	25th	50th	75th	95th
Winter	-13.30	-5.95	-1.79	2.49	9.01
Spring	-21.07	-11.04	-5.04	1.80	10.17
Summer	-36.10	-17.77	-6.39	6.25	26.24
Autumn	-8.20	0.46	5.97	12.05	21.50

Table 3.6. Scenarios of Temperature Change (°C) from an Ensemble of GCMs for the 5th, 25th, 50th, 75th, and 95th Percentiles for the A2 Scenario for 2050 Relative to 1971-2000 Means

	5th	25th	50th	75th	95th
Winter	0.2	1.0	1.5	2.0	2.9
Spring	0.8	1.3	1.7	2.0	2.6
Summer	1.1	1.5	1.8	2.1	2.5
Autumn	1.0	1.5	1.8	2.1	2.6

Table 3.7. Scenarios of Precipitation Change (Percent) from an Ensemble of GCMs for the 5th, 25th, 50th, 75th, and 95th Percentiles for the A2 Scenario for 2050 Relative to 1971-2000 Means

	5th	25th	50th	75th	95th
Winter	-12.7	-5.7	0.4	5.6	13.6
Spring	-22.9	-12.8	-6.0	0.5	10.3
Summer	-31.2	-15.0	-5.2	5.9	21.3
Autumn	-7.3	1.3	7.0	12.7	22.1

The 2-, 5-, 10-, 25-, 50-, and 100-year return periods for mean monthly precipitation show only modest differences between the current climate and the projected climate in 2050 at the three PDF percentiles (figure 3.10). As expected, there is a decrease in monthly precipitation extremes at the 5th percentile for the less rare return periods (2- to 25-year), relative to the current climate, which would be expected given the reduction in precipitation by up to 36 percent in summer. However, given the shape of the beta-p distribution, the 100-

year precipitation event is slightly larger than the baseline. Results for the 50th percentile indicate that the less rare return periods are on par with current climate patterns, but that the rare return periods may have modestly larger storms. At the 95th percentile, storms are generally larger across the board.

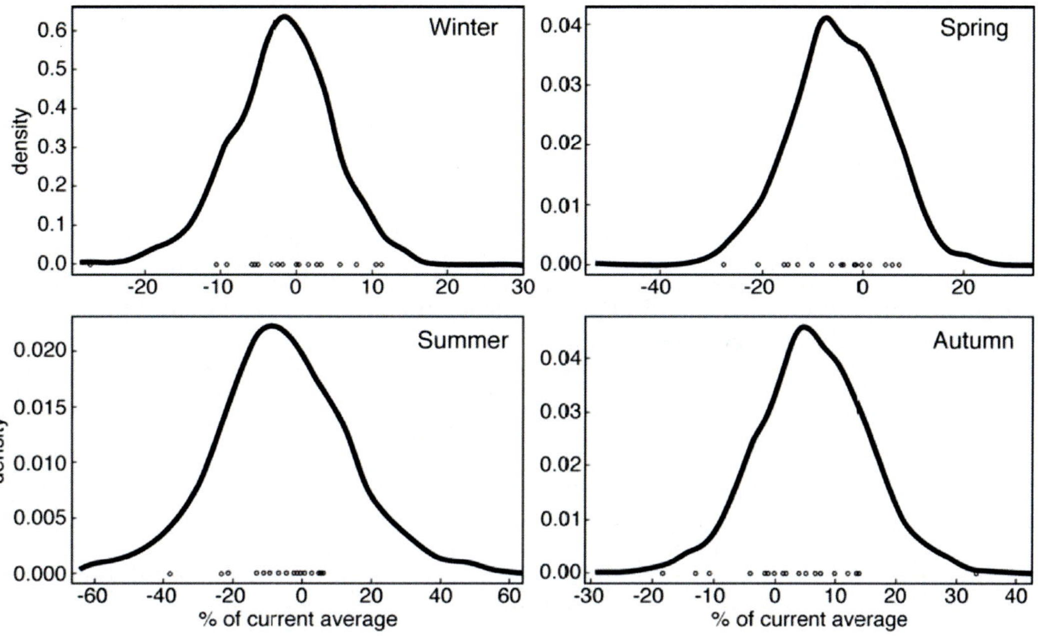

Figure 3.9. Probability density functions for seasonal precipitation change in the Gulf Coast study area for 2050 created by using the A1B emissions scenario

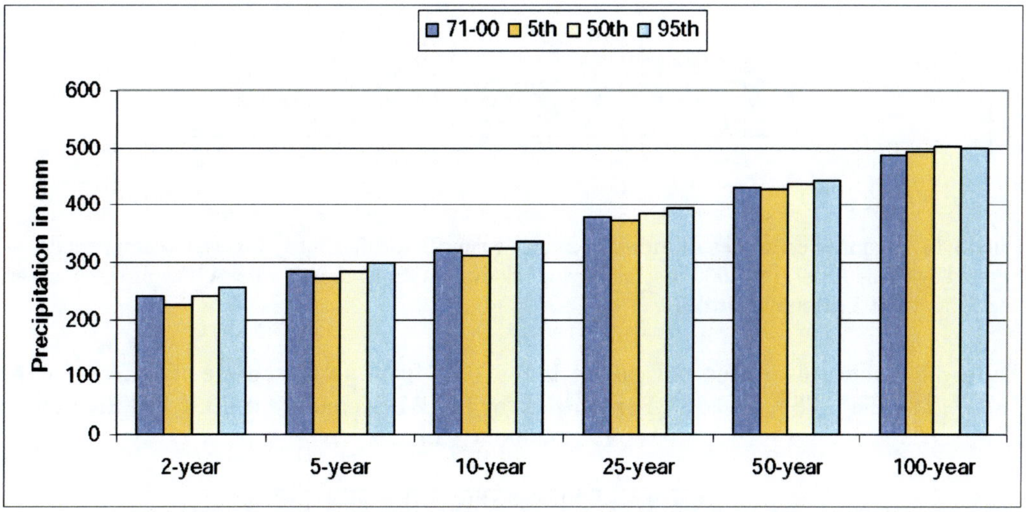

Figure 3.10. Quantile estimates of monthly precipitation for the 2- to 100-year return period generated by using the 1971 to 2000 baseline period relative to GCM output for the A1B emissions scenario at the 5, 50, and 95 percent quartiles

Monthly runoff extremes show a very different relationship to the current climate. At both the 5th and 50th percentiles, there is a dramatic reduction in projected runoff (figure 3.11). The mid-range of the GCMs suggests a decline in runoff relative to the 1971-2000 baseline period. Runoff rates are lower because precipitation is somewhat reduced, but perhaps more importantly, the projected increases in temperature also lead to increases in potential and actual evapotranspiration, and evapotranspiration rates are highest in the Gulf and southeastern United States compared to other regions (Hanson, 1991). An increase in actual evapotranspiration, without any increase in precipitation, translates into a reduction in runoff rates. However, at the 95th percentile, precipitation increases anywhere from 9 to 26 percent, depending on season.

Extremes in monthly deficit show a more complex pattern between the quartiles and over the various return periods (figure 3.12). The 5th percentile shows much larger deficits occurring relative to the 1971-2000 baseline. This is especially relevant at the two-year return period, which is nearly 30 percent larger in magnitude/intensity than in the current climate. This observation makes sense because as temperatures become somewhat warmer, thereby increasing potential evapotranspiration, there also are substantial reductions in precipitation. The net effect of this combination would be an increase in deficits (and drought intensity). Smaller reductions in precipitation at the 50th percentile dampen the increases in deficits. At the 95th percentile, increases in temperature are more than offset by the dramatic increases in precipitation, with deficits substantially reduced in their intensity.

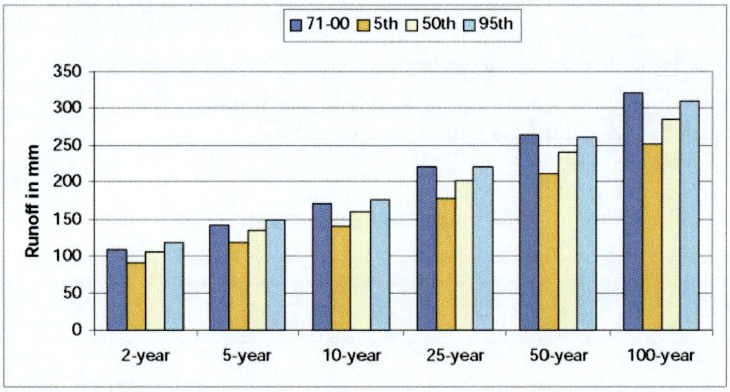

Figure 3.11. Quantile estimates of monthly average runoff for the 2- to 100- year return period generated by using the 1971 to 2000 baseline period relative to GCM output for the A1B emissions scenario at the 5, 50, and 95 percent quartiles

Table 3.8. Scenarios of Temperature Change (°C) from an Ensemble of GCMs for the 5th, 25th, 50th, 75th, and 95th Percentiles for the B1 Scenario for 2050 Relative to 1971-2000 Means

	5th	25th	50th	75th	95th
Winter	-0.31	0.44	1.02	1.53	2.32
Spring	0.67	1.05	1.32	1.62	2.03
Summer	0.64	1.09	1.35	1.63	2.03
Autumn	0.62	1.04	1.33	1.62	2.07

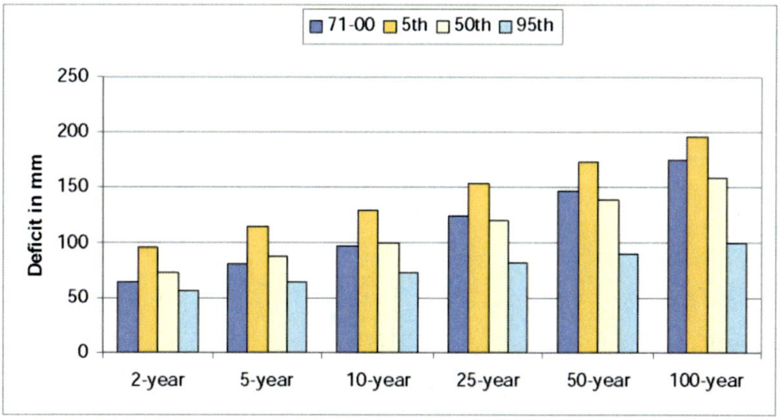

Figure 3.12. Quantile estimates of monthly average deficit for the 2- to 100- year return period generated by using the 1971 to 2000 baseline period relative to GCM output for the A1B emissions scenario at the 5, 50, and 95 percent quartiles

Table 3.9. Scenarios of Precipitation Change (Percent) from an Ensemble of GCMs for the 5th, 25th, 50th, 75th, and 95th Percentiles for the B1 Scenario for 2050 Relative to 1971-2000 Means

	5th	25th	50th	75th	95th
Winter	-9.77	-4.37	-0.52	3.36	9.51
Spring	-16.94	-7.96	-2.94	2.41	11.38
Summer	-27.06	-14.16	-3.36	7.43	24.19
Autumn	-7.83	-0.06	5.63	11.13	19.40

3.1.6. Changes in Daily Temperature

To examine trends in extreme temperature for the study area, daily maximum temperature, and minimum temperature were analyzed from 1950 through 2005. The historical analysis presented uses a dataset and tools developed for an analysis of North American extremes based on the daily data set from the USHCN (National Oceanic and Atmospheric Administration [NOAA], 2006). Temperature indices of transportation sensitive parameters were created on a station basis and then averaged together. For localized analyses, anomalies of the indices for all stations within 500 km of the target location were averaged together. For the U.S. time series, anomalies of station-level indices were first averaged into 2.5° latitude by 2.5° longitude grid boxes. Where a grid box did not have any stations, the values of the indices from neighboring grid boxes were interpolated into that grid box in order to make the averaging area more spatially representative. The grid box values were then averaged on an area-weighted basis to create U.S. time series. The time series figures show the annual values and a smoothed line derived from a locally weighted regression (Lowess filter; Cleveland et al., 1988). An advantage of a Lowess filter is that it is not impacted very much by one extreme annual value that might occur in an El Niño year, and therefore it depicts the underlying long-term changes quite well.

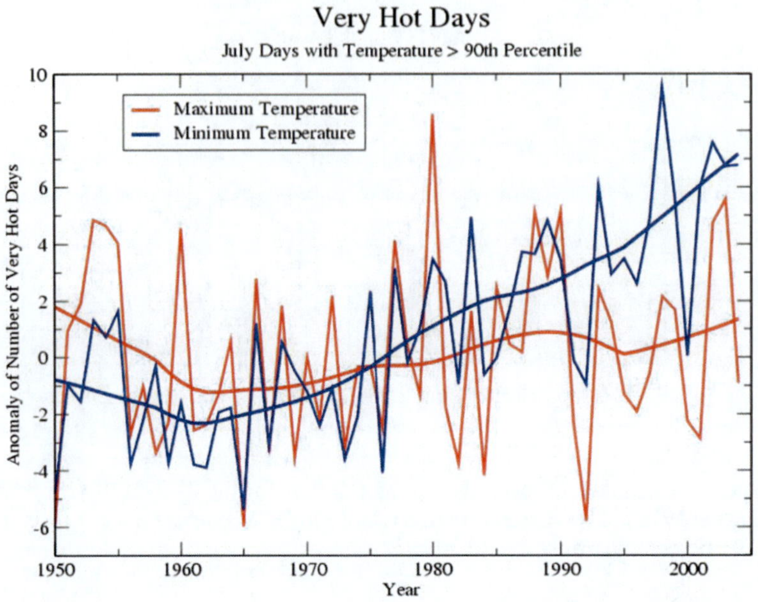

Figure 3.13. The change in the warmest 10 percent of July maximum and minimum temperatures at each station across the entire United States, for 1950-2004. Note that the number of days above the 90th percentile in minimum temperature is rising faster than it is in maximum temperature

The number of very hot days has been increasing on average across the United States. Figure 3.13 shows the average change since 1950 in the warmest 10 percent of July maximum and minimum temperatures at each station. The positive trend in minimum temperatures implies significantly warmer nights. The maximum temperature decreased after a period of south-central U.S. droughts in the 1950s and has been increasing ever since.

Temperature trends across the Gulf States region are not as pronounced as they are nationally due to the moderating effect of the proximate Gulf of Mexico waters. Figure 3.14 shows the anomaly in the number of days above 100 °F averaged over stations within 500 km of Dallas, TX. Although centered outside the Gulf States region considered in this report, many of the stations are well within it, and this figure is certainly representative of the behavior of Gulf States extreme temperatures in the recent past. Note the cooling following the 1 950s droughts. Also, note that the magnitude of interannual variations is considerably larger than any trend.

Notwithstanding this absence of a detectible trend in the number of days exceeding a high threshold temperature, it is very likely that in the future the number of very hot days will substantially increase. Figure 3.15 shows a prediction of the average number of days exceeding 37.8 °C (100 °F) in the June-July-August (JJA) season 25 years, 50 years, and 90 years into the future under the SRES A2 scenario for Houston, TX, the closest station to Galveston, TX, with available data. The algorithm used for this prediction exploits current observations as well as predictions of the JJA average temperature from 17 of the climate models contained in the WGNE-CMIP3 database prepared for the IPCC Fourth Assessment Report. Twenty-five years from now, the probability of a week (although not necessarily continuous) of 37.8 °C temperatures in this region is greater than 50 percent. Fifty years from now, the overall heating is such that the probability of three weeks of 37.8 °C temperatures is greater than 50 percent. Note that results obtained under either the B 1 or A1B forcing scenarios would be statistically indistinguishable from these results until well after the mid-century mark.

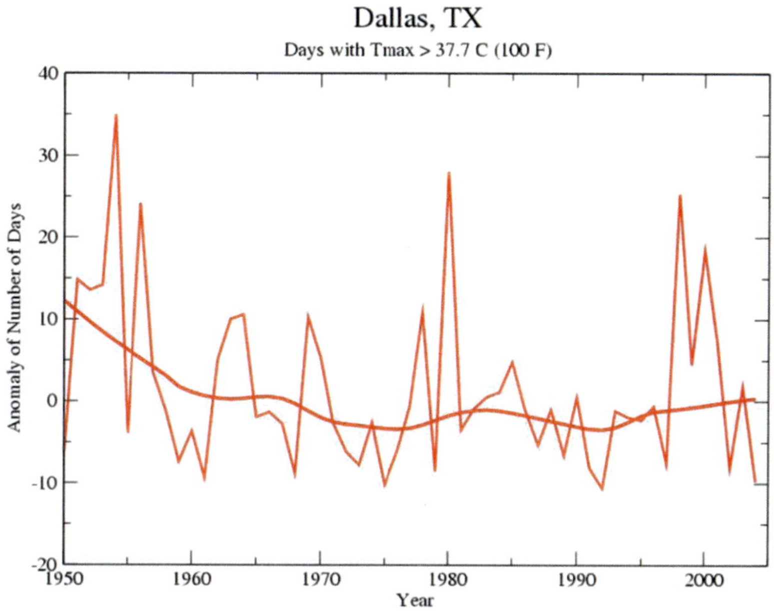

Figure 3.14. Historical time series from stations within 500 km of Dallas, TX, showing anomalies of the number of days above 37.7 °C (100 °F), for 1950-2004

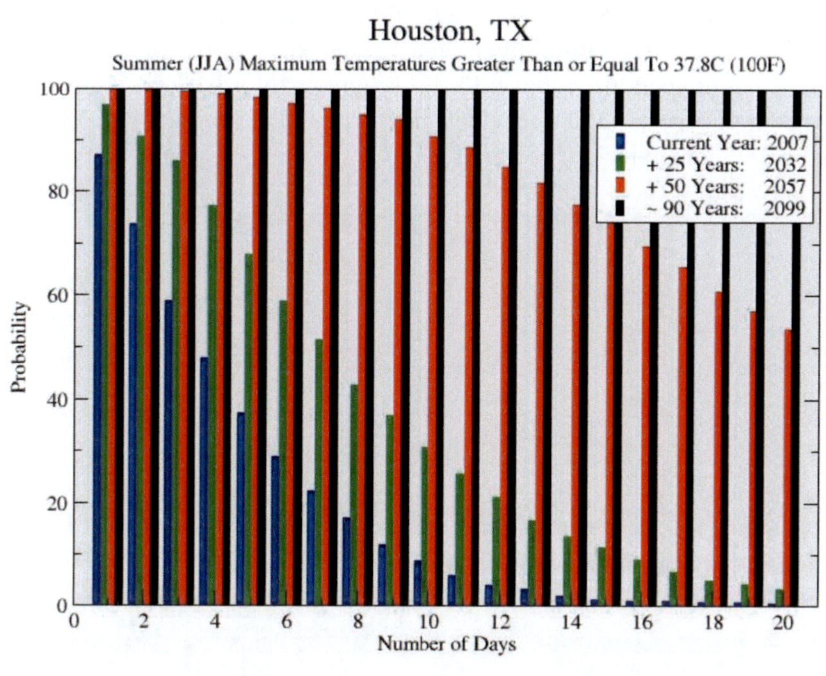

Figure 3.15. The current and future probabilities of having 1 to 20 days during the summer at or above 37.8 °C (100 °F) in or near Houston, TX, under the A2 emissions scenario

Climate models predict that the extreme temperature events could change more than the average climate over the course of the next century (IPCC, 2007). One way of quantifying this is to consider 20-year return values of the annual maximum of the daily average temperature. The 20-year return value is that value that is exceeded by a random variable once every 20 years on average over a long period of time. Such an event is truly rare, occurring only three or four times over the course of a typical human lifetime. Generalized extreme value theory provides a robust statistical framework to perform these calculations (Zwiers and Kharin, 1998; Wehner, 2005). Figure 3.16 shows the predicted change in this quantity at the end of 21st century under the SRES A1B scenario from a mean model constructed from 10 models from the WGNE-CMIP3[1] database. Over the Gulf States region, this extreme value change is about 1°C greater than the change in the average temperature. Another way to put this in perspective is to consider the frequency with which currently considered rare events will be encountered in the future. Figure 3.17 shows the number of times in a 20-year period that the 1990-1999 return value would be reached near the end of the century. The purple shaded regions exceed 10 times; hence, currently considered rare events are likely to happen every other year or more frequently.

3.1.7. Changes in Specific Temperature Maxima Affecting Transportation

Transportation analysts have identified several specific attributes of temperature change of concern in transportation planning. Changes in annual days above 32.2 °C (90 °F) and maximum high temperature, for example, will impact the ability to construct and maintain transportation facilities. Concrete loses strength if it is set at air temperatures greater than 32.2 °C and the ability of construction workers and maintenance staff to perform their duties is severely curtailed at temperatures above 32.2 °C degrees. In order to properly design for the thermal expansion of concrete and steel elements of transportation facilities, knowledge of the maximum expected temperatures is required.

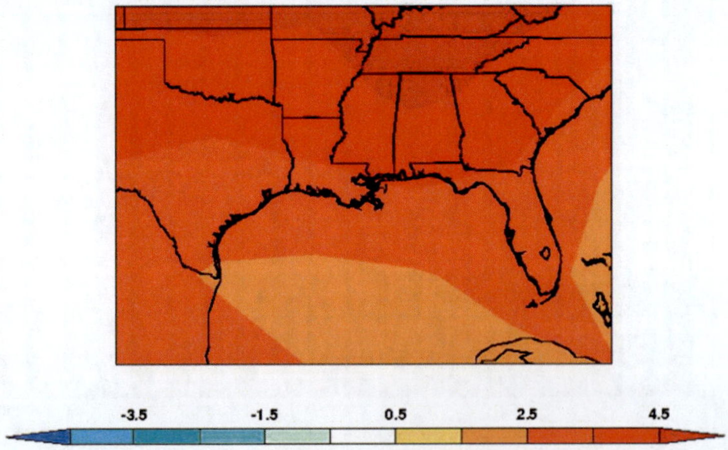

Figure 3.16. Mean model predicted change (°C) of the 20-year return value of the annual maximum daily averaged surface air temperature under the A1B emissions scenario in the Gulf States region. This analysis compares the 1990-1999 period to the 2090-2099 period

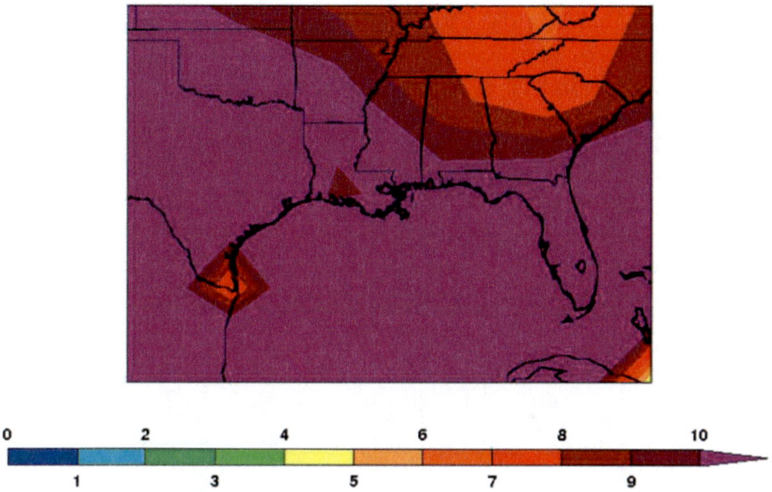

Figure 3.17. Number of times on average, over a 20-year period, that the 1990-1999 annual maximum daily averaged surface air temperature 20- year return value levels would be reached under the SRES A1B 2090-2099 forcing conditions. Under 1990-1999 forcing conditions, this value is defined to be one

Since global climate models are integrated at spatial scales around 200 km, a linear regression analysis was used to downscale relationships between the three variables of greatest concern at the localized scale of a weather station to the transportation sector. Historical data from the USHCN for eight observation stations in the Gulf Coast study area were analyzed to determine highest temperature of record, mean number of days at minimum temperature 32.2 °C or higher, and mean daily temperature. Table 3.10 shows the reported observations for the eight weather stations for days above 32.2 °C and the associated annual and July mean daily temperatures.

Based on the relationship established in the regression analysis of the historical data, changes in mean and extreme temperatures were calculated for the study area relevant to the temperatures in 2050 and 2100, as predicted by the global climate models used in this study. The analysis focused on the relationship between mean daily temperature, output from the climate models at 200-km scales, and the desired values downscaled to local spatial scales: number of days above 32.2 °C and the highest temperature of record. Comparisons were made to each of the annual mean daily temperatures and mean daily temperatures for the month of July to determine which relationship better provided the desired forecast variables.

A linear regression of days above 32.2 °C (90 °F) as an independent variable for the stations shown was undertaken for each of the annual mean daily and the July mean daily temperatures as the dependent variables. The regression of observed days above 32.2 °C versus annual mean daily temperature showed that for each rise by 0.6 °C (1 °F) in annual mean daily temperature there is an associated 3.9-day increase in the annual days above 32.2 °C. However, the data for New Orleans falls outside the trend line for this relationship.[2] The regression of days above 32.2 °C versus July mean daily temperature showed that for each rise by 0.6 °C (1 °F) in July mean daily temperature there is an associated 10-day increase in the annual days above 32.2 °C.[3]

Table 3.10. Days above 32.2 °C (90 °F) and Mean Daily Temperature in the Study Area for Datasets Running Through 2004. The Start Date Varies by Location (Note the Number of Years of Observed Data)

Station	Years of Observed Data	Annual Days Above 90 °F	Normal Mean Daily (°F)	
			Annual	July
Mobile, AL	42	74	66.8	81.5
Baton Rouge, LA	45	84	67.0	81.7
Lake Charles, LA	40	76	67.9	82.6
New Orleans, LA	58	72	68.8	82.7
Meridian, MS	40	80	64.7	81.7
Houston, TX	35	99	68.8	83.6
Port Arthur, TX	44	83	68.6	82.7
Victoria, TX	43	106	70.0	84.2

The regression of observed high temperature versus annual mean daily temperature suggested that for each rise by 0.6 °C (1 °F) in annual mean daily temperature there is an associated 0.3 °C (0.6° F) rise in high temperature. However, this relationship only has an R-squared of 0.10, largely because the data for New Orleans falls outside the trend line. The regression of high temperature versus July mean daily temperature showed that for each rise by 0.6 °C (1 °F) in July mean daily temperature there is an associated 1.2 °C (2 °F) rise in the high temperature.[4]

The mean daily temperature for the study area is 27.6 °C (81.7 °F). Based on the relationships established above, this implies that the existing high temperature should be approximately 40.6 °C (105 °F). For each additional 0.6 °C (1° F) degree increase in July mean daily temperature that is forecasted by the GCMs, this high temperature can be expected to increase by 1.2 °C (2 °F). Using the relationship developed, this implies that the baseline/historical number of days above 32.2 °C (90 °F) is approximately 77 days. For each additional increase of 0.6 °C (1 °F) in July mean daily temperature that is forecasted by the GCMs, the number of days above 32.2 °C (90 °F) can be expected to increase approximately 10 days..

Airport runway length in the United States is generally calculated based on the mean maximum temperature (that is, the average of the daily high temperatures) during the hottest month of the year during the prior 30-year record. August is the month with the highest monthly mean maximum temperature in the Gulf Coast study area. Mean maximum temperature is reported by NOAA for 283 NOAA stations across the United States, six of which are located in the study area. The average mean maximum temperature for the hottest month of the year from these six stations is 33.1 °C (91.6 °F). To verify this, we determined the 30-year mean maximum temperature data (1972 to 2002) from the Carbon Dioxide Information Analysis Center (CDIAC), which encompasses 12 reporting stations located in the study area. CDIAC data provides station elevation data as well as latitude and longitude data. The average mean maximum temperature from the 12 CDIAC stations is 33.0 °C (91.4 °F) (table 3.11). The airport section in the subsequent chapter deals more specifically with this dataset in an analysis of how runway length may be impacted by changes in temperature during the next 50 to 100 years.

3.1.8. Increasing Daily Precipitation Extremes

As mentioned above, current generation climate models are limited in their ability to simulate individual storms by a lack of horizontal resolution. From a simple theoretical argument (Allen and Ingram, 2003), it is expected that extreme precipitation events should become more intense as the climate warms. The IPCC (2007) concluded that the frequency of heavy precipitation events had increased over most areas during the past century and that a continued increase in heavy precipitation events is very likely during the 21st century. The largest rainfall rates occur when a column of air is completely saturated and precipitates out nearly completely. The Clausius-Claperyon relationship dictates that as the air temperature increases, the atmosphere has the ability to hold more water vapor. Hence, under a warmer climate, it is very likely that specific humidity will increase both on average and in extreme saturation conditions. Extreme-value analysis of model output for daily precipitation in the Gulf States region, similar to the analysis discussed above with daily surface air temperatures, reveals a predicted increase of around 10 percent in the 20- year return value of the annual maximum of daily averaged precipitation, as shown in figure 3.18. The coarse horizontal resolution of the climate models used in this analysis results in an underestimation of extreme precipitation events (Wehner, 2005). Furthermore, these models lack the resolution to simulate tropical cyclones, a further source of extreme precipitation events. However, these deficiencies likely cause the prediction errors to be conservative, and it is likely that daily mean precipitation levels that are currently rare will become more commonplace in the future.

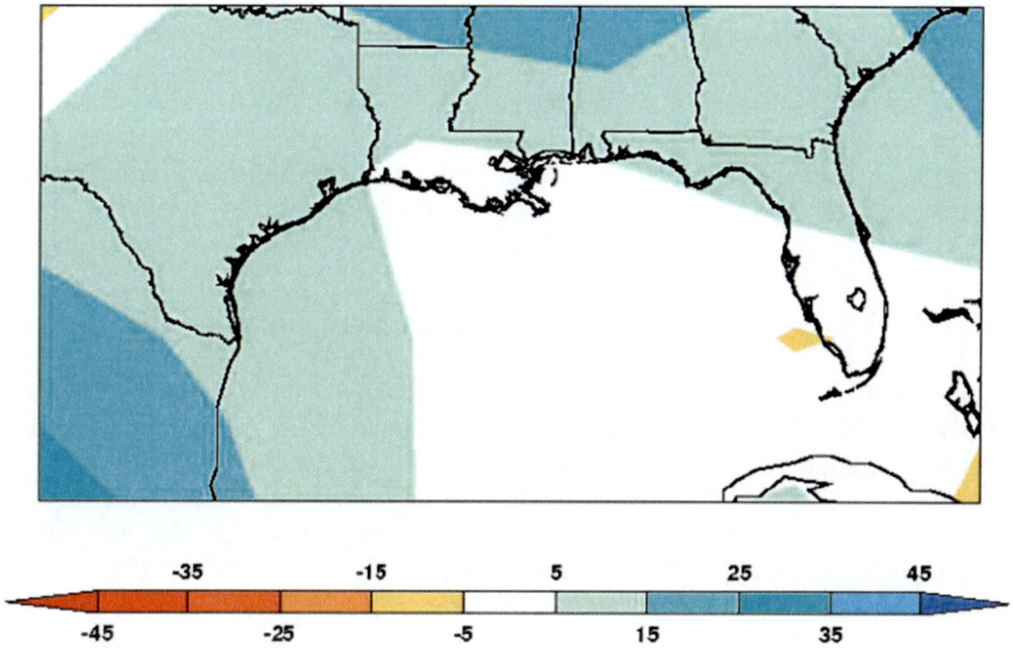

Figure 3.18. Mean model-predicted fractional change of the 20-year return value of the annual maximum daily averaged precipitation under the SRES A1B in the Gulf States region. This analysis compares the 1990-1999 period to the 2090-2099 period

Table 3.11. Modeled Outputs of Potential Temperature Increase (°C [°F]) Scenarios for August

Mid-Term Potential (2050 Scenarios)				Long-Term Potential (2100 Scenarios)			
Temperature Increase by Scenario Percentile: °C (°F)				**Temperature Increase by Scenario Percentile: °C (°F)**			
Scenario	**5th**	**50th**	**95th**	**Scenario**	**5th**	**50th**	**95th**
A1B	1.6 (2.9)	2.5 (4.5)	**3.4 (6.1)**	**A1B**	3.0 (5.4)	3.9 (7.0)	5.0 (9.0)
B1	**0.9 (1.6)**	1.8 (3.2)	2.6 (4.7)	**B1**	**1.8 (3.2)**	2.7 (4.9)	3.6 (6.5)
A2	1.1 (2.0)	2.3 (4.1)	**3.4 (6.1)**	**A2**	3.3 (5.9)	4.7 (8.5)	**6.0 (10.8)**

Note: Lowest/highest changes in bold.

3.2. Hurricanes and Less Intense Tropical Storms

Tropical cyclones (called hurricanes in the Atlantic and Gulf Coast regions) pose a severe risk to natural systems, personal property, and public infrastructure in the Gulf Coast region, and this risk will likely be exacerbated as the temperature of atmosphere and sea surface increase. Whereas loss of life from hurricanes has decreased in recent decades, property losses due to rapid population growth and economic development of coastal areas has increased (Herbert et al., 1997; Pielke and Pielke, 1997; Pielke and Landsea, 1998). Hurricanes have their greatest impact at the coastal margin where they make landfall and sustain their greatest strength. Severe beach erosion, surge overwash, inland flooding, and windfall casualties are exacted on both cultural and natural resources. Transportation facilities – roads, rails, pipelines, airports, ports – in coastal counties will likely be subjected to increasing hurricane intensity in the coming decades. Changes in Atlantic Basin hurricane formation and the behavior of hurricanes that make landfall in the Gulf Coast region have important implications for transportation planning, design, and maintenance in the short and long term.

3.2.1. Assessing Trends in Historical Hurricane Frequency and Intensity

Understanding hurricane frequency and landfall patterns is an important process in calculating insurance liabilities and rates for coastal communities as well as forecasting future risk under a changing climate. Several studies have shown that landfalling hurricanes are more or less frequent for given coastal reaches of the United States (see figure 3.19) and within given decades over the recorded history of North Atlantic storms (Simpson and Lawrence, 1971; Ho et al., 1987; Neumann, 1991; Jarrell et al., 1992; Gray et al., 1997; Pielke and Pielke, 1997; Neumann et al., 1999; Vickery et al., 2000). While different methods have been employed to calculate landfall probabilities at the state and county levels, there is general agreement that south Florida, the Carolinas, and the western Gulf Coast are most frequently impacted by major hurricanes (figure 3.19).

Studies of multidecadal hurricane variability and cycles have been complicated by the relatively short period of available and reliable data. Landfall counts of tropical storms and hurricanes at Grand Isle, LA, produced with a hurricane simulation model, HURASIM,

(Doyle and Girod, 1997) (appendix E) for five-year periods from 1951 through 2005 show periods of greater and lesser hurricane history with short- and long-term variability (figure 3.20). If there is any pattern, historical records exhibit episodic hurricane activity rather than trends toward more frequent or stronger hurricanes, despite the most recent period of intense hurricane activity. While the long-term frequency trend of named storms within the Atlantic Basin has remained fairly constant, interannual variability is prominent particularly among major hurricanes (Gray, 1990; Landsea et al., 1992; Gray et al., 1997; Goldenberg et al., 2001; Bell and Chelliah, 2006). Hurricane spawning patterns have been linked to regional oscillation cycles, Atlantic thermohaline circulation, and African West Sahel rainfall patterns that have improved our understanding and forecasting of hurricane activity in the North Atlantic Basin (Gray et al., 1997; Landsea et al., 1999).

Increased tropical storm activity is likely to accompany global warming as a function of higher sea surface temperatures, which have been observed globally (figure 3.21). The kinetic energy of tropical storms and hurricanes is fueled from the heat exchange in warm tropical waters. An increase in sea surface temperature (SST) from global climate change is likely to increase the probability of higher sustained winds per tropical storm circulation (Emanuel, 1987; Holland, 1997; Knutson et al., 1998). Sea surface temperature has increased significantly in the main hurricane development region of the North Atlantic during the past century (Bell et al., 2007) (figure 3.22) as well as in the Gulf of Mexico (Smith and Reynolds, 2004) (figure 3.23).

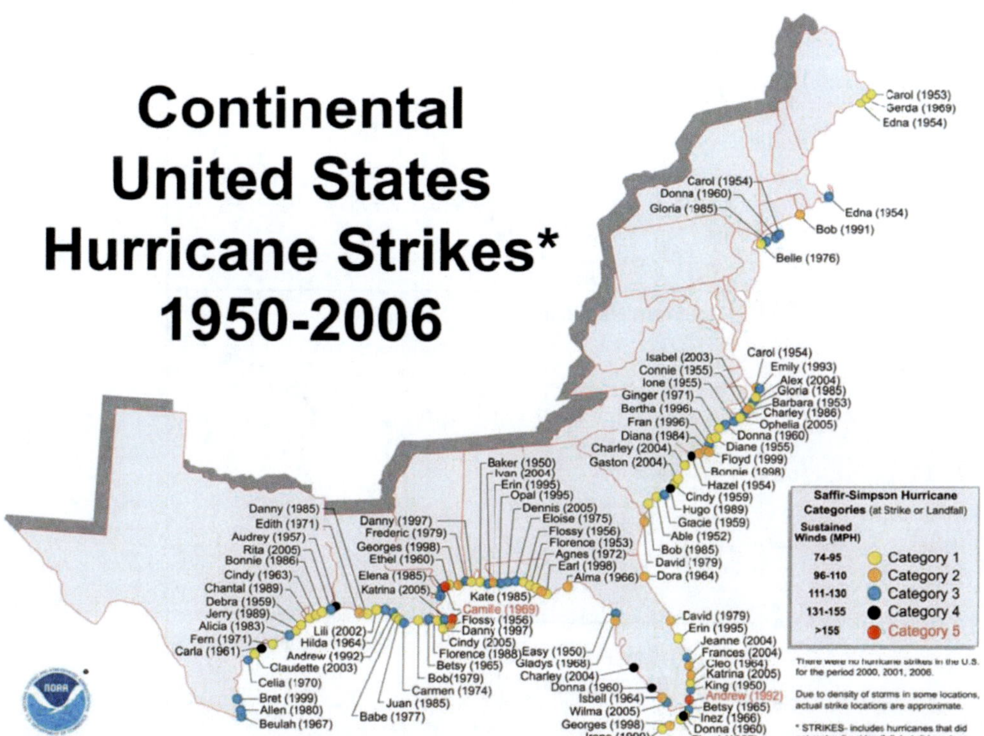

Source: NOAA, National Climate Data Center, Asheville, NC.

Figure 3.19. Geographic distribution of hurricane landfalls along the Atlantic and Gulf Coast regions of the United States, from 1950 to 2006

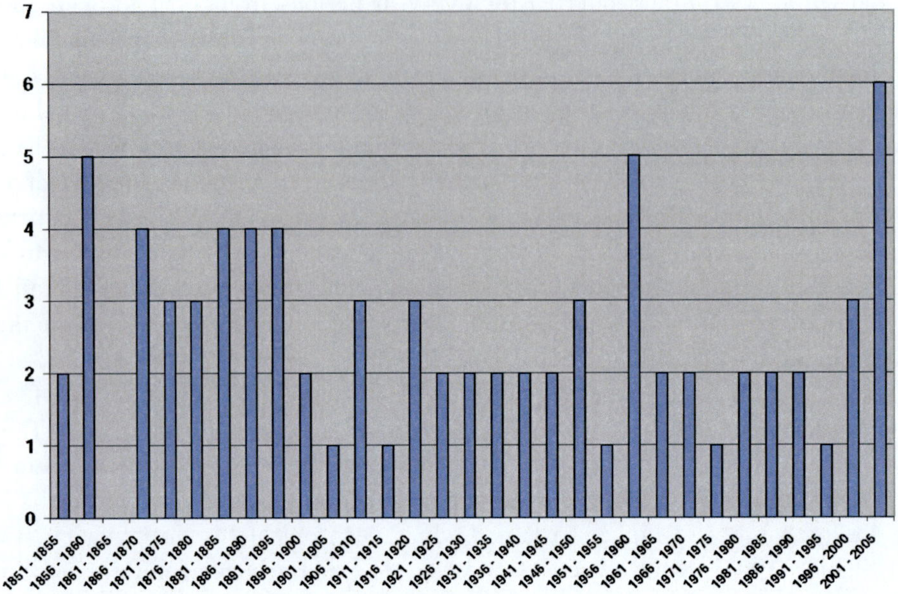

Source: NOAA National Hurricane Research Division.

Figure 3.20. Frequency histogram of landfalling storms of tropical storm strength or greater in Grand Isle, LA, summarized on a 5-year basis, for the period 1851-2005

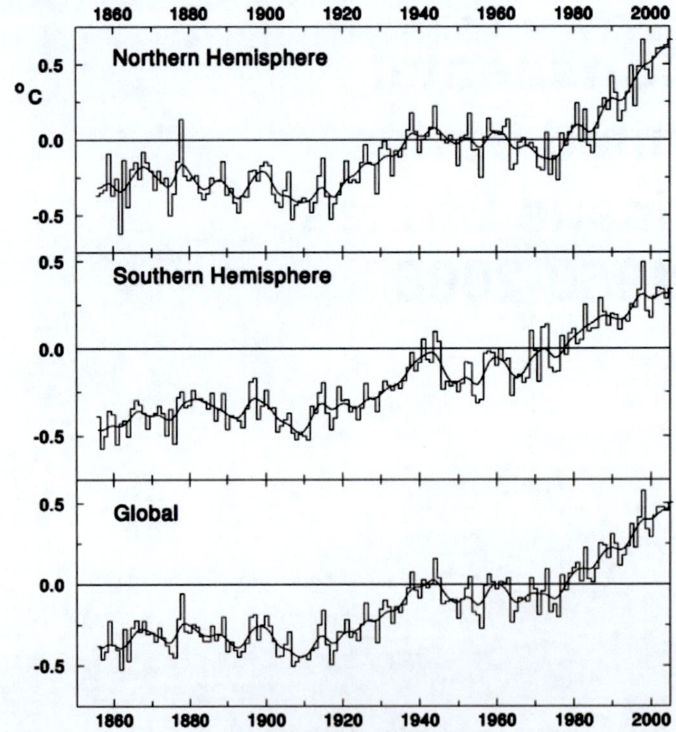

Source: NOAA, National Climate Data Center, Asheville, NC.

Figure 3.21. Hemispherical and global mean sea surface temperatures for the period of record from 1855 to 2000

Many scientists have evaluated the relationships between 20th-century warming and hurricane intensity, with some suggesting that the incidence of intense hurricanes over the past decade for the Atlantic basin could signal the beginning of an El Nino-Southern Oscillation (ENSO) related cycle of increased hurricane activity (Gray, 1984; O'Brien et al., 1996; Saunders et al., 2000). Henderson-Sellers et al. (1998) found no discernible trends in global hurricane trends with respect to number, intensity, or location during the past century. More contemporary analysis of the upswing in intense hurricane activity since the 1990s demonstrates that the proportion of intense, more destructive hurricanes has increased in some ocean basins, including the North Atlantic, concomitant with rising sea surface temperature (Emanuel, 2005; Hoyos et al., 2006; Mann and Emanuel, 2006; Trenberth and Shea, 2006; Webster et al., 2005). Some studies conclude that the increase in recent decades is due to the combination of natural cyclical events (such as the North Atlantic Oscillation) and human-induced increases in sea surface temperature (Elsner, 2006).

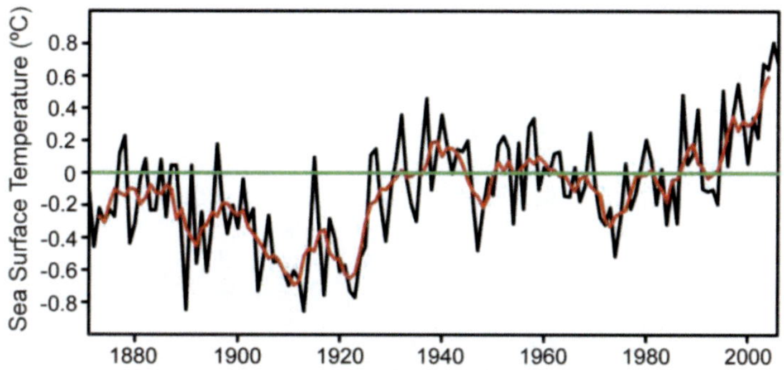

Source: Bell et al., 2007.

Figure 3.22. Sea surface temperature trend in the main hurricane development region of the North Atlantic during the past century. Red line shows the corresponding 5-yr running mean. Anomalies are departures from the 1971–2000 period monthly means

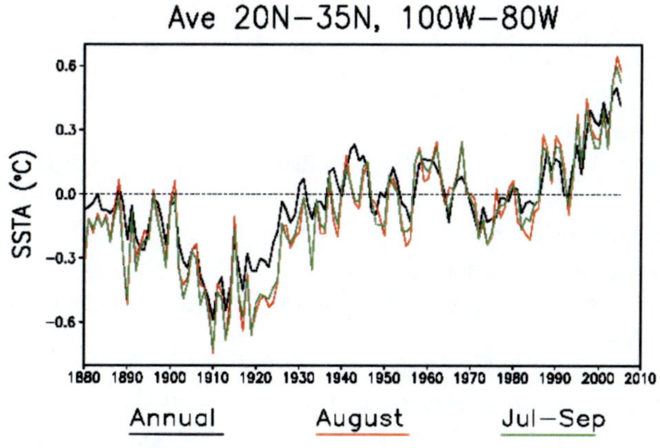

ERSST v.2: Extended Reconstructed Sea Surface Temperatures, version 2.
Source: Smith and Reynolds, 2004.

Figure 3.23. Sea surface temperature trends in the Gulf of Mexico region produced by using the ERSST v.2 database. The plot includes the SST anomalies averaged annually, as well as the anomalies determined from the averages for August only and the July- September peak of the hurricane season

Ocean currents that regulate heat content also appear to play an important role in the intensity of hurricanes when atmospheric conditions are favorable (Shay, 2006). In the Gulf of Mexico, the Loop Current is a heat conveyor that can build a heat reservoir spanning 200-300 km in diameter and 80-150 m in depth that is generally oriented towards the central Gulf Coast (figure 3.24) (Jaimes et al., 2006). Satellite-based and *in situ* measurements support the hypothesis that the warm water brought into the Gulf of Mexico by the Loop Current played an important role in the rapid intensification of Hurricanes Katrina, Rita, and Wilma (Jaimes et al., 2006).

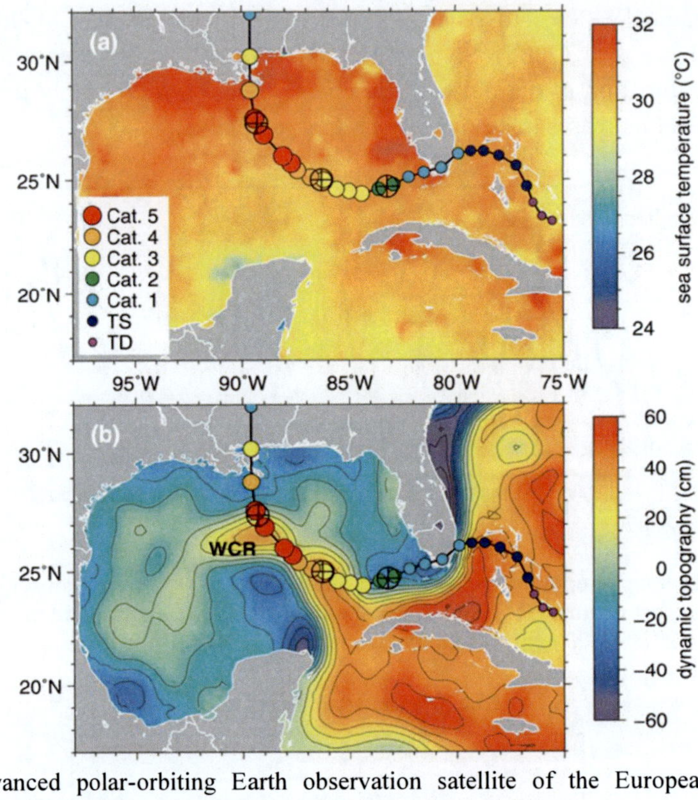

Envisat: An advanced polar-orbiting Earth observation satellite of the European Space Agency that\provides measurements of the atmosphere, ocean, land, and ice

GFO: Satellite program of U.S. Navy GEOSAT Follow-On (GFO) program to maintain continuous ocean observations

Jason-1: NASA satellite monitoring global ocean circulation

POES: Polar Observational Environmental Satellite

TOPEX: TOPEX/Poseidon is a CNES and NASA satellite monitoring ocean surface topography.

Source: Scharroo et al., 2005.

Figure 3.24 The location and intensity of Hurricane Katrina at intervals of 6 hours show two intensification events. Circles indicate data from National Hurricane Center advisories showing storm intensity (T S and TD stand for tropical storm and tropical depression, respectively): (a) Intensification is not correlated with sea surface temperature (from POES high-resolution infrared data); (b) In contrast, the intensifications correlate well with highs in the ocean dynamic topography (from Jason 1,TOPEX, Envisat, and GFO sea surface height data). The Loop Current can be seen entering the Gulf south of Cuba and exiting south of Florida; the warm-core ring (WCR) is the prominent high shedding from the Loop Current in the center of the Gulf

Santer et al. (2006) used 22 climate models to study the possible causes of increased SST changes in the Atlantic and Pacific tropical cyclogenesis region, where SST increased from 0.32 °C to 0.67 °C (0.57 – 1.21 °F) over the 20th century. Their analysis suggests that century-timescale SST changes of this magnitude cannot be explained solely by unforced variability of the climate system. In experiments in which forcing factors are varied individually rather than jointly, human-caused changes in greenhouse gases are the main driver of the 20th-century SST increases in both tropical cyclogenesis regions. Ouuchi et al. (2006) used an atmospheric general circulation model at 20-km horizontal resolution to directly simulate the relationship between the tropical storm cycle and SST. This hurricane resolving model produced seasonal tropical storm statistics under present day conditions and was capable of hurricane force winds. When driven with the SST anomalies taken from A1B scenario experiments, the model produced fewer tropical storms everywhere except the North Atlantic Basin, where an increase was predicted. Tropical storms were more intense on average in all basins in these modeling experiments.

These results and those from similar studies suggest that as radiative forcing and SST continue to increase, hurricanes will be more likely to form in the Atlantic and Pacific Basins and more likely to intensify in their destructive capacity. In its Fourth Assessment Report, the IPCC (2007) concludes that:

- There is observational evidence for an increase of intense tropical cyclone activity in the North Atlantic since about 1970, correlated with increases of tropical sea surface temperatures;
- Multidecadal variability and the quality of the tropical cyclone records prior to the beginning of routine satellite observations in about 1970 complicate the detection of long-term trends in tropical cyclone activity; and
- There is no clear trend in the annual numbers of tropical cyclones.

3.2.2. Gulf Coast Hurricane History

Gulf coast ecosystems are exposed to varying degrees of hurricane disturbance as influenced by storm frequency, periodicity, and duration. Figure 3.25 shows that tropical storm landfall across the Gulf of Mexico Basin increases geometrically from west to east. Because most storms spawn in tropical waters in the eastern Atlantic there is a greater probability for eastern landmasses on the same latitude to incur tropical storms (Elsner, 1999). Temporal patterns of the past century reveal periods of relatively frequent hurricanes as well as inactive periods for most of the Gulf Coast region. The relatively calm period of record for hurricanes from the 1950s through the 1970s has some hurricane specialists purporting an increase in North Atlantic storms over the past decade related to ENSO oscillations and general warming trends (Elsner and Kara, 1999). Palynological and geological studies offer another means to reconstruct the regional history of hurricane activity over several centuries coincident with species changes and sedimentary overwash indicative of surge heights and storm intensity. One study of lake sediments in coastal Alabama suggests that major hurricanes of a Category 4 or 5 struck the Alabama coast with a frequency of about 600 years during the past three millennia (Liu and Fearn, 1993).

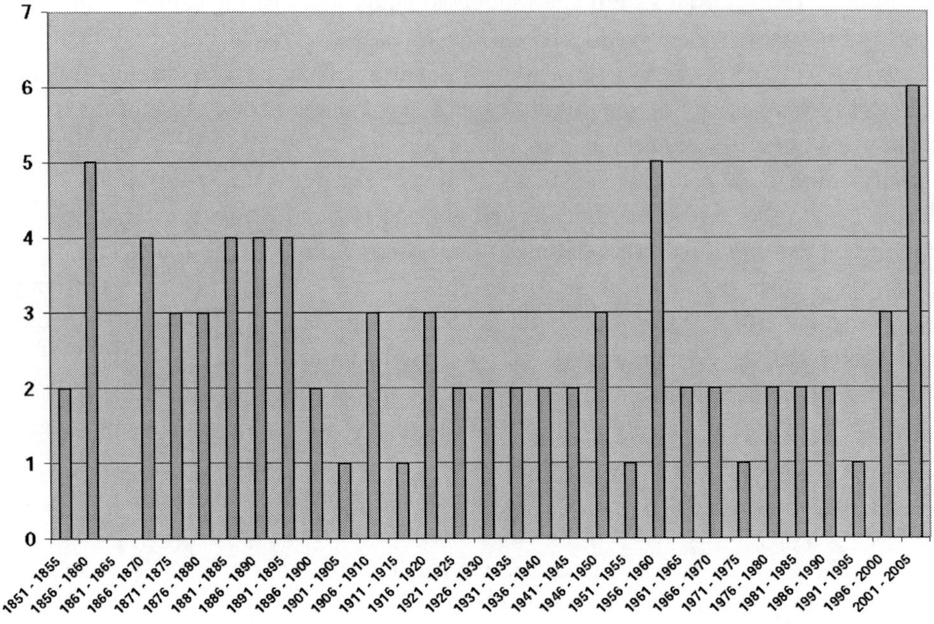

Figure 3.25. Frequency histogram of tropical storm events for coastal cities across the Gulf of Mexico region of the United States over the period of record from 1851 to 2006

3.2.3. HURASIM: Model Application

HURASIM is a spatial simulation model of hurricane structure and circulation for reconstructing estimated windforce and vectors of past hurricanes. The model uses historical tracking and meteorological data of dated North Atlantic tropical storms from 1851 to present. A description of the HURASIM model is presented in appendix E.

The HURASIM model was applied in a hindcast mode to reconstruct hurricane windfields across the Gulf Coast region from Galveston, TX, to Pensacola, FL, on a 10-km grid basis for the period of record from 1851 to 2003. The model calculated windspeed and direction for every 15 minutes of storm movement retaining only wind events of 30 mph or greater for all proximal storms and grid cells within the study region. Storm tracking for calendar years 2004 and 2005 have not been added to the Hurricane Database (HURDAT) dataset as yet and, therefore, have been omitted from this analysis despite record storm activity that may be associated with multidecadal cycles and/or current global warming trends.

3.2.4. Historical Storm Frequency Across the Northern Gulf Coast Study Region

HURASIM model results were categorized by storm class based on the commonly used Saffir-Simpson scale over a 153-year period from 1851 to 2003 to gain an historical perspective of recurrence potential and spatial distribution of storm events along the northern Gulf Coast between Galveston, TX, and Pensacola, FL. Table 3.12 outlines the Saffir-Simpson scale for categorizing storms by intensity associated with range of windspeed. Storms on the Saffir-Simpson scale also have been ascribed typical storm surge levels based on observations during the 20th century. For example, NOAA states that storm surge during landfall of a Category 1 hurricane is "generally 4 to 5 feet above normal," and a Category 3 hurricane storm surge is "generally 9 to 12 feet above normal" (NOAA, 2007). In the Gulf Coast region, however, storm surge is highly variable for a given class of storm on the Saffir-Simpson scale in the Gulf Coast region. For example, Hurricane Camille, a Category 5 hurricane at landfall, had a peak storm surge in coastal Mississippi of 7.6 m (25 ft), while the storm surge associated with Hurricane Katrina (a Category 3 hurricane at landfall) had peak storm surge of 8.5 m (28 ft) (Graumann et al., 2005).

Figure 3.26 shows the frequency patterns of storm events with Category 1, 2, and 3 winds or higher across the study region. Results show that storm frequency by storm class is highest for southeastern coastal Louisiana than elsewhere and lowest in inland locations, decreasing with increasing latitude. Secondary locations with high hurricane incidence include Galveston, TX, and the Mississippi coast. Coastal reaches west of Galveston, TX, the chenier plain of southwest Louisiana, and northwest Florida have experienced low to moderate hurricane frequency respectively. The highest frequency of Category 3 storm winds or greater for the entire region are seven storms over the 153-year period, equivalent to four to five storms per century. Based on the historical perspective alone, transportation planners should expect at least one major hurricane of Category 3 or greater to strike the northern Gulf Coast every 20 years. Over the same 20 years, planners can expect another Category 2 hurricane and two Category 1 hurricanes for a combined incidence rate of at least one hurricane every five years. While this rate is indicative of the worst-case grid location coastwide and over the entire historical record, the chance for storm track convergence elsewhere within the region is expected to be similar. However, storm frequency may be influenced by multidecadal variability such that some sites may experience higher incidence depending on the timeframe and whether it spans periods of on and off cycles.

3.2.5. Temporal and Spatial Analysis of Hurricane Landfall

The northern Gulf Coast exhibits spatially disjunct patterns of storm strikes related to the landfall tracks and storm categories (figure 3.27). Of storms exceeding Category 3-level winds between 1851 and 2003, the HURASIM model counted a maximum of seven storms equal to a recurrence interval of one major hurricane every 22 years for southeastern Louisiana. Hurricane tracking records are available from 1851 to present, but data accuracy was greatly improved at the turn of the century with expanded and instrumented weather stations and since 1944 when aircraft reconnaissance of tropical storms was instituted. HURASIM model output was analyzed by segmented time periods to determine short-term return frequencies of tropical storms to account for cyclical behavior and data accuracy for successive intervals of 15, 30, and 50 years of the longer 153-year record from 1851 to 2003. Data analysis focuses on the maximum potential return interval of storms by category according to the Saffir-Simpson scale. Given the prospect of questionable data accuracy of storm history and multidecadal storm cycling, it was deemed prudent to report storm frequencies for different time intervals to establish upward bounds of storm recurrence probabilities for catastrophe planning and assessment akin to worst-case scenarios. Shorter time windows are likely to exhibit a wide range of storm recurrence probabilities (both high and low) relative to longer periods.

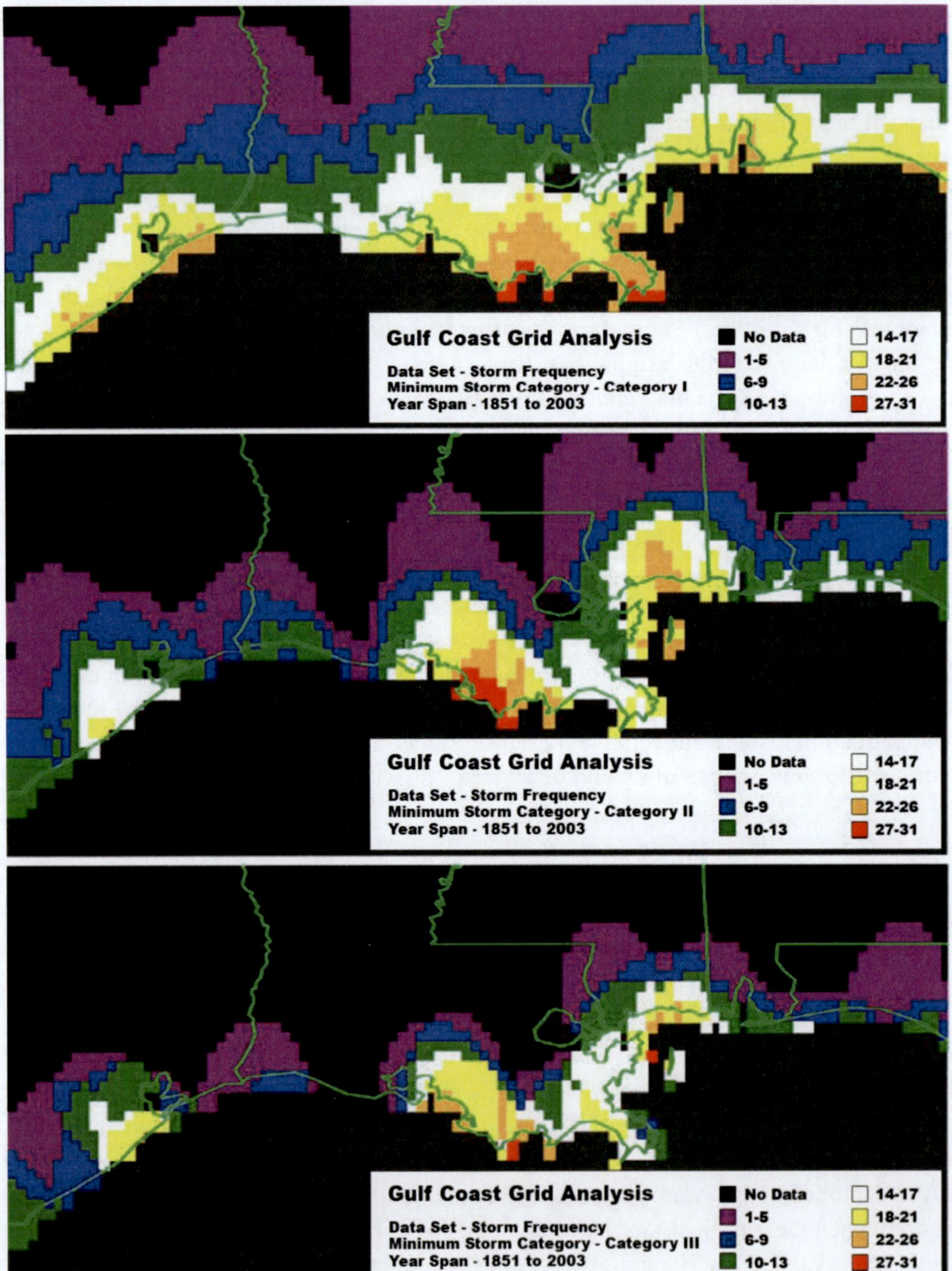

Figure 3.26. Frequency analysis of storm events exhibiting Category 1, 2, and 3 winds or higher across the Gulf Coast study area

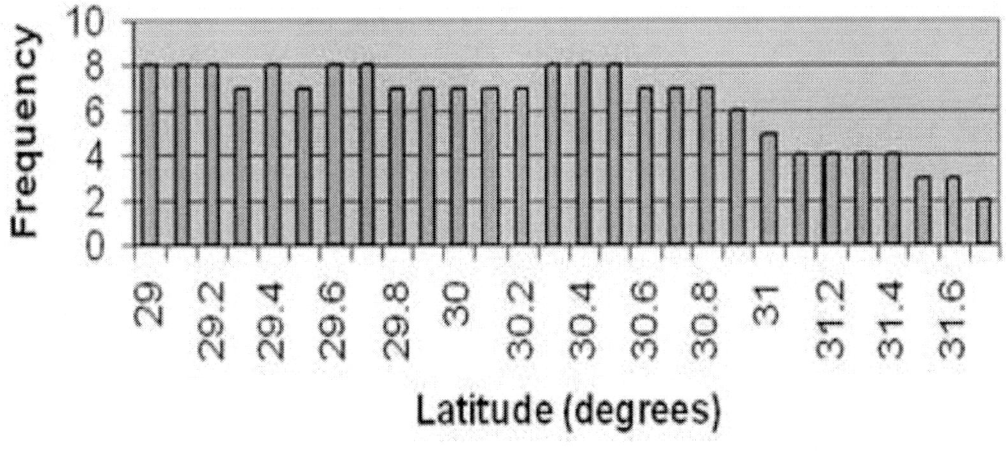

Figure 3.27. Latitudinal gradient of declining storm frequency of Category 1 hurricanes or greater from Grand Isle, LA, inland, illustrating the reduction of storm strength over land away from the coast, for the period 1951-2000

Table 3.12. Saffir-Simpson Scale for Categorizing Hurricane Intensity and Damage Potential. Note that Maximum Sustained Wind Speed is the Only Characteristic Used for Categorizing Hurricanes

Saffir-Simpson Scale and Storm Category	Central Pressure (mbar)	Maximum Sustained Wind Speed (mph)	Damage Potential
1	980	74-95	Minimal
2	965-979	96-110	Moderate
3	945-964	111-130	Extensive
4	920-944	131-155	Extreme
5	< 920	>155	Catastrophic

The shorter the period of observation, the greater the probability of inflating the calculated return interval. Figure 3.28 shows the storm frequency for 15-, 30-, and 50-year intervals for Category 1 storms or greater for the most active grid location across the study area. The most active time period historically for all time intervals was the latter 19th century, despite concerns of data accuracy for this period. These data show a potential maximum of storm incidence of three to five hurricanes every 10 years, nearly twice the strike frequency for the entire 153-year record. The lowest incidence of hurricane activity within the Gulf Coast study region for all time intervals spans the 1970s and 1980s with two to three hurricanes for every 10 years. These historical hurricane return intervals provide an expected range of .2 and .5 probability that a hurricane may strike a given coastal county within the study region that can be used to guide coastal planning and preparation. Recent hurricane studies spurred by the upswing in hurricane activity of the 1990s and early 21st century reveal the highly variable and cyclical nature of hurricane activity in the Northern Gulf of Mexico, as well as the need for reliable datasets that can be used to quantify longterm trends and relationships with sea surface temperature (Goldenberg et al., 2001).

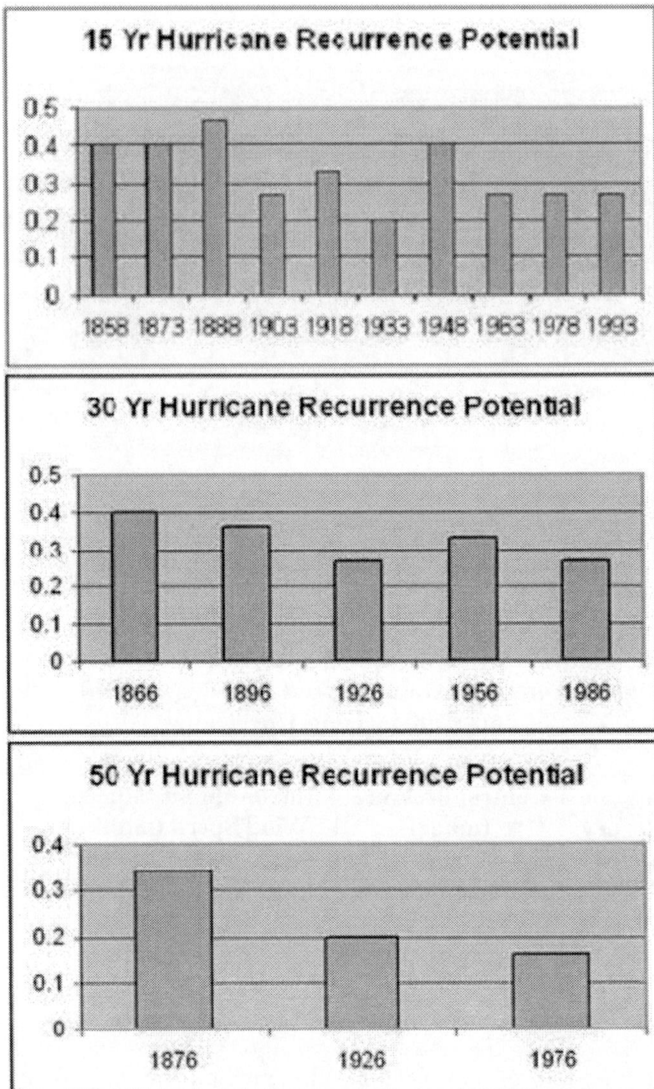

Figure 3.28. Storm frequency variation for 15-, 30-, and 50-year intervals for Category 1 storms or greater for the most active grid location across the Gulf Coast study region

3.2.6. Patterns of Hurricane Wind Direction

The HURASIM model outputs wind direction during storm landfall, which often relates to storm impact based on exposure to direct wind force. Road signs, for example, may be more prone to damage or destruction depending on their orientation to circulating storm winds. Because most storms approach the coast from the Gulf of Mexico on a northerly track, approaching storm winds are easterly and northeasterly on account of the counterclockwise rotation of North Atlantic tropical storms. Figure 3.29 displays simulated wind rows and direction of wind force derived for one of the most active grid cell locations in the study region at Grand Isle, LA, for tropical storm and hurricane conditions over the 153-year period of record. The concentration of wind rows is westerly and southerly for tropical storm events in accordance with prevailing storm approach from the south. Hurricane-force winds and direction at Grand Isle demonstrate a distinct shift to southwesterly and southeasterly directions as a result of major hurricanes passing to the east. As hurricanes pass inland of a given site, yet sustain their strength, backside winds in the opposite direction can occur. The length of each wind row is a function of the total number of 15-minute intervals of storm track interpolation and passage extracted from the HURASIM model. Longer wind rows are indicative of more frequent occurrences. Wind row data and polargrams have been generated for each grid cell within the Gulf Coast study region so that local and regional characterization of wind direction can be determined.

3.2.7. Modeling Climate Change Effects on Tropical Cyclones into the 21st Century

Early theoretical work on hurricanes suggested an increase of about 10 percent in wind speed for a 2 °C (4 °F) increase in tropical sea surface temperature (Emanuel, 1987). A 2004 study from the Geophysical Fluid Dynamics Laboratory in Princeton, New Jersey, that utilized a mesoscale model, downscaled from coupled global climate model runs, indicated the possibility of a 5 percent increase in the wind speeds of hurricanes by 2080 (cf. IPCC, 2001). To explore how climate change could affect 21st century hurricane intensity, windspeeds of hurricanes during 1904 to 2000 were modeled and then projected to increase from 5 to 20 percent over the equivalent forecast period of 2004 to 2100. Storm tracking for calendar years 2004 and 2005 have not been added to the HURDAT (NOAA/NCDC) dataset as yet and, therefore, have been omitted from this analysis despite record storm activity in 2005 that may be associated with multidecadal cycles and/or current global warming trends. Future storm intensities were calculated by multiplying the historical wind reconstructions with the proportional increase based on the forecast year relative to a ramping increase to 5, 10, 15, and 20 percent by the year 2100. The theoretical and empirical limits of maximum hurricane intensity appear to be highly correlated with SSTs (Miller, 1958; Emanuel, 1986, 1988; Holland, 1997). While climatologists debate the weight of contributing factors, including SST, modeling and recent empirical evidence suggest that a 10 percent or more increase in potential intensity is plausible under warming conditions predicted for the 21st century (Emanuel, 1987; Camp and Montgomery, 2001; Knutson and Tuleya, 2004).

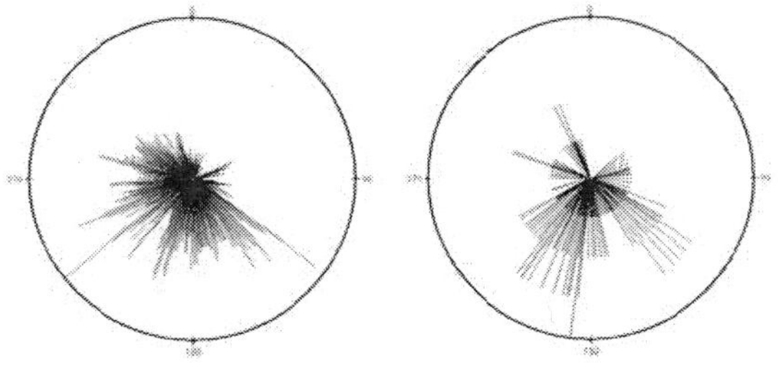

Figure 3.29. Simulated wind rows and direction of wind force derived from the HURASIM model for one of the most active grid cell locations in the study area at Grand Isle, LA, for tropical storm and hurricane conditions over the 153-year period of record, 1851 - 2003

Due to the differences in multidecadal hurricane activity over the 20th century, it was appropriate to evaluate the potential increase in storm frequency relative to the period of record. Figure 3.30 shows the potential increase in storm frequency by years 2050 and 2100 under a climate change scenario that supposes increased ramping of hurricane intensity concomitant with warming sea surface temperatures projected at 5, 10, 15, and 20 percent over the 21st century. Results show that an increase of one to two hurricanes above the historical frequency can be expected by year 2050 and up to four added hurricanes by year 2100. The potential gain of four hurricanes over the next century from a 20 percent increase in storm intensities nearly doubles the strike probability of the historical record. Not only will hurricane incidence increase under these assumptions, but individual storms will be stronger such that more catastrophic storms are likely to develop regardless of landfall location. These models and simulated data provide transportation planners with discrete and generalized probabilities of potential hurricane impact based on past and future climate.

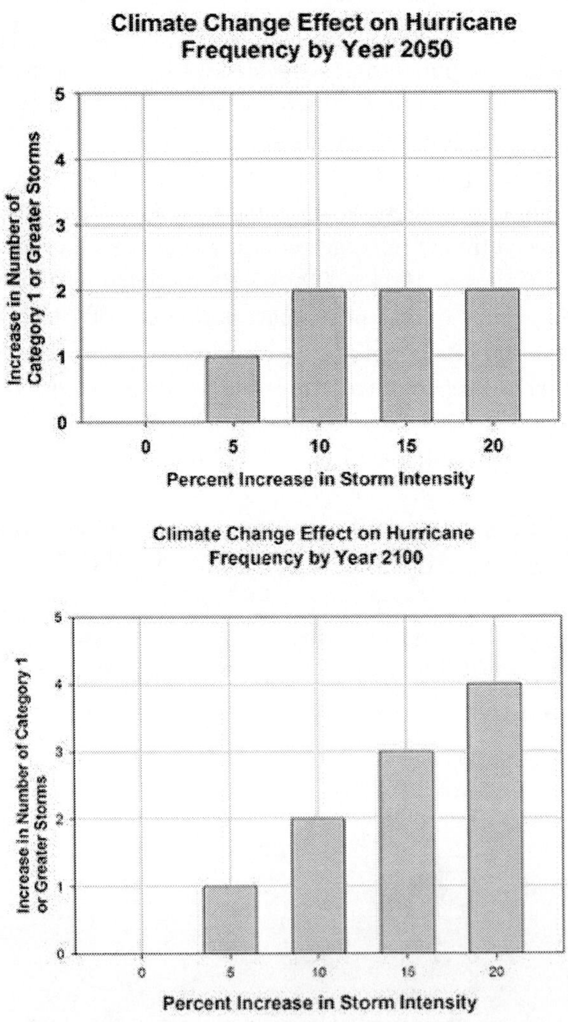

Figure 3.30. Potential increase in the number of hurricanes by the years 2050 and 2100, assuming an increase in hurricane intensity at 5, 10, 15, and 20 percent concomitant with warming sea surface temperatures projected

3.3. SEA LEVEL RISE AND SUBSIDENCE

Changes in climate during ice ages and warming periods have affected sea levels and coastal extent, as evidenced from geologic records. Currently, global sea level is on the rise and is likely to accelerate with continued fossil fuel consumption from modernization and population growth (IPCC, 2001, 2007). As sea level rises, coastal shorelines will retreat, and low-lying areas will tend to be inundated more frequently, if not permanently, by the advancing sea. Subsidence (or sinking) of the land surface already is contributing to the flooding of transportation infrastructure in many Gulf Coast counties. In order to assess the vulnerability of transportation systems to inundation due to sea level rise, an integrated assessment of all important influences on coastal flooding must be considered. Relative sea level rise (RSLR) is the combined effect of an increase in ocean volume resulting from thermal expansion, the melting of land ice ("eustatic" sea level rise), and the projected changes in land surface elevation at a given location.

In this section, global sea level trends are first reviewed, including a comparison of IPCC findings in the Third and Fourth Assessments. This is followed by an examination of sea level rise and subsidence in the study region. The application of two different models to project RSLR in the region is then discussed, and a summary of the modeled range of projected RSLR to 2100 is presented.

3.3.1. Historical and Projected Global Sea Level Trends

Sea level has risen more than 120 m since the peak of the last ice age (about 20,000 B.P.) and over the 20th century by 1-2 mm/year (Douglas, 1991, 1997; Gornitz, 1995; IPCC, 2001). The rate of global sea level rise since 1963 is estimated at 1.8 mm/year (IPCC, 2007). More recent analysis of satellite altimetry data for the period from 1993 to 2003 shows a global average rate of sea level rise of about 3.1 (2.4-3.8) mm per year. Whether the faster rate since 1993 reflects decadal variability or a long-term acceleration over the 20th-century rate is unclear. There is high confidence, however, that the rate of observed sea level rise was greater in the 20th century compared to the 19th century (IPCC, 2007).

The rate of sea level rise in the world ocean basins varied significantly during the 20th century. Sea level rise during the 21st century is projected to have substantial geographical variability as well. The historical rate of sea level rise calculated from tide gauge records, and satellite altimetry is much higher in the Gulf of Mexico than in many other ocean basins (see IPCC, 2007, Working Group I, page 412).

The IPCC Third Assessment Report (TAR) (2001) projected an increase of 0.09-0.88 m in average global sea level by year 2100 with a mid-range estimate of 0.45 m. The range of projected sea level rise through 2100 is slightly lower and narrower in the IPCC Fourth Assessment Report (AR4) (see table 3.1). The midpoint of the projections in sea level rise differs by roughly 10 percent, and the ranges in the two assessment reports would have been similar if they had treated uncertainties in the same way (IPCC, 2007). As noted earlier, the IPCC 2007 sea level rise projections do not include rapid dynamical changes in ice flow from Greenland or Antarctica. If realized, some of the model-based projections could more than double the rate of sea level rise over the past century.

3.3.2. Tide Records, Sea Level Trends, and Subsidence Rates along the Central Gulf Coast

Changes in mean water level at a given coastal location are affected by a combination of changes in sea level in an ocean basin and by local factors such as land subsidence. Gulf Coastal Plain environments, particularly in the central and western parts of the Gulf Coast study area, are prone to high rates of land surface subsidence attributed to soil decomposition and compaction, deep fluid extraction (Morton et al., 2001, 2002; White and Morton, 1997), and the lack of sediment deposition. For example, the Mississippi River delta region demonstrates relative sea level rates of 10 mm/year, tenfold greater than current eustatic sea level rise (Penland and Ramsay, 1990; Gornitz, 1995). Cahoon et al. (1998) measured subsidence rates for several Gulf Coast sites ranging from a low of 2.7 mm/year in the Big Bend region of northwest Florida up to 23.9 mm/year for coastal Louisiana. Some of the forces driving shallow subsidence apparently included seasonal changes in water levels and a periodic occurrences of major storms.

The National Ocean Service (NOS), a division of NOAA, validates and reposits historical water level records at primary tide stations along the coasts and Great Lakes of the United States. Historic data from tide stations located within the Gulf Coast study region have been downloaded from the NOS web site at <http://tidesandcurrents.noaa.gov> in graphical and digital formats to be used in model simulations for projecting future sea level rise. Three tide stations at Pensacola, FL, Grand Isle, LA, and Galveston, TX, comprise the most reliable long-term tide records corresponding with the eastern, central, and western coverage of the study area (figure 3.31). The mean sea level trend for these gauges shows Grand Isle, LA, with the highest rate at 9.85 mm/year, followed by Galveston, TX, at 6.5 mm/year, and Pensacola, FL, at the lowest rate of 2.14 mm/year. These trend values are indicative of the high rates of local subsidence in Louisiana and Texas relative to the more stable geology underlying the Florida Panhandle. Multiple studies have extracted subsidence rates from these and other tide gauges within the Gulf Coast sector with some variability in rate estimates and methodology that mostly reaffirm regional patterns of generally high or low subsidence trends (Swanson and Thurlow, 1973; Penland and Ramsay, 1990; Zervas, 2001; Shinkle and Dokka, 2004).

Long-term tide gauge records are among the most reliable measures of local and regional subsidence. However, tide records also include the long-term trend of eustatic sea level change, which over the last century has been estimated at 1.7-1.8 mm/year on a global basis (Douglas, 1991, 1997, 2001; IPCC, 2001, 2007; Holgate and Woodworth, 2004). Accounting for historical eustatic change in accord with the global average equates to regional subsidence rates of 8.05 mm/year for Grand Isle, LA, and the Mississippi River deltaic plain; 4.7 mm/year for Galveston, TX, and the chenier plain; and 0.34 mm/year for Pensacola, FL, and Mississippi/Alabama Sound of the central Gulf Coast. The high subsidence rate of the Mississippi River Delta region at Grand Isle, LA, is more than four times greater than the historical eustatic trend of the last century and will account for a relative rise in sea level approaching 0.81 m by the year 2100, apart from future eustatic changes. Some areas within the coastal zone of Louisiana have subsidence rates exceeding 20 mm/year, demonstrating the potential range and variability within a subregion (Shinkle and Dokka, 2004).

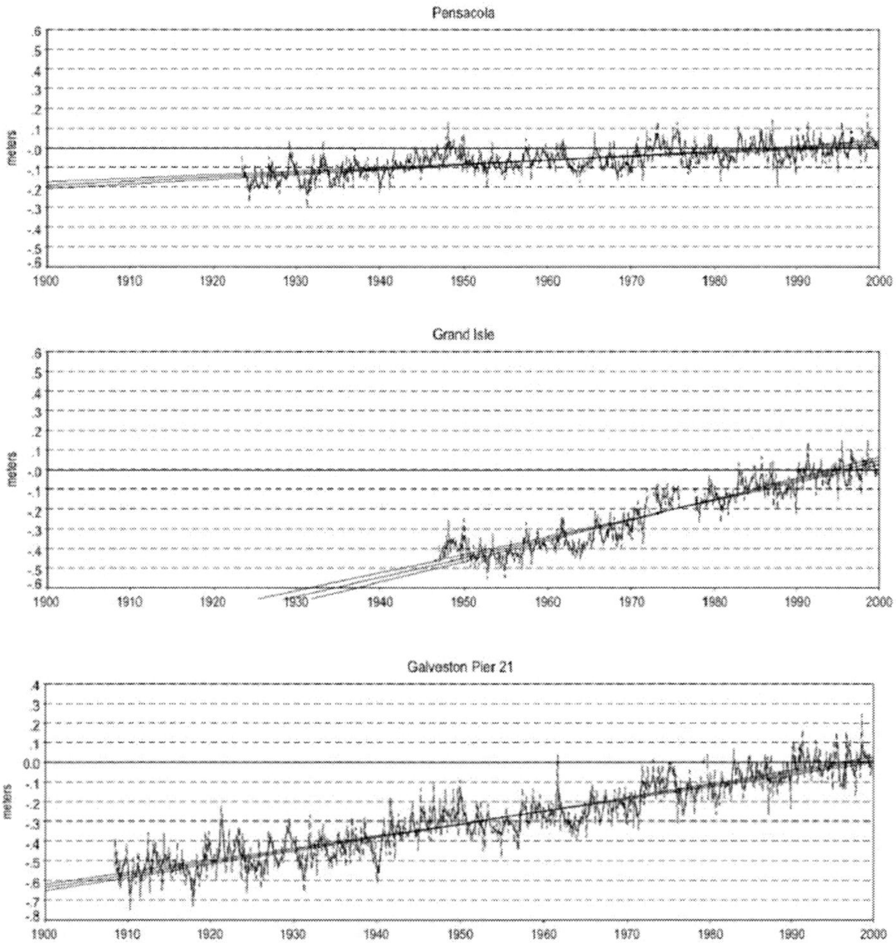

Figure 3.31. Tide gauge records and mean sea level trend line for three northern Gulf Coast tide stations at Pensacola, FL, Grand Isle, LA, and Galveston, TX, corresponding with the eastern, central, and western coverage of the study area (1900-2000)

Subsidence rates across a broad region like the Gulf Coast are highly variable on a local scale even within a representative coastal landform such as the Mississippi River deltaic plain or chenier plain. Many factors contribute to the rate and process of subsidence at a given locale by natural compaction, dewatering, and subsurface mineral extractions. Releveling surveys of benchmark monuments and well heads provide additional evidence and rates of rapid subsidence (Morton et al., 2001, 2002; Shinkle and Dokka, 2004). An extensive releveling project of the Lower Mississippi River coastal plain of first-order benchmarks along major highway corridors provides an expansive network of measured subsidence rates (Shinkle and Dokka, 2004). Oil and gas extractions in coastal Louisiana and southeastern Texas have accelerated local subsidence and wetland loss concomitant with production (Morton et al., 2001, 2002). Releveling projects in large cities such as New Orleans and Houston-Galveston have demonstrated high subsidence rates related to sediment dewatering and groundwater pumping, increasing the vulnerability to local flooding (Gabrysch, 1984; Zilkowski and Reese, 1986; Gabrysch and Coplin, 1990; Holzschuh, 1991; Paine, 1993; Galloway et al., 1999; Burkett et al., 2002).

3.3.3. Sea Level Rise Scenarios for the Central Gulf Coast Region

Two different sea level rise models were used to assess the range of sea level change that could be expected in the study area during the next 50 to 100 years. The Sea Level Rise Rectification Program (SLRRP) (see appendix F) is a model developed by the U.S. Geological Survey to explore the combined effects of future sea level change and local subsidence on coastal flooding patterns. CoastClim is a commercially available model that allows users to select GCM and emissions scenarios to predict sea level change within GCM grid cells over oceans. Table 3.13 outlines the selection list of GCM models that were available for use with SLRRP and the CoastClim models at the time of this study.

SLRRP projects future sea level rise for select tide gauge locations by rectifying the historical tide record of monthly means for the period of record and adding the predicted global mean eustatic sea level change obtained from IPCC (2001).[5] The tidal data input for the SLRRP model is composed of mean monthly water levels, which captures both short-term seasonal deviations and long-term trends of sea level change. Monthly values are derived from averaged hourly recordings for each month. A mean sea level trend is calculated for each tide gauge station, which includes both the local subsidence rate of vertical land movement and the eustatic rate of global sea level change for the period of record. Data records are given in stage heights for different tidal datums such as mean low water, mean tide level, and mean high water, which were rectified to the North American Vertical Datum of 1988 (NAVD88) to readily compare with land-based elevations of roads and other transportation infrastructure. Monthly extremes data also were used in this study to show that daily highs within a month can exceed the monthly average by as much 0.2 84 m and 0.196 m for Galveston, TX, and Pensacola, FL, respectively. (SLRRP model procedures and inputs are explained in further detail in appendix F.)

Table 3.13. GCM Model-Selection Options Based on Data Availability for the USGS SLRRP and CoastClim Models for Generating Future Sea Level Rise Projections. There are 3 GCM Model Datasets Shared Between SLRRP and CoastClim and a Total of 11 GCM Models and Datasets Altogether

SLRRP GCM Listing	CoastClim GCM Listing
CSIRO_Mk2	CGCM1
CSM 1.3	CGCM2
ECHAM4/OPYC3	CSIRO_Mk2
GFDL_R15_a	GFDL_R15_b
HadCM2	GFDL_R30_c
HadCM3	HadCM2
PCM	HadCM3

Notes: CGMC1, CGCM2: Canadian Global Coupled Model .
CSIRO_Mk2: Commonwealth Scientific and Industrial Research Organisation [Australia] –.
CSM 1.3: Climate Simulation Model (NCAR)
ECHAM4/OPYC3: A coupled global model developed by the Max-Planck-Institute for Meteorology (MPI) and Deutsches Klimarechenzentrum (DKRZ) in Hamburg, Germany
GFDL_R15a, R15b, R30c: Geophysical Fluid Dynamics Laboratory
HADCM2, HADCM3: Hadley Centre Coupled Model.
PCM: Parallel Climate Model – DOE/NCAR

The SLRRP model indicates that surface elevations between 47.8 cm and 119.6 cm (NAVD88) will be inundated by sea level rise through 2050, dependent on geographic location, emissions scenario, and GCM forecast. The SLRRP model suggests that surface elevations between 70.1 cm and 199.6 cm (NAVD88) will be inundated by sea level rise through 2100, again dependent on geographic location, emissions scenario, and GCM forecast. Table 3.14 provides SLRRP model results showing the mean land surface elevations (cm, NAVD88) subject to coastal flooding for Galveston, TX, Grand Isle, LA, and Pensacola, FL, by 2050 and 2100 based on averaged output for all seven GCM models for the A1F1, B1, A1B, and A2 emissions scenarios.

The CoastClim V.1. model is another database tool for extracting predicted sea level for a given location, GCM, and emissions scenario much like the SLRRP model. CoastClim has a global database to predict regional patterns of sea level change associated with grid cell output of inclusive GCM models. CoastClim's user-friendly interface allows the user to select the region of interest from a global map. With a mouse click on the shoreline map, CoastClim picks the closest GCM grid cell and extracts a normalized index of regional sea level change relative to the global mean sea level. The normalized index is derived as a ratio or scaling factor for the average pattern of sea level change for the region or grid cell resolution divided by the global mean sea level change for the forecast period of 2071 to 2100. Table 3.15 shows the equivalent normalized index for each of seven GCM model selections for Galveston, TX, Grande Isle, LA, and Pensacola, FL. The different models display a variable range of grid cell resolution and projected sea level response above and below the global mean from 0.88 to 1.04 for the northern Gulf Coast region. The user also can select from six SRES emissions scenarios (A1B, A1F1, A1T, A2, B1, and B2) to run for a given GCM application. CoastClim displays the predicted outcome in relative sea level rise above zero in tabular and graphical format from 1990 to 2100.

Table 3.14. USGS SLRRP Model Results Showing the Mean Land Surface Elevations (cm [NAVD88]) Subject to Coastal Flooding for the Gulf Coast Region by 2050 and 2100 under a High, Mid, and Low Scenario Based on Combined Output for All Seven GCM Models for the A1F1, B1, A1B, and A2 Emissions Scenarios

Year 2050	Low				Year 2100	Low			
	A1FI	B1	A1B	A2		A1FI	B1	A1B	A2
Galveston, TX	83.0	80.9	83.4	83.4	Galveston, TX	130.7	117.0	124.9	127.0
Grand Isle, LA	107.5	106.0	108.8	106.3	Grand Isle, LA	171.2	159.7	168.7	167.6
Pensacola, FL	48.0	47.8	48.4	53.7	Pensacola, FL	83.9	70.1	78.2	75.2
Year 2050	Mid				Year 2100	Mid			
	A1FI	B1	A1B	A2		A1FI	B1	A1B	A2
Galveston, TX	88.9	86.7	88.7	88.8	Galveston, TX	146.0	129.5	137.1	140.8
Grand Isle, LA	113.6	111.8	114.2	111.8	Grand Isle, LA	185.3	171.4	180.2	181.3
Pensacola, FL	53.9	53.6	53.7	60.0	Pensacola, FL	99.2	82.6	90.3	89.3
Year 2050	High				Year 2100	High			
	A1FI	B1	A1B	A2		A1FI	B1	A1B	A2
Galveston, TX	94.8	92.5	93.9	94.3	Galveston, TX	161.3	142.0	149.3	154.5
Grand Isle, LA	119.6	117.6	119.6	117.3	Grand Isle, LA	199.6	183.1	191.7	195.1
Pensacola, FL	59.8	59.4	58.9	66.3	Pensacola, FL	114.5	95.0	102.5	103.5

Table 3.15. Regional Grid Cell Counts and Normalized Indices of Sea Level Rise Relative to Global Mean Sea Level Projections for Northern Gulf Coast Tide Gage Locations by Different GCM Models Used in CoastClim Simulations

CoastClim Models	Gulf Coast Grid Cell Count	Normalized SLR Index		
		Galveston, TX	Grand Isle, LA	Pensacola, FL
CGCM1	5	0.89	0.89	0.89
CGCM2	5	1.04	1.04	0.95
CSIRO_Mk2	3	0.90	0.94	0.94
GFDL_R15_b	2	0.94	0.88	0.89
GFDL_R30_c	6	0.98	1.01	1.01
HadCM2	2	1.02	1.02	1.02
HadCM3	7	1.03	1.00	0.96

Table 3.16. CoastClim Model Results Showing the Mean Sea Level Rise (CM) for the Gulf Coast Region by 2050 and 2100 under a High, Mid, and Low Scenario Based on Combined Output for All Seven GCM Models for the A1F1, B1, A1B, and A2 Emission Scenarios

Year 2050	Low				Year 2100	Low			
	A1FI	B1	A1B	A2		A1FI	B1	A1B	A2
Galveston, TX	40.5	39.2	40.2	39.6	Galveston, TX	81.8	72.4	76.3	78.6
Grand Isle, LA	60.6	59.3	60.3	59.8	Grand Isle, LA	118.8	109.3	113.3	115.6
Pensacola, FL	14.2	13.0	14.0	14.2	Pensacola, FL	33.6	24.3	28.2	32.0
Year 2050	Mid				Year 2100	Mid			
	A1FI	B1	A1B	A2		A1FI	B1	A1B	A2
Galveston, TX	46.2	44.3	45.8	44.8	Galveston, TX	101.8	84.9	92.2	95.4
Grand Isle, LA	66.4	64.4	66.0	64.9	Grand Isle, LA	138.9	121.8	129.3	132.4
Pensacola, FL	20.0	18.1	19.6	19.8	Pensacola, FL	53.5	36.8	44.1	49.3
Year 2050	High				Year 2100	High			
	A1FI	B1	A1B	A2		A1FI	B1	A1B	A2
Galveston, TX	54.3	51.6	53.8	52.1	Galveston, TX	130	103.7	115.5	119.3
Grand Isle, LA	74.5	71.7	73.9	72.3	Grand Isle, LA	167.3	140.7	152.5	156.4
Pensacola, FL	28.1	25.3	27.5	27.5	Pensacola, FL	81.6	55.6	67.2	73.9

CoastClim was used to generate predicted outcomes for seven different GCM models, six SRES scenarios, and three greenhouse gas forcing conditions of low, mid, and high for a total of 126 individual sea level rise curves for the 21st century. Results indicate that sea level rise will vary with both the selected model and emissions scenario. The high emissions A1 F 1 outcome for all GCM models predicts the highest rates of sea level change among SRES options, with a minimum eustatic sea level rise of 0.67 m by 2100, maximum potential rise of

1.55 m, and a mid-range around 1 m depending on model selection. The CoastClim model shows that relative sea level will rise between 12.68 cm and 75.42 cm by 2050, dependent on-site location, emissions scenario, and GCM forecast. By 2100, CoastClim predicts a potential sea level rise between 23.64 cm and 172.06 cm depending on-site location, emission scenario, and GCM forecast. Table 3.16 displays the CoastClim model results of the mean predicted sea level rise (cm) for the Gulf Coast region by 2050 and 2100 under high, mid-, and low IPCC (2001) scenarios based on combined output for all seven GCM models for the A1F1, B1, A1B, and A2 emissions scenarios. However, these same eustatic rates are captured in the SLRRP model but rectified to a geodetic datum and local tidal conditions that more accurately reflect the potential for coastal flooding.

3.4. Storm Surge

Storm surge is a wave of water that is pushed onshore by the force of the winds in the right quadrant of hurricane approach that can often inundate shoreline and inland areas up to many miles, length, and width. The added wave energy from advancing storms combines with normal tides to create the hurricane storm tide, which increases mean water levels to record heights, usually inundating roadways and flooding homes and businesses. The level of surge in a particular area is determined by the slope of the offshore continental shelf and hurricane intensity. The stronger the hurricane and the shallower the offshore water, the higher the surge will be. This advancing surge combines with the normal tides to create the hurricane storm tide, which can increase the mean water level 15 ft or more. In addition, wind-driven waves are superimposed on the storm tide. This rise in water level can cause severe flooding in coastal areas, particularly when the storm tide coincides with the normal high tides.

3.4.1. Predicting Storm Surge with the SLOSH Model

NOAA's National Weather Service forecasters model storm surge using the SLOSH (Sea, Lake, and Overland Surges from Hurricanes) model. NOAA and the Federal Emergency Management Agency (FEMA) use SLOSH to predict potential height of storm surge so as to evaluate which coastal areas are most threatened and must evacuate during an advancing storm. The SLOSH model is a computerized model run by NOAA's National Hurricane Center (NHC) to estimate storm surge heights and winds resulting from historical, hypothetical, or predicted hurricanes by taking into account storm barometric pressure, size, forward speed, track, and wind force. The model accounts for astronomical tides by specifying an initial tide level but does not include rainfall amounts, riverflow, or wind-driven waves. SLOSH also considers the approach or angle of hurricane landfall, which can effectively enhance surge height of westerly and northwesterly approaching storms along the northern Gulf Coast. Graphical output from the model displays color-coded storm surge heights for a particular area in feet above the model's reference level, the National Geodetic Vertical Datum (NGVD), which is the elevation reference for most maps. Emergency managers use output data and maps from SLOSH to determine which areas must be evacuated for storm surge.

Modeling, theory, and recent empirical evidence suggest that hurricane intensity is likely to increase in the Gulf Coast region (see prior section on hurricanes). Even if hurricanes do not become more intense, however, sea level rise alone will increase the propensity for flooding that will occur when hurricanes make landfall in the Gulf Coast region. To assess the combined potential effects of hurricanes and sea level rise on the Gulf Coast transportation sector, a database of storm surge heights for Category 3 and 5 hurricanes was developed by using NOAA's SLOSH model for all coastal counties (extending inland from coastal counties along the Gulf of Mexico to those counties incorporating I-10) for the study area. Resulting surge elevations were overlaid on ArcView™ representations of each study area, enabling views of the study area in its entirety and minimum graphic representations at the county/parish level.

The NHC developed the SLOSH model to predict storm surge potential from tropical cyclones for comprehensive hurricane evacuation planning. The SLOSH models requires grid-based configurations of near-shore bathymetry and topography on a basin level. The NHC has defined 38 basins in the Atlantic and Pacific Oceans of which there are 14 subbasins that define the offshore and onshore geomorphology of the Gulf Coast shoreline from the Florida Keys to the Laguna Madre of Texas. SLOSH model simulations were performed for a merged suite of SLOSH basins (n=7) that covers the central Gulf Coast between Galveston, TX, and Mobile, AL, (table 3.17). SLOSH output were compiled for 28 simulation trials to extract surge levels for varying storm intensities (Categories 2-5) and landfall approaches. A sample simulation of surge height predictions are shown based on combined output for storms of Category 2, 3, 4, and 5 approaching the eastern half of the study area (Louisiana, Mississippi, and Alabama) on different azimuths (figure 3.32). Storm intensity, speed, and direction produces different storm surge predictions. Model simulation trials conducted for the SLOSH basin that covers New Orleans involved calibration and validation checks with historical storms and flood data.

Study area SLOSH applications involved the collection, synthesis, and integration of various geospatial information and baseline data for the central Gulf Coast region relevant to storm surge model implementation and predictions, with the following objectives:

- To derive a database of storm surge heights for Category 3 and 5 hurricanes by using NOAA's SLOSH model for all coastal counties (extending inland from coastal counties along the Gulf of Mexico to those counties incorporating I-10), for the study area spanning Galveston, TX, to Mobile, AL;
- To overlay the resulting surge elevations on ArcView™ representations of each study area, enabling views of the study area in its entirety and minimum graphic representations at the county level;
- To add topographic contours at 1 -m intervals to the study area datasets; and
- To color code storm surge heights based on surge elevation in meters.

The integration of SLOSH output with local geospatial data will be particularly useful in phase II of the study, which will involve an assessment of transportation impacts for a particular county or Metropolitan Planning Organization (MPO) within the study area.

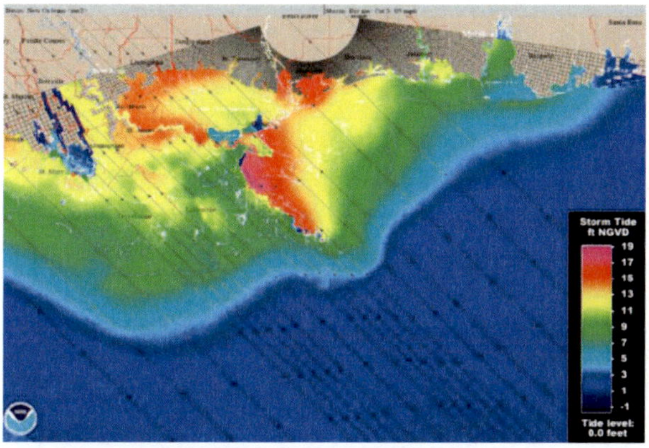

Figure 3.32. Merged results of Category 2 through 5 hurricane surge simulations of a slow-moving storm approaching from the southeast (toward northwest in database), generated by using SLOSH model simulations

Table 3.17. Seven SLOSH Basin Codes, Name Descriptions, and Storm Categories Included in the Central Gulf Coast Study Region and Simulation Trials from Mobile, AL, to Galveston, TX

Basin Code	Basin Name	Storm Category
EMOB	Elliptical Mobile Bay	Cat2, Cat3, Cat4, Cat5
NBIX	MS – Gulf Coast	Cat2, Cat3, Cat4, Cat5
MS2	New Orleans	Cat2, Cat3, Cat4, Cat5
LFT	Vermilion Bay	Cat2, Cat3, Cat4, Cat5
EBPT	Elliptical Sabine Lake	Cat2, Cat3, Cat4, Cat5
EGL2	Elliptical Galveston Bay (2002)	Cat2, Cat3, Cat4, Cat5
PSX	Matagorda Bay	Cat2, Cat3, Cat4, Cat5

3.4.2. Future Sea Level Rise and Storm Surge Height

Sea level rise can be incorporated into surge height predictions from SLOSH simulations for future years by elevating surge levels in proportion to the amount of rise for any given scenario (figure 3.33). Sea level change will be particularly important in influencing this coastal area, since the land already is subject to flooding with supranormal tides and surge and rainfall events of even smaller, less powerful, tropical storms. Improved spatial detail and vertical accuracy of coastal elevations will greatly enhance predictions of the spatial extent of flooding from projected sea level rise and storm surges. Lidar imagery used in this project for coastal Louisiana offers distinct advantages for modeling purposes and graphical representation over other available Digital Elevation Model (DEM) data sources such as the National Elevation Dataset (figure 3.34). Also, it is expected that storm surges superimposed on higher mean sea levels will tend to exacerbate coastal erosion and land loss. During Hurricanes Rita and Katrina, for example, 562 km^2 (217 mi^2) of land in coastal Louisiana was converted to open water (Barras, 2006), and the Chandeleur Island chain was reduced in size by roughly 85 percent (USGS, 2007). The implications of the loss of these natural storm buffers on transportation infrastructure have not been quantified.

Surge analyses were conducted for the Gulf Coast study area by reviewing historical tide records and simulated hurricane scenarios based on the NOAA SLOSH model. Highest tide records for over 70 coastal tide stations were obtained from historical records within the study area, with the highest recorded surge of 6.2 m (20.42 ft) (NAVD88) at Bay St. Louis, MS, in the wake of a northerly approaching Category 5 storm, Hurricane Camille (1969). After Hurricane Katrina (2005), high watermark surveys in New Orleans proper and east along the Gulf Coast in Mississippi revealed storm surge heights approaching 8.5 m (28 ft) mean sea level (m.s.l.). Simulated storm surge from NOAA SLOSH model runs across the central Gulf Coast region demonstrate a 6.7-7.3-m (22-24-ft) potential surge with major hurricanes of Category 3 or greater without considering a future sea level rise effect. Storm approach from the east on a northwesterly track can elevate storm surge 0.3- 1.0 m (1-3 ft) in comparison to a storm of equal strength approaching on a northeasterly track. The combined conditions of a slow churning Category 5 hurricane making landfall on a westerly track along the central Gulf Coast under climate change and elevated sea levels indicate that transportation assets and facilities at or below 9 m (30 ft) m.s.l. are subject to direct impacts of projected storm surge.

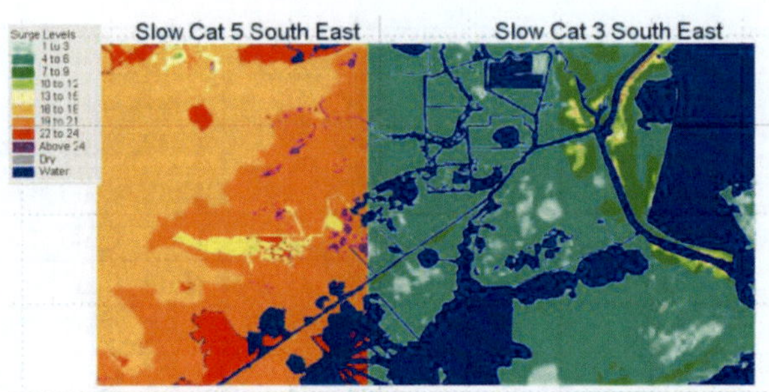

Figure 3.33. Color schemes illustrate the difference in surge inundation between a Category 3 and Category 5 storm approaching the southeastern Louisiana coast from the southeast

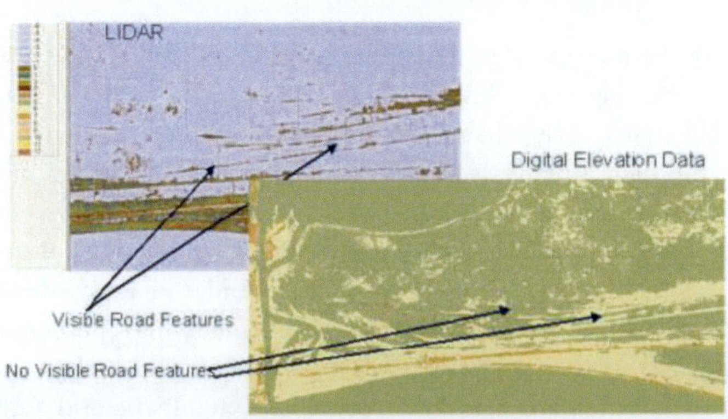

Figure 3.34. Comparison of lidar and National Digital Elevation Data (DEM) for eastern Cameron Parish, LA. The advantages of using a lidarderived topography are many, particularly as the effects of climate change are likely to be subtle in the short term but significant for this low-lying coast where 1 ft of added flooding will impact a large land area

3.5. Other Aspects of Climate Change with Implications for Gulf Coast Transportation

Temperature, precipitation, runoff, sea level rise, and tropical storms are not the only components of Gulf Coast climate that have the potential to change as the temperature of the atmosphere and the sea surface increase. Changes in wind and wave regime, cloudiness, and convective activity could possibly be affected by climate change and would have implications for some modes of transportation in the Gulf Coast region.

3.5.1. Wind and Wave Regime

There have been very few long-term assessments of near surface winds in the United States. Groisman and Barker (2002) found a decline in near surface winds of about -5 percent during the second half of the 20th century for the United States, but they suggest that a stepwise increase in the number of wind-reporting stations noticeably reduced the variance of the regionally averaged time series. They note that most reporting stations are located near airports and other developed areas. They did not attribute the decrease to climate change or land use change. Warming trends can be expected to generate more frequent calm weather conditions typical of summer months and generally characterized by lower winds than more windy conditions typical of cold-season months (Groisman et al., 2004).

Few studies have been made of potential changes in prevailing ocean wave heights and directions as a consequence of climate change, even though such changes can be expected (Schubert et al., 1998; McLean et al., 2001). In the North Atlantic, a multidecadal trend of increased wave height has been observed, but the cause is poorly understood (Guley and Hasse, 1999; Mclean et al., 2001). Wolf (2003) attributes the increasing North Atlantic wave height in recent decades to the positive phase of the North Atlantic Oscillation, which appears to have intensified commensurate with the slow warming of the tropical ocean (Hoerling et al., 2001; Wang et al., 2004). Changes in wave regime will not likely be uniform among ocean basins, however, and no published assessments have focused specifically on how climate change may affect wind and wave regime in the Gulf of Mexico. One 3-year study of wave and wind climatologies for the Gulf of Mexico (Teague et al., 1997) indicates that wave heights and wind speeds increase from east to west across the Gulf. This particular study, which is based on TOPEX/POSEIDEN satellite altimetry and moored surface buoy data, also indicates seasonality with the highest wind speeds and wave heights in the fall and winter.

Scenarios of future changes in seasonal wave heights constructed by using climate model projections for the northeastern Atlantic indicated increases in both winter and fall seasonal means in the 21st century under three forcing scenarios (Wang et al., 2004). The IPCC (2007) concludes that an increase in peak winds associated with hurricanes will accompany an increase in tropical storm intensity. Increasing average summer wave heights along the U.S. Atlantic coastline are attributed to a progressive increase in hurricane activity between 1975 and 2005 (Komar and Allan, 2007). Wave heights greater than 3 m increased by 0.7 to 1.8 m during the study period, with hourly averaged wave heights during major hurricanes increasing significantly from about 7 m to more than 10 m since 1995 (Komar and Allan, 2007). A more recent study of wave heights in the central Gulf of Mexico between 1978 and

2005 suggests a slight increase, but the trend is not statistically significant (Komar and Allan, 2008) (figure 3.35).

If tropical storm windspeed increases as anticipated (see section 3.2.8), this will tend to have a positive effect on mean wave height during the coming decades. Wave heights in coastal bays also will tend to increase due to the combined erosional effects of sea level rise and storms on coastal barrier islands and wetlands (Stone and McBride, 1998; Stone et al., 2003).

3.5.2. Humidity and Cloudiness

As the climate warms, the amount of moisture in the atmosphere is expected to rise much faster than the total precipitation amount (Trenberth et al., 2003). The IPCC (2007) has concluded that tropospheric water vapor increased over the global oceans by 1.2 ± 0.3 percent per decade from 1988 to 2004, consistent in pattern and amount with changes in SST and a fairly constant relative humidity. Several studies have reported an increase in the near surface specific humidity (the mass of water vapor per unit mass of moist air) over the United States during the second half of the past century (Sun et al., 2000; Ross and Elliot, 1996). Sun et al. (2000) found that during 1948 to 1993, the mean annual specific humidity under clear skies steadily increased at a mean rate of 7.4 percent per 100 years.

Gaffen and Ross (1999) analyzed annual and seasonal dewpoint temperature, specific humidity, and relative humidity at 188 first-order weather stations in the United States for the period from 1961 to 1995. (Relative humidity is a measure of comfort based on temperature and specific humidity.) Coastal stations in the southeastern United States were moister than inland stations at comparable latitude, and stations in the eastern half of the country had specific humidity values about twice those at interior western stations. This dataset also shows increases in specific humidity of several percent per decade and increases in dewpoint of several tenths of a degree per decade over most of the country in winter, spring, and summer, with nighttime humidity trends larger than daytime trends (Gaffen and Ross, 1999). In the southeastern United States, specific humidity increased 2 to 3 percent per decade between 1973 and 1993 (Ross and Elliot, 1996), and this trend is expected to continue.

3.5.3. Convective Activity

Sun et al. (2001) documented a significant increase in total, low, cumulonimbus, and stratocumulus cloudiness across the United States during 1948 to 1993. The largest changes in the frequency of cumulonimbus cloudiness occurred in the intermediate seasons, especially in the spring. The increase in the frequency of cumulonimbus cloud development is consistent with the nationwide increase in the intensity of heavy and very heavy precipitation observed by Karl and Knight (1998) and Groisman et al. (2004). Cumulonimbus clouds are commonly associated with afternoon thunderstorms in the Gulf Coast region. The historical and projected increase in summer minimum temperatures for the study area suggest an increase in the probability of severe convective weather (Dessens, 1995, Groisman et al., 2004).

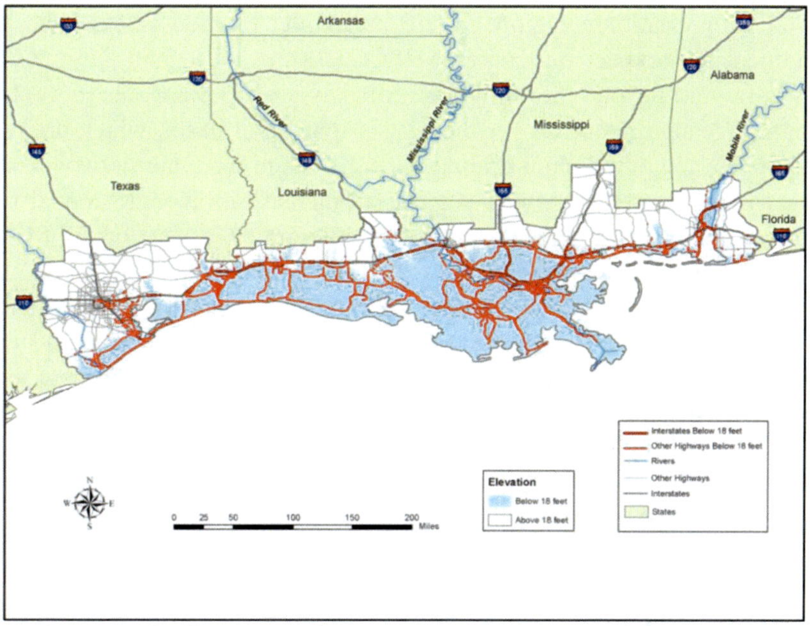

Figure source: Komar and Allan, 2008; data source: NOAA National Buoy Data Center, Stennis, MS.

Figure 3.35. Trend in summer wave height (1978-2005) in the mid-Gulf of Mexico

3.6. CONCLUSIONS

The empirical climate record of the past century, in addition to climate change scenarios, was examined to assess the past and future temperature and hydrology of the central Gulf Coast region. The empirical record of the region shows an annual temperature pattern with high values in the 1920s-1940s, with a drop in annual temperatures in late 1950s, which persisted through the 1970s. Annual temperatures then began to climb over the past three decades but still have not reached the highs of previous decades. The timing of the increase in Gulf Coast temperatures is consistent with the global "climate shift" since the late 1970s (Karl et al., 2000 and Lanzante, 2006) when the rate of temperature change increased in most land areas.

Annual precipitation in the study area shows a suggestion toward increasing values, with some climate divisions, especially those in Mississippi and Alabama, having significant long-term trends. There also is a modeled long-term trend of increasing annual runoff regionwide. Over the entire record since 1919, there was an increase in rainfall that, combined with relatively cool temperatures, led to an estimated 36 percent increase in runoff. Modeled future water balance, however, suggests that runoff is expected to either decline slightly or remain relatively unchanged, depending upon the balance of precipitation and evaporation. Moisture deficits and drought appear likely to increase across the study area, though model results are mixed. These findings are consistent with the IPCC (2007), which concludes that it is very likely that heat waves, heat extremes, and heavy precipitation events over land will increase during this century and that the number of dry days (or spacing between rainfall events) will increase. Even in mid-latitude regions where mean precipitation is expected to decrease, precipitation intensity is expected to increase (IPCC, 2007).

Changes in rainfall beyond the study area can play an important role in the hydrology of the coastal zone. Weather patterns over the Mississippi River basin, which drains 41 percent of the United States, and other major drainages contribute to the total runoff in the Gulf Coast region. Several recent modeling efforts suggest an increase in average annual runoff in the eastern half of the Mississippi River watershed, while drainage west of the Mississippi and along the southern tier of states is generally predicted to decrease (Milly et al., 2005; IPCC, 2007). In the case of the Mississippi River, drainage to the coast is not presently a major factor in terms of flooding of infrastructure, because the river is levied and only a small portion of its flow reaches the marshes and shallow waters of the Louisiana coastal zone. Drainage of the Mississippi River and other rivers to the coast, however, is important in maintaining coastal soil moisture and water quality. The decline of approximately 150,000 acres of coastal marsh in southern Louisiana in 2000 was attributed to extreme drought, high salinities, heat and evaporation, and low river discharge (State of Louisiana, 2000).

As stated earlier, climate models currently lack the spatial and temporal detail needed to make confident projections or forecasts for a number of variables, especially on small spatial scales, so plausible "scenarios" are often used to provide input to decision making. Output from an ensemble of 21 general circulation models (GCMs) run with the three emissions scenarios indicate a wide range of possible changes in temperature and precipitation out to the year 2050. The models agree to a warmer Gulf Coast region of about 1.5 °C ± 1 °C, with the greatest increase in temperature occurring in the summer. Based on historical trends and model projections, we conclude that it is very likely that in the future the number of very hot days will substantially increase across the study area. Due to the non-normality of temperature distributions over the five Gulf States, extreme high temperatures could be about 1°C greater than the change in the average temperature simulated by the GCMs.

Scenarios of future precipitation are more convoluted, with indications of increases or decreases by the various models, but the models lean slightly toward a decrease in annual rainfall across the Gulf Coast. However, by compounding changing seasonal precipitation with increasing temperatures, average runoff is likely to remain the same or decrease, while deficits (or droughts) are more likely to become more severe.

Each of the climate model and emissions scenarios analyzed in this report represent plausible future world conditions. As stated earlier, GCMs currently lack the spatial and temporal detail needed to make projections or forecasts, so plausible "scenarios" are often used to provide input to decision making. These models also lack the capacity for simulating small-scale phenomena such as thunderstorms, tornadoes, hail, and lightning. However, climate models do an excellent job of simulating temperature means and extremes. Hourly and daily precipitation and runoff extremes are much more difficult to simulate due to horizontal resolution constraints. However, based on observational and modeling studies the IPCC (2007) and numerous independent climate researchers have concluded that more intense precipitation events are very likely during this century over continental land masses in the Northern Hemisphere.

Recent empirical evidence suggests a trend towards more intense hurricanes formed in the North Atlantic Basin, and this trend is likely to intensify during the next century (IPCC, 2007). In the Gulf region, there is presently no compelling evidence to suggest that the number or paths of tropical storms have changed or are likely to change in the future.

Convective activity, heavy precipitation events, and cloudiness all appear likely to increase in the Gulf Coast region as the climate warms.

Change in the rate of sea level rise is dependent on a host of interacting factors that are best evaluated on decadal to centennial time scales. Two complimentary modeling approaches were applied in this study to assess the potential rise in sea level and coastal submergence over the next century. Both models were used to estimate RSLR by 2050 and 2100 under a range of greenhouse gas emissions scenarios. Both models account for eustatic sea level change as estimated by the global climate models and also incorporate values for land subsidence in the region based on the historical record. One model, CoastClim, produces results that are closer to a simple measure of future sea level change under the scenarios of future climate. A similar model, SLRRP, also incorporates values for high and low tidal variation attributed to astronomical and meteorological causes, which are pulled from the historical record. The SLRRP model is rectified to the NAVD88 that is commonly used by surveyors to calculate the elevations of roads, bridges, levees, and other infrastructure. The tide data used in the SLRRP model is based on a monthly average of the mean high tide (called mean high higher water) for each day of the month. The SLRRP results capture seasonal variability and interannual trends in relative sea level change, while the CoastClim results do not.

The three long-term tide gauge locations analyzed in this study represent three subregions of the study area: Galveston, TX (the chenier plain); Grand Isle, LA (the Mississippi River deltaic plain); and Pensacola, FL (Mississippi/Alabama Sound). For each of these gauges, we examined potential range of relative sea level rise through 2050 and 2100 using the SRES B1, A1B, A2, and A1F1 emissions scenarios based on the combined output of 7 GCMs (table 3.14). Results for the year 2100 generated with CoastClim range from 24 cm (0.8 ft) in Pensacola to 167 cm (5.5 ft) in Grand Isle. Results for the year 2100 from SLRRP, which as noted above accounts for historical tidal variation, are somewhat higher: predicted relative sea level ranges from 70 cm (2.3 ft, NAVD88) in Pensacola to 199 cm (6.5 ft, NAVD88) in Grand Isle.

Storm surge simulations accomplished basin-specific surge height predictions for a combination of storm categories, track speeds, and angled approach on landfall that can be summarized by worst-case conditions to exceed 6 to 9 m (20 to 30 ft) along the central Gulf Coast. Storm attributes and meteorological conditions at the time of actual landfall of any storm or hurricane will dictate actual surge heights. Transportation officials and planners within the defined study area can expect that transportation facilities and infrastructure at or below 9 m of elevation along the coast are subject to direct and indirect surge impacts. Sea level rise of 1 to 2 m (3-6 ft) along this coast could effectively raise the cautionary height of these surge predictions to 10 m (33 ft) or more by the end of the next century.

Changes in climate can have widespread effects on physical and biological systems of low- lying, sedimentary coasts. However, the large and growing pressures of development are responsible for most of the current stresses on Gulf Coast natural resources, which include: water quality and sediment pollution, increased flooding, loss of barrier islands and wetlands, and other factors that are altering the resilience of coastal ecosystems (U. S. Environmental Protection Agency, 1999). Human alterations to freshwater inflows through upstream dams and impoundments, dredging of natural rivers and engineered waterways, and flood-control levees also have affected the amount of sediment delivered to the Gulf coastal zone. Roughly 80 percent of U.S. coastal wetland losses have occurred in the Gulf Coast region since 1940, and predictions of future population growth portend increasing pressure on Gulf Coast communities and their environment. Sea level rise will generally increase marine transgression on coastal shorelines (Pethick, 2001) and the frequency of barrier island overwash during storms, with effects most severe in coastal systems that already are stressed and deteriorating. An increase in tropical storm intensity or a decrease in fresh water and sediment delivery to the coast would tend to amplify the effects of sea level rise on Gulf Coast landforms.

Our assessment of historical and potential future changes in Gulf Coast climate section draws on publications, analyses of instrumental records, and models that simulate how climate may change in the future. Model results, climatic trends during the past century, and climate theory all suggest that extrapolation of the 20th-century temperature record would likely underestimate the range of change that could occur in the next few decades. The global near-surface air temperature increase of the past 100 years is approaching levels not observed in the past several hundred years (IPCC, 2001); nor do current climate models span the range of responses consistent with recent warming trends (Allen and Ingram, 2002). Regional "surprises" are increasingly possible in the complex, nonlinear Earth climate system (Groisman et al., 2004), which is characterized by thresholds in physical processes that are not completely understood or incorporated into climate model simulations; e.g., interactive chemistry, interactive land and ocean carbon emissions, etc. While there is still considerable uncertainty about the *rates* of change that can be expected (Karl and Trenberth, 2003), there is a fairly strong consensus regarding the direction of change for most of the climate variables that affect transportation in the Gulf Coast region. Key findings from this analysis and other published studies for the study region include:

Warming Temperatures – An ensemble of GCMs indicate that the average annual temperature is likely to increase by 1-2 °C (2-4 °F) in the region by 2050. Extreme high temperatures also are expected to increase, and within 50 years the probability of experiencing 21 days a year with temperatures of 37.8 °C (100 °F) is greater than 50 percent.

Changes in Precipitation Patterns – While average annual rainfall may increase or decrease slightly, the intensity of individual rainfall events is likely to increase during the 21st century. It is possible that average soil moisture and runoff could decline, however, due to increasing temperature, evapotranspiration rates, and spacing between rainfall events.

Rising Sea Levels – Relative sea level is likely to rise between 1 and 6 ft by the end of the 21st century, depending upon model assumption and geographic location. The highest rate of relative sea level rise will very likely be in the central and western parts of the study area (Louisiana and East Texas) where subsidence rates are highest.

Storm Activity – Hurricanes are more likely to form and increase in their destructive potential as the sea surface temperature of the Atlantic and Gulf of Mexico continue to increase. Rising relative sea level will exacerbate exposure to storm surge and flooding. Depending on the trajectory and scale of individual storms, facilities at or below 9 m (30 ft) could be subject to direct storm surge impacts.

Table 3.18. SLRRP Model Parameters and Results Showing the Mean Sea Level Rise Projections for the Gulf Coast Region by 2050 and 2100 Under a High, Mid, and Low Scenario Based on Combined Output for All Seven GCM Models for the A1F1 Emission Scenario

Model Parameters	Scenarios	Louisiana-Texas Chenier Plain	Louisiana Deltaic Plain	Mississippi-Alabama Sound
Tide Gauge		Galveston, TX	Grand Isle, LA	Pensacola, FL
Sea Level Trend (mm/yr)		6.5	9.85	2.14
Subsidence (mm/yr)		4.7	8.05	0.34
Sea Level Rise by 2050 (cm, NAVD88)	High	94.8	119.6	59.8
	Mid	88.9	113.6	53.9
	Low	83.0	107.5	48.0
Sea Level Rise by 2100 (cm, NAVD88)	High	161.3	199.6	114.5
	Mid	146.0	185.3	99.2
	Low	130.7	171.2	83.9

3.7. REFERENCES

Ahijevych, D. A., Carbone, R. E. & Davis, C. A. (2003). Regional Scale Aspects of the Diurnal Precipitation Cycle. Preprints, *31st International Conference on Radar Meteorology,* Seattle, Washington, American Meteorological Society, 349-352B.

Allen, R. G. (2003). *REF-ET User's Guide*. University of Idaho Kimberly Research Stations: Kimberly, Idaho.

Allen, R. G., Pereira, L. S., Raes, D. & Smith, M. (1998). *Crop Evapotranspiration-Guidelines for Computing Crop Water Requirements.* FAO Irrigation and Drainage Paper 56, Food and Agriculture Organization, Rome.

Allen, M. R. & Ingram, W. J. (2002): Constraints on future changes in climate and the hydrologic cycle. *Nature*, *419*, 224-232.

Bell, G. D. & Chelliah, M. (2006). Leading tropical modes associated with interannual and multidecadal fluctuations in North Atlantic hurricane activity, *Journal of Climatology*, *19*, 590– 612.

Bell, G. D., Blake, E., Landsea, C. W., Chelliah, M., Pasch, R., Mo, K. C. & Goldenberg, S. B. (2007). Tropical Cyclones: Atlantic Basin. In: "Chapter 4: The Tropics", from the *State of the Climate in 2006*, A. Arguez (ed.). *Bulletin of the American Meteorological Society, 88*, S1-S135.

Benjamin, J., Shay, L. K., Uhlhorn, E., Cook, T. M., Brewster, J., Halliwell, G. & Black, P. G. (2006). Influence of Loop Current ocean heat content on Hurricanes Katrina, Rita, and Wilma. In: *27th Conference on Hurricanes and Tropical Meteorology*, Paper C3.4, American Meteriological Society, 24-28 April 2006, Monterey, California. 4 pages.

Burkett, V. B., Zilkowski, D. B. & Hart, D. A. (2002). Sea Level Rise and Subsidence: Implications for Flooding in New Orleans, Louisiana. In: Prince, K.R., Galloway, D.L. (Eds.) *Subsidence Interest Group Conference, Proceedings of the Technical Meeting*, Galveston, Texas, 27-29 November 2001. USGS Water Resources Division Open-File Report Series 03-308, U.S. Geological Survey, Austin, Texas.

Cahoon, D. R., Day, J. W., Jr., Reed, D. & Young, R. (1998). Global climate change and sea level rise: estimating the potential for submergence of coastal wetlands. In: *Vulnerability of coastal wetlands in the Southeastern United States: climate change research results, 1992-1997.* [G.R. Guntenspergen and B.A. Vairin, eds.] U.S. Geological Survey Biological Resources Division Biological Science Report, USGS/BRD/BSR-1998-0002, pp 2 1-34.

Christopherson, R. W. (2000). *Geosystems*. Prentice-Hall, Upper Saddle River, New Jersey.

Dessens, J. (1995). Severe convective weather in the context of a nighttime global warming. *Geophysical Research Letters, 22*, 124 1-1244.

Dingman, S. L., (2002). *Physical Hydrology, 2nd Ed.* Prentice Hall, Upper Saddle River, New Jersey.

Douglas, B. C. (1991). Global sea level rise. *Journal of Geophysical Research, 96*, 6981-92.

Douglas, B. C. (1997). Global sea rise: A redetermination. *Surveys in Geophysics*, Volume 18, pages 279-292.

Douglas, B. C. (2001). Sea level change in the era of the recording tide gauge. In: *Sea level Rise: History and Consequences*, International Geophysics Series, Volume 75, edited by B. Douglas, M. Kearney, and S. Leatherman, Chapter 3, pages 37-64. Academic Press, San Diego, California.

Easterling, D. R., Karl, T. R., Mason, E. H., Hughes, P. Y., Bowman, D. P., Daniels, R. C. & Boden, T. A. (Eds.), *United States Historical Climatology Network (U.S. HCN) Monthly.*

Elsner, J. B. (2006). Evidence in support of the climate change-Atlantic hurricane hypothesis. *Geophysical Research Letters, 33*, L1 6705.

Emanuel, K. (1987). The dependence of hurricane intensity on climate. *Nature, 326*, 483-485.

Emanuel, K. (2005). Increasing destructiveness of tropical cyclones over the past 30 years. *Nature, 436*, 686-688.

Fontenot, R. (2004). *An Evaluation of Reference Evapotranspiration Models in Louisiana.* M.N.S. Thesis, Department of Geography and Anthropology, Louisiana State University.

Gabrysch, R. K. (1984). *Ground-water withdrawals and land-surface subsidence in the Houston-Galveston region, Texas, 1906-1980.* Texas Department of Water Resources Report, *287*, Austin, Texas, 64 pages.

Gabrysch, R. K. & Coplin, L. S. (1990). *Land-surface subsidence resulting from groundwater withdrawals in the Houston-Galveston Region, Texas, through 1987.* U.S. Geological Survey, Report of Investigations No. 90-01, 53 pages.

Gaffen, D. J. & Ross, R. J. (1999). Climatology and trends in U.S. surface humidity and temperature, *Journal of Climate, 12.*

Galloway, D., Jones, D. R. & Ingebritsen, S. E. (1999). *Land Subsidence in the United States.* U.S. Geological Survey Circular 1182.

Gornitz, V. (1995). Sea level rise: a review of recent past and near-future trends. *Earth Surface Processes and Landforms, 20*, 7-20.

Goldenberg, S. B., Landsea, C. W., Mestas-Nunez, A. M. & Gray, W. M., (2001). The recent increase in Atlantic hurricane activity: Causes and implications. *Science, 293*, 474-479.

Graumann, A., Houston, T., Lawrimore, J., Levinson, D., Lott, N., McCown, S., Stephens, S. & Wuertz, D. (2005). *Hurricane Katrina – a climatological perspective.* October 2005, Updated August 2006. Technical Report 2005-01. 28 pages. NOAA National Climate Data Center, available at http://www.ncdc.noaa.gov/oa/reports/tech-report200501 z.pdf.

Groisman, P. Y., Knight, R. W., Karl, T. R., Easterling, D. R., Sun, B. & Lawrimore, J. H. (2004). Contemporary changes of the hydrological cycle over the contiguous United States: Trends derived from in situ observations. *Journal of Hydrometeorology, 5*, 64-85.

Groisman, P. Y. & Barker, H. P. (2002). Homogeneous blended wind data over the contiguous United States. Preprints, *13th Conference on Applied Climatology,* Portland, Oregon, American Metrological Society, J114–J117.

Guley, S. K. & Hasse, L. (1999). Changes in wind waves in the North Atlantic over the last 30 years. *International Journal of Climatology*, *19*, 1,091-1,117.

Guttman, N. B. & Quayle, R. G. (1996). A Historical Perspective of U.S. Climate Divisions. *Bulletin of the American Meteorological Society, 77*, 293-303.

Hanson, R. L. (1991). *Evapotranspiration and Droughts*, In: R.W., Paulson, E.B., Chase, R. S. Roberts, & D. W. Moody, Compilers, National Water Summary 1988-89- *Hydrologic Events and Floods and Droughts*: U.S. Geological Survey Water-Supply Paper 2375, pages 99-104.

Hoerling, M. P., Hurrell, J. W. & Taiyi, X. (2001). Tropical origins for recent North Atlantic climate change. *Science, 292*, 90-92.

Holgate, S. J. & Woodworth, P. L. (2004). Evidence for enhanced coastal sea level rise during the 1 990s. *Geophysical Research Letters*, Volume, *31*, L07305.

Holzschuh, J. C. (1991). *Land subsidence in Houston, Texas U.S.A.: Field-trip guidebook for the Fourth International Symposium on Land Subsidence*, May 12-17, 1991, Houston, Texas, 22 pages.

Hoyos, C. D., Agudelo, P. A., Webster, P. J., Curry, J. A. (2006). Deconvolution of the Factors Contributing to the Increase in Global Hurricane Intensity. *Science, 312,* 94-97.

IPCC (Intergovernmental Panel on Climate Change), 1996: *Climate Change 1995: Impacts, Adaptations and Mitigation of Climate Change: Scientific-Technical Analyses.* Cambridge University Press, New York, New York, 880 pages.

IPCC, (2001). *Climate Change 2001: Impacts, Adaptation, and Vulnerability.* In: J. J., McCarthy, O. F., Canziani, N. A., Leary, et al. (Eds.). Contribution of Working Group II to the Third Assessment Report of the Intergovernmental Panel on Climate Change. New York, New York: Cambridge University Press. 944 pages. (Available on-line at http://www.ipcc.ch/).

IPCC, (2007). *Climate Change 2007: The Physical Science Basis, Summary for Policy-Makers.* Contribution of Working Group I to the Fourth Assessment Report of the Intergovernmental Panel on Climate Change. Geneva, Switzerland, 21 pages.

Jensen, D. T., Hargreaves, G. H., Temesgen, B. & Allen, R. G. (1997). Computation of ETo under Nonideal Conditions. *Journal of Irrigation and Drainage Engineering, 123(5)*, 394-400.

Karl, T., Knight, R. & Plummer, N. (1995). Trends in High-Frequency Climate Variability in the Twentieth Century. *Nature, 377*, 217-220.

Karl, T. R., Williams, C. N., Jr., & Quinlan, F. T. (1990):. *United States Historical Climatology Network (HCN) Serial Temperature and Precipitation Data,* ORNL/CDIAC-30, NDP-019/R1, Carbon Dioxide Information and Analysis Center, Oak Ridge National Laboratory, Oak Ridge, Tennessee.

Karl, T. R. & Williams, C. W. Jr., (1987). An Approach to Adjusting Climatological Time Series for Discontinuous Inhomogeneities, *Journal of Applied Meteorology,* Volume, *26*, 1,744-1,763.

Karl, T. R., Williams, C. N. Jr., Young, P. J. & Wendland, W. M. (1986). A Model to Estimate the Time of Observation Bias with Monthly Mean Maximum, Minimum, and Mean Temperatures for the United States, *Journal of Applied Meteorology,* Volume, *25*, 145-160, 1986.

Karl, T. R., Diaz, H. F. & Kukla, G. (1988). Urbanization: Its Detection and Effect in the United States Climate Record. *Journal of Climate*, 1,099-1,123.

Karl, T. R. & Knight, R. W. (1998). Secular trends of precipitation amount, frequency, and intensity in the United States. *Bulletin of the American Metrological Society, 79*, 231-241.

Karl, T. R., Knight, R. W. & Baker, B. (2000). The record breaking global temperatures of 1997 and 1998: evidence for an increase in the rate of global warming? *Geophysical Research Letters*, *27(5)*, 719-722.

Karl, T. R. & Trenberth, K. E. (2003). Modern global climate change. *Science*, *302*, (5651), 1, 719-1,723.

Keim, B. D. (1997). Preliminary Analysis of the Temporal Patterns of Heavy Rainfall across the Southeastern United States. *Professional Geographer, 49(1)*, 94-104.

Keim, B. D., Wilson, A. M., Wake, C. P. & Huntington, T. G. (2003). Are there spurious temperature trends in the United States Climate Division database? *Geophysical Research Letters* 30(7): 1404, doi: 10.1 029/2002GL0 16295, 2003.

Keim, B. D., Fisher, M. R. & Wilson, A. M. (2005). Are there spurious precipitation trends in the United States Climate Division database? *Geophysical Research Letters 32*, L04702, doi: 10. 1029/2004GL021985, 2005.

Komar, P. D. & Allan, J. C. (2007). Higher Waves Along U.S. East Coast Linked to Hurricanes. *EOS*, *Transactions American Geophysical Union*, Volume *88*, Issue 30, page 301.

Komar, P. D. & Allan, J. C. (2008). Increasing hurricane-generated wave heights along the U.S. east coast and their climate controls. *Journal of Coastal Research*, March 2008, Volume *24*, Issue 2.

Lanzante, J. R., Peterson, T. C., Wentz, F. J. & Vinnikov, K. Y. (2006). What do observations indicate about the changes of temperature in the atmosphere and at the surface since the advent of measuring temperatures vertically? In: *Temperature Trends in the Lower Atmosphere*, Edited by T.R. Karl, S.J. Hassol, C.D. Miller and W.L. Murray. U.S. Climate Change Science Program, Synthesis and Assessment Product 1.1, Washington, D.C. pages 47-70.

Liu, K. & Fearn, M. L. (1993). Lake-sediment record of late Holocene hurricane activity from coastal Alabama. *Geology, 21*, 973-976.

Mann, M. E. & Emanuel, K. A. (2006). Atlantic hurricane trends linked to climate change. *EOS* 87, *24*, 233-244.

McLean, R. F., Tsyban, A., Burkett, V., Codignotto, J., Forbes, D., Ittekkot, V., Mimura, N., and R. J. Beamish, (2001). Coastal zones and marine ecosystems. In: *Climate Change: Impacts, Adaptation, and Vulnerability.* Third Assessment Report, Working Group II report of the Intergovernmental Panel on Climate Change (IPCC). IPCC Secretariat, Geneva, Switzerland, pages 343-379.

Milly, P. C. D., Dunne, K. A. & Vecchia, A. V. (2005). Global pattern of trends in streamflow and water availability in a changing climate. *Nature, 438*, 347-3 50.

Morton, R. A., Buster, N. A. & Krohn, M. D. (2002). Subsurface Controls on Historical Subsidence Rates and Associated Wetland Loss in Southcentral Louisiana. *Transactions Gulf Coast Association of Geological Societies*. Volume *52*, pages 767-778.

Morton, R. A., Purcell, N. A. & Peterson, R. L. (2001). Field evidence of subsidence and faulting induced by hydrocarbon production in coastal southeast Texas. *Transactions Gulf Coast Association of Geological Society*, Volume, *51,* 239-248.

Nakicenovic, N. & Swart, R. (Eds.), (2000). *Special Report on Emissions Scenarios*. Cambridge University Press. Cambridge, United Kingdom, 612 pages http://www.grida.no/climate/ipcc/emission/See Special Report on Emissions Scenarios.

NOAA, (2007). *The Saffir Simpson Scale*. http://www.nhc.noaa.gov/aboutsshs.shtml.

Nerem, R. S., Chambers, D. P., Leuliette, E. W., Mitchum, G. T. & Giese, B. S. (1999). Variations in Global Mean Sea Level Associated with the 1997-1998 ENSO Event: Implications for Measuring Long-Term Sea Level Change. *Geophysical Research Letters, 26(19)*, 3,005-3,008.

Nicholls, R. J., Wong, P. P., Burkett, V., Codignotto, J., Hay, J., McLean, R., Ragoonaden, S. & Woodroffe, C. (2007). Coastal Systems and Low-Lying Areas. In: *Climate Change Impacts, Adaptations, and Vulnerability.* Intergovernmental Panel on Climate Change Fourth Assessment Report. IPCC Secretariat, Geneva, Switzerland.

Oouchi, K., Yoshimura, J., Yoshimura, H., Mizuta, R., Kusunoki, S., and A. Noda, (2006). Tropical Cyclone Climatology in a Global-Warming Climate as Simulated in a 20km-Mesh Global Atmospheric Model: Frequency and Wind Intensity Analyses. *Journal of the Meteorological Society of Japan*, *84*, 259-276,10.

Parson, E., Burkett, V., Fisher-Vanden, K., Keith, D., Mearns, L., Pitcher, H., Rosenzweig, C. & Webster, M. (2007). *Global Change Scenarios: Their Development and Use.* Subreport 2.1B of Synthesis and Assessment Product 2.1 by the U.S. Climate Change Science Program and the Subcommittee on Global Change Research. Department of Energy, Office of Biological & Environmental Research, Washington, D.C., U.S.A., 106 pages.

Paine, J. G. (1993). Subsidence of the Texas coast – Inferences from historical and late Pleistocene sea levels: *Tectonophysics,* Volume *222*, pages 445-458.

Penland, S. & Ramsey, K. (1990). Relative sea level rise in Louisiana and the Gulf of Mexico: 1908-1988. *Journal of Coastal Research, 6*, 323-342.

Pethick, J. (2001). Coastal management and sea-level rise. *Catena*, *42*, 307-322.

Quayle, R.G., Easterling, D. R., Karl, T. R. & Hughes, P. Y. (1991). Effects of Recent Thermometer Changes in the Cooperative Station Network, *Bulletin of the American Meteorological Society*, *Volume 72*, 1,718-1,724.

Ross, R. J. & Elliott, W. P. (1996). Tropospheric water vapor climatology and trends over North America: 1973–93. *Journal of Climate, 9,* 3,561-3,574.

Ruosteenoja, K., Carter, T. R., Jylha, K. & Tuomenvirta, H. (2003). *Future Climate in World Regions: An Intercomparison of Model-Based Projections for the New IPCC Emissions Scenarios*. Finnish Environment Institute, Helsinki, 83 pages.

Santer, B. D., Wigley, T. M. L., Gleckler, P. J., Bonfils, C., Wehner, M. F., AchutaRao, K., Barnett, T. P., Boyle, J. S., Brüggemann, W., Fiorino, M., Gillett, N., Hansen, J. E., Jones, P. D., Klein, S. A., Meehl, G. A., Raper, S. C. B., Reynolds, R. W., Taylor, K. E. & Washington, W. M. (2006). Forced and unforced ocean temperature changes in

Atlantic and Pacific tropical cyclogenesis regions. *Proceedings of the National Academy of Sciences*, *103*, 13, 905-13,910.

Scharroo, R., Smith, W. H. F. & Lillibridge, J. L. (2005). Satellite Altimetry and the Intensification of Hurricane Katrina. *Eos*, Volume *86*, No. 40, 4 October, page 366.

Shay, L. K. (2006). Positive feedback regimes during tropical cyclone passage. In: *14th Conference on Sea-Air Interactions*, AMS Annual Meeting, 30 January-3 February, Atlanta, Georgia.

Shell International, (2003). *Scenarios: an Explorer's Guide*. Global Business Environment. [Online at: www-static.shell.com/static/royal-en/downloads/scenarios_explorersguide.pdf.

Shingle, K. D. & Dokka, R. K. (2004). *Rates of vertical displacement at benchmarks in the Lower Mississippi valley and the northern Gulf Coast*. NOAA Technical Report NOS/NGS 50, U.S. Department of Commerce, 135 pages.

Smith, T. M. & Reynolds, R. W. (2004). Improved Extended Reconstruction of SST (1854-1997). *Journal of Climate, 17*, 2,466-2,477.

Stone, G. W. & McBride, R. A. (1998). Louisiana barrier islands and their importance in wetland protection: Forecasting shoreline change and subsequent response of wave climate. *Journal of Coastal Research*, *14*, 900-14,916.

Stone, G. W., Morgan, J. P., Sheremet, A. & Zhang, X. (2003). *Coastal land loss and wave-surge predictions during hurricanes in Coastal Louisiana: implications for the oil and gas industry*. Coastal Studies Institute, Louisiana State University, 67 pages.

State of Louisiana, (2000). *Brown Marsh Questions and Answers*. Coastal Wetlands Planning, Protection, and Restoration Task Force, Baton Rouge, Louisiana Department of Natural Resources, Fact Sheet #5. [Online at: http://www.brownmarsh.net/factSheets/brown%20marsh%20Q%20and%20A.pdf].

Sun, B., Ya. Groisman, P., Bradley, R. S. & Keimig, F. T. (2000). Temporal changes in the observed relationship between cloud cover and surface air temperature. *Journal of Climate, 13*, 4,341-4,357.

Sun, B., Ya. Groisman, P. & Mokhov, I. I. (2001). Recent changes in cloud type frequency and inferred increases in convection over the United States and the former USSR. *Journal of Climate, 14*, 1,864-1,880.

Swanson, R. L. & Thurlow, C. I. (1973). Recent Subsidence Rates along the Texas and Louisiana Coasts as Determined from Tide Measurements. *Journal of Geophysical Research*: *Volume*, *78,* 2,665-2,671.

Tebaldi, C., Mearns, L. O., Smith, R. L. & Nychka, D. (2004). Regional Probabilities of Precipitation Change: A Bayesian Analysis of multimodel simulations. *Geophysical Research Letters*, Volume *31*.

Tebaldi, C., Smith, R. L., Nychka, D. & Mearns, L. O. (2005). Quantifying Uncertainty in Projections of Regional Climate Change: A Bayesian Approach to the Analysis of Multimodel Ensembles. *Journal of Climate*, Volume *18*, Number 10, 1,524-1,540.

Teague, W. J., Hwang, P. A., Jacobs, G. A., Thompson, E. F. & Wang, D. W. (1997). *A three- year climatology of waves and winds in the Gulf of Mexico*. U.S. Naval Research Laboratory, Stennis, Mississippi. NRL/MR/7332-97-8068. 20 pages.

Trenberth, K. E., Dai, A., Rasmussen, R. M. & Parsons, D. B. (2003). The changing character of precipitation. *Bulletin of the American Meteorological Society, 84*, 1,205-1,217.

Trenberth, K. E. & Shea, D. J. (2006). Atlantic hurricanes and natural variability in 2005. *Geophysical Research Letters, 33*, L12704.

Turc, L. (1961). Evaluation des besoins en eau d'irrigation, evapotranspiration potentielle, formule climatique simplifee et mise a jour. (In French). *Annales Agronomiques, 12(1)*, 13-49.

Twilley, R. R., Barron, E., Gholz, H. L., Harwell, M. A., Miller, R. L., Reed, D. J., Rose, J. B., Siemann, E., Welzel, R. G. & Zimmerman, R. J. (2001). *Confronting Climate Change in the Gulf Coast Region: Prospects for Sustaining Our Ecological Heritage*. Union of Concerned Scientists, Cambridge, Massachusetts and Ecological Society of America, Washington, D.C., 82 pages.

U.S. Environmental Protection Agency (EPA), 1999: *Ecological Condition of Estuaries in the Gulf of Mexico*. Office of Research and Development, National Health and Environmental Effects Research Laboratory, Gulf Ecology Division, Gulf Breeze, Florida. EPA-620-R-98-004. 80 pages.

U.S. Geological Survey (USGS), 2007. Predicting flooding and coastal hazards. U.S. Geological Survey, Coastal and Marine Geology Program. Sound Waves Monthly Newsletter, June 2007. page 1.

Wang, X. L., Zweirs, F. W. & Swail, V. R. (2004). North Atlantic Ocean Wave Climate Change Scenarios for the Twenty-First Century, *Journal of Climate,* pages 2,368– 2,383.

Webster, P. J., Holland, G. J., Curry, J. A. & Chang, H. (2005). Changes in tropical cyclone number, duration, and intensity in a warming environment. *Science,*309,1,844-1,846.

Wehner, M. (2005). Changes in daily precipitation and surface air temperature extremes in the IPCC AR4 models. *U.S. CLI VAR Variations*, 3, (2005) pages 5-9. Lawrence Berkeley National Laboratory, LBNL-61594.

White, W. A. & Morton, R. A. (1997). Wetland losses related to fault movement and hydrocarbon production, southeastern Texas coast. *Journal of Coastal Research, 13*, 1,305-1,320.

Wigley, T. M. L., Raper, S. C. B., Hulme, M. & Smith, S. (2000). *The MAGICC/SCENGEN Climate Scenario Generator: Version 2.4.* Technical Manual, Climatic Research Unit, UEA, Norwich, United Kingdom, 48 pages.

Wilks, D. S. (1993). Comparison of three-parameter probability distributions for representing annual and partial duration precipitation series. *Water Resources Research, 29*, 3,543-3,549.

Wolf, J. (2003). Effects of Climate Change on Wave Height at the Coast. *Geophysical Research Abstracts*, European Geophysical Society, *5*, 13,351.

Zervas, C., (2001). Sea Level Variations of the United States 1854-1999; NOAA Technical Report NOS CO-OPS 36, NOAA, Silver Spring, Maryland, 66 pages.

Zilkoski, D. B. & Reese, S. M. (1986). *Subsidence in the Vicinity of New Orleans as Indicated by Analysis of Geodetic Leveling Data*. NOAA Technical Report NOS 120 NGS 38, National Oceanic and Atmospherics Administration, Rockville, Maryland.

Zwiers, F. W. & Kharin, V. V. (1998). Changes in the extremes of the climate simulated by CCC GCM2 under CO_2 doubling. *Journal of Climate*, *11*, 2,200-2,222.

End Notes

[1] Working Group on Numerical Experimentation Coupled Model Intercomparison Project phase 3.
[2] As a result, this relationship only has an R-squared of 0.27.
[3] With an R-squared of 0.61 if New Orleans is included and 0.77 if New Orleans is excluded from the analysis.
[4] The R-squared associated with this data is 0.42 if New Orleans is included and 0.89 if New Orleans is excluded from the analysis.
[5] The sea level rise estimates from the IPCC Fourth Assessment Report were not available when the sea level rise simulations were run for this study. The projected range of sea level change in the IPCC Fourth Assessment Report (2007) has an upper limit that is slightly lower and a lower limit that is slightly greater than the projections contained in the IPCC Third Assessment Report (2001). The IPCC Fourth Assessment Report also indicates, however, that the rate of historical sea level rise was greater in the Gulf of Mexico than in most other ocean basins, so the global average rate may tend to underestimate the rate of change in the study area.

In: Impacts of Climate Change on Transportation and... ISBN: 978-1-60741-424-7
Editor: Iason Pavlopoulos

Chapter 4

WHAT ARE THE IMPLICATIONS OF CLIMATE CHANGE AND VARIABILITY FOR GULF COAST TRANSPORTATION?

Lead Authors: Robert S. Kafalenos and Kenneth J. Leonard

Contributing Authors: Daniel M. Beagan, Virginia R. Burkett, Barry D. Keim, Alan Meyers, David T. Hunt, Robert C. Hyman, Michael K. Maynard, Barbara Fritsche, Russell H. Henk, Edward J. Seymour, Leslie E. Olson, Joanne R. Potter and Michael J. Savonis

The major climate drivers discussed in chapter 3.0 have significant implications for the transportation system in the Gulf Coast region. This chapter provides an overview, in section 4.1, of the impacts of climate change on the region's transportation infrastructure. It starts with a summary organized around the primary climate effects addressed in chapter 3.0 (temperature, precipitation, sea level rise, and storm activity) and continues with a discussion of freight and private sector concerns. In section 4.2, it shifts to a more detailed discussion organized by transportation mode; this subsection ends by summarizing and discussing freight and private sector concerns involving multiple modes. Finally, we use a series of case studies in section 4.3 to illustrate some of the effects of the 2005 hurricanes on transportation.

Based on the analysis of the climate scenarios relayed in chapter 3.0, climate change is likely to have the largest impact on highways, ports, and rail, particularly through sea level rise and storm surge. Temperature increases, particularly temperature extremes, are likely to increase energy consumption for refrigerated storage as well as rail and highway maintenance. Bridges, included in multiple modes, also could be affected by changes in precipitation, particularly through changes in peak stream flow. Changes in severe weather patterns (thunderstorms) or cloud cover could affect flight operations. See tables 4.3 through 4.6 for summary statistics.

4.1. Climate Drivers and Their Impacts on the Transportation System

This section focuses on the main impacts on transportation facilities and features (e.g., bridges) resulting from the primary climate drivers: temperature, precipitation, sea level rise, and storm activity, and summarizes some of the issues that affect multiple modes.[1] While each climate factor has implications for the transportation network, relative sea level rise (RSLR) and storm activity have the potential to cause the most serious damage to transportation infrastructure in this study region. The relative significance of different climate factors will vary from region to region. The section closes with a look at key cross- modal issues, particularly private sector involvement and the potential for climate impacts in the Gulf Coast region to disrupt freight movements outside the study region.

As noted in Chapter 3.0, the climate impacts on transportation infrastructure assessed in this study rely on the combination of an understanding of historical climate trends and future projections from general circulation models (GCM). While model results imply that change will be gradual and linear, it should be noted that regional "surprises" are increasingly possible in the complex, nonlinear Earth climate system (Groisman et al., 2004), which is characterized by thresholds in physical processes that are not completely understood or incorporated into climate model simulations; e.g., interactive chemistry, interactive land and ocean carbon emissions, etc. While there is still considerable uncertainty about the rates of change that can be expected (Karl and Trenberth, 2003), there is a fairly strong consensus concerning the direction of change for most of the climate variables that affect transportation in the Gulf Coast region.

4.1.1. Effects of Warming Temperatures

Based on the results presented in chapter 3.0 for the Gulf Coast subset of the GCM runs performed for the IPCC Fourth Assessment Report (2007), the average temperature in the Gulf Coast region appears likely to increase by at least 1.5 °C ± 1 °C (2.7 °F ± 1.8 °F) during the next 50 years. While changes in average temperatures have some implications for transportation infrastructure and services, the more significant consideration is the potential change in temperature extremes. As the number of days that the temperature is above 32 °C (90 °F) increases – rising in the next century to as much as 115 days (plus or minus 16 days) per year from the current level of 77 days – stress will increase on both the infrastructure itself and on the people who use and provide transportation services. Temperature extremes are most likely to cause the greatest maintenance problems. The greater frequency of very hot days will lead to greater need for maintenance of roads and asphalt pavement (although some paving materials may handle temperature extremes better than others), rail tracks and freight facilities, some vehicles, and facility buildings and structures due to degradation in materials. Further, construction and maintenance schedules may be affected, as work crews may be unable to work during extreme heat events. For aviation, longer runways may be required, although this will probably be offset by advancements in engine technology and airframe materials.

Increases in temperatures also are likely to increase energy consumption for cooling. This applies particularly to freight operations, including ports where energy is required to provide for refrigeration, as well as to trains and truck operations. Air conditioning requirements for passengers also can be expected to increase, which may lead to a need for additional infrastructure at terminal facilities. This has both environmental and economic costs and may pose a public health concern to vulnerable populations during emergency situations.

4.1.2. Effects of Precipitation Levels and Patterns

Precipitation and Runoff

In this study, annual and monthly (January and July) precipitation totals are examined. Changes in mean precipitation levels appear to have a less significant effect on transportation than do sea level rise, storm surge, and temperature extremes. However, the potential exists for increased intensity in individual precipitation events, which would likely affect transportation network operations, safety, and storm water management infrastructure. Runoff resulting from such events could lead to increased peak streamflow, which could affect the sizing requirement for bridges and culverts.

As reported in chapter 3.0, the climate models show relatively wide variance in average precipitation projections, with plausible scenarios showing annual rainfall potentially increasing or decreasing by as much as 13 percent by 2050 and by ± 15 percent by 2100. However, regardless of whether average precipitation rises or falls, higher temperatures are expected to result in more rapid evaporation. This would result in declining soil moisture and decreased runoff to rivers and streams. The size and extent of natural habitats adjacent to highways may be altered, resulting in changes in some plant and animal communities. These ecological changes may have implications for environmental mitigation strategies and commitments.

While changes in annual average precipitation may have some effects, change in the intensity of individual rainfall events is likely to be the more significant implication for the transportation system. An increase in the intensity or frequency of heavy downpours may require redesign of storm water management facilities for highway, bridges and culverts, ports, aviation, and rail. Severe weather events are correlated to higher incidence of crashes and delays, affecting both safety and mobility. Further, aviation services can be disrupted by intense rainfall events as well as an increase in the probability of severe convective weather. No attempt is made in this study to quantify potential changes in intensity under the climate scenarios presented in chapter 3.0.

4.1.3. Relative Sea Level Rise

Background

Scenarios of 61 cm and 122 cm (2 and 4 ft) of relative sea level rise were selected as inputs to our analysis of potential transportation impacts in the study area. These scenarios were selected based on the range of projected relative sea level rise (discussed in chapter 3.0) of 24-199 cm (about 1-7 ft, depending on location, GCM, and a given emission scenario from the Special Report on Emissions Scenarios (SRES). Even the lowest end of the range of increase in relative sea level has the potential to threaten a considerable proportion of the transportation infrastructure in the region. Future planning, construction, and maintenance activities should be informed by an understanding of the potential vulnerabilities. This subsection begins with a summary of the relative sea level rise analysis conducted for this study (see chapter 3.0 for the full discussion) and continues by summarizing the potential effects of relative sea level rise on the transportation modes.

As noted in chapter 3.0, RSLR is the combined effect of the projected increase in the volume of the world's oceans (eustatic sea level change), which results from increases in temperature and melting of ice, and the projected changes in land surface elevation at a given location. In the Gulf Coast region, change in land surface elevation change is dominated by subsidence, or sinking, of the land surface. While sea level may continue to rise incrementally, the potential for abrupt increases in relative levels cannot be dismissed. Gradual and relatively consistent rates of sea level increases will be more easily addressed by transportation planners and designers than would more abrupt or discontinuous changes in water levels. No analysis is conducted regarding the implications of a catastrophic degree of sea level change that would result from major changes in the rate of land ice decline (e.g., a rapid collapse of the Greenland Ice Sheet).

Two different sea level rise models were used to estimate potential RSLR in the study area. Both models were used to estimate relative sea level rise by 2050 and 2100 under the greenhouse gas emissions scenarios considered in this study (see chapter 3.0 for more on the scenarios). Both models account for eustatic sea level change and land subsidence in the region based on the historical record. One model, CoastClim, produces results that approximate future change in RSLR under the climate scenarios. A similar model, SLRRP, also incorporates values for high and low tidal variation attributed to astronomical and meteorological causes, which are pulled from the historical record. The tide data used is based on a monthly average of the mean high tide (mean higher high water) for each day of the month. The SLRRP results presented in the study are the highest predicted monthly sea level elevations by 2050 and 2100. Thus, the SLRRP results capture seasonal variability and interannual trends in sea level change, while the CoastClim results do not.

Results for the low- and high-range RSLR cases are summarized in tables 4.1 and 4.2. (See tables 3.14 and 3.16 for the full range of results.) Analysis was conducted for three long-term tide gage locations, as subsidence rates vary substantially across the region: regional subsidence rates are 4.7 mm/year (0.19 in/year) for Galveston, TX, and the chenier plain; 8.05 mm/year (0.32 in/year) for Grand Isle, LA, and the Mississippi River deltaic plain; and 0.34 mm/year (0.013 in/year) for Pensacola, FL, and the Mississippi/Alabama Sound of the central Gulf Coast. Results generated using CoastClim range from 24 cm (0.8 ft) in Pensacola to 167 cm (5.5 ft) in Grand Isle. Results from SLRRP, which as noted above accounts for historical tidal variation, are somewhat higher, with predicted sea level ranging from 70 cm (2.3 ft) in Pensacola to 199 cm (6.5 ft) in Grand Isle (North American Vertical Datum 88 [NAVD88]).

This phase I analysis broadly examines the potential effects of sea level rise on the region as a whole; the results related in this study should not be used to predict specific impacts on any single location at a specific point in time. Impacts were analyzed assuming two different levels of relative sea level rise; 61 cm (2 ft) and 122 cm (4 ft). From a regional perspective, the selection of this range for analysis is clearly supported by the model results. In fact, given that the results range from 24 cm to 199 cm (0.8 to 6.5 ft), analyzing for 61- and 122-cm (2-

and 4-ft) increases in RSLR may be overly conservative from a regional perspective. For both Galveston and especially Grand Isle, analyzing at the 122-cm (4-ft)level is conservative, given that the high-range scenario results modeled to the year 2100 range from 130 cm (4.3 ft) to 199 cm (6.5 ft) for these two areas. In the case of Pensacola, given that three of the four values that define the range of the results are above 61 cm (2 ft), this level should be considered a conservative input value. The 122-cm (4-ft) level, however, is representative of the high-range scenario results (114 cm or 3.8 ft) for Pensacola.

The effect of existing flood control works has not been addressed in this study. Many existing facilities at lower elevations are protected by levees and other physical structures, which are intended to provide resistance to storm surge. The present land-based elevation data allows us to identify general geographic zones of potential risk and to identify areas that merit further study. More detailed future assessments of specific sites and facilities should consider the presence and viability of protective structures as part of an analysis of risk and vulnerability at those locations.

As discussed in chapter 3.0, RSLR will not be uniform across the region. This study's results are meant to give a broad indication of where relative sea levels could be by the year 2100 and what infrastructure could be affected as a result of the analysis under the 61- and 122-cm (2- and 4-ft) RSLR scenarios. This analysis provides a first approximation of potential vulnerabilities and provides insights for transportation planners; more detailed analyses can then be conducted to further assess specific locations and facilities that may be at risk. Phase II of this study will examine specific sublocations within the region and incorporate location-specific projections of future RSLR.

Impact on Transportation

Relative sea level rise poses the greatest danger to the dense network of ports, highways, and rail lines across the region. An increase in relative sea level of 61 cm (2 ft) has the potential to affect 64 percent of the region's port facilities, while a 122-cm (4-ft) rise in relative sea level would affect nearly three-quarters of port facilities. This is not surprising given that port facilities are adjacent to a navigable water body. For highways and rail, while the percentages are lower, the effect also is quite large. About a quarter of arterials and interstates, nearly half of the region's intermodal connector miles, and 10 percent of its rail miles would be affected by a 122-cm (4-ft) rise. Because goods are transferred to and from ports by both trucks and rail, service interruptions on selected segments of infrastructure are likely to affect much more than these percentages imply due to the disruption to network connectivity. For example, an increase in relative sea level of 61 cm (2 ft) would affect 220 km (137 mi) of I-10 east of New Orleans, which could affect on-road transport of both people and goods into and out of New Orleans and, to a lesser extent, Houston. Similarly, while less than 10 percent of rail miles would be affected, most of the rail lines linking New Orleans to the rail system could be affected. This could hinder freight movements in the region, especially since New Orleans is the main east-west link for rail located in the region, one of four in the United States. While airports in the region are less directly vulnerable to sea level rise, the vulnerability of roads and rail lines serving them affects the passenger and freight services these facilities provide as well. See table 4.3 for a summary of this information.

Table 4.1. Relative Sea Level Rise Modeled by Using SLRRP

	Low Range	High Range
Galveston, TX	117 cm (3.8 ft)	161 cm (5.3 ft)
Grand Isle, LA	160 cm (5.2 ft)	199 cm (6.5 ft)
Pensacola, FL	70 cm (2.3 ft)	114 cm (3.8 ft)

Table 4.2. Relative sea level rise modeled by using CoastClim

	Projected Subsidence by 2100	RSLR, B1-Low Range	Subsidence, Percent of Low Range	RSLR, A1F1-High Range	Subsidence, Percent of High Range
Galveston, TX	51.7 cm (1.7 ft)	72 cm (2.4 ft)	71.8%	130 cm (4.3 ft)	39.7%
Grand Isle, LA	88.6 cm (2.9 ft)	109 cm (3.5 ft)	81.3%	167 cm (5.5 ft)	53.0%
Pensacola, FL	3.7 cm (0.12 ft)	24 cm (0.8 ft)	15.4%	82 cm (2.7 ft)	4.5%

Table 4.3. Relative Sea Level Rise Impacts on Gulf Coast Transportation Modes: Percentage of Facilities Vulnerable

Relative Sea Level Rise	Interstate Highways	Ports (Freight)	Rail Lines	Airports
61 cm (2 ft)	19%	64%	5%	1 airport
122 cm (4 ft)	24%	72%	9%	3 airports

Relative sea level rise is likely to have an impact on the other modes as well. While bus routes can be adjusted over time should facilities no longer be of use, light rail facilities are not so easily moved; some of the light rail routes in Galveston and New Orleans would be affected by a 61-cm (2-ft) rise. Airports would not escape the direct and indirect effects of relative sea level rise; New Orleans International airport, at 122-cm (4 ft), and two other smaller airports could be affected directly by higher sea levels. Others could be affected indirectly if the roads and connectors leading to them are flooded.

The data and analysis for both relative sea level rise and storm surge are based on land area elevations, rather than facility elevations. Facility elevations generally were not readily available for this phase of the study in a consolidated and geospatial format. The elevation of land areas was determined from the National Elevation Dataset (NED) maintained by the United States Geological Survey (USGS) (USGS, 2004). Mapping data for transportation infrastructure was obtained from the U.S Department of Ttransportation's (DOT) Bureau of Transportation Statistics (BTS, 2004).

The NED has a horizontal resolution of 30 m (98 ft). Since the positional accuracy of the transportation facilities is plus or minus 80 m (262 ft), the elevation data is sufficient only to make general conclusions about transportation facilities that are vulnerable to flooding. While some sections of the transportation network – particularly roads and rail lines – may be elevated, it is important to note that inundation of even short segments of the system can shut down significant portions of the broader network due to the essential connectivity provided by these segments. Furthermore, such inundation can undermine infrastructure's foundations and substructures.

4.1.4. Storm Activity

As discussed in chapter 3.0, the intensity of hurricanes making landfall in the Gulf Coast study area is likely to increase. In addition, the climate analysis indicates that the number of hurricanes may increase as the temperature of the sea surface continues to warm. Simulated storm surge from model runs across the central Gulf Coast at today's elevations and sea levels demonstrated a 6.7- to 7.3-m (22- to 24-ft) potential surge for major hurricanes of Category 3 or greater. Based on recent experience, even these levels may be conservative; surge levels during Hurricane Katrina (rated a Category 3 at landfall) exceeded these heights in some locations.

Many of the region's major roads, railroads, and airports have been constructed on land surfaces at elevations below 5 m (16.4 ft). Storm surge poses significant risk to transportation facilities[2] due to the immediate flooding of infrastructure, the damage caused by the force of the water, and secondary damage caused by collisions with debris. While surges at varying heights may disrupt operations and damage infrastructure, the effects of storm surges of 5.5 and 7 m (18 and 23 ft) were assessed for the purposes of this analysis.

This assessment does not take into account the possible dampening of surge effects due to distance inland from coastal areas and the buffering qualities of both ecological systems (barrier islands, wetlands, marshes) and the built environment. The analysis identifies portions of the transportation network that are at land elevations below 5.5 and 7 m (18 and 23 ft) as an initial indication of areas and facilities that may be at risk and warrant more detailed analysis. Areas significantly inland from the coast or protected by buffering systems may be less vulnerable, depending on site-specific coastal geomorphology and the characteristics of individual storm events.

As shown clearly by Hurricanes Katrina and Rita, storm surge has the potential to cause serious damage and loss of life in low-lying areas. As considered in this study, much of the region's infrastructure is vulnerable to storm surges on the order of 5.5 to 7 m (18 to 23 ft), though the specific infrastructure that would be flooded depends on the characteristics of a given storm, including its landfall location, wind speed, direction, and tidal conditions.

As in the case of relative sea level rise, ports, highway, and rail are the transportation facilities that would be most directly affected by storm surge. Ports have the most exposure, because 98 percent of port facilities are vulnerable to a storm surge of 5.5 m (18 ft). Fifty-one percent of arterials and 56 percent of interstates are located in areas that are vulnerable to a surge of 5.5 m (18 ft), and the proportions rise to 57 and 64 percent, respectively, for a surge of 7 m (23 ft). Some 73 percent of intermodal connector miles are vulnerable to surges of 5.5 or 7 m (18 ft or 23 ft). Thirty-three percent of rail lines are vulnerable to a storm surge of 5.5 m (18 ft); this proportion climbs to 41 percent vulnerable at 7 m (23 ft). Twenty-nine airports are vulnerable to a surge of 7 m (23 ft), and one major commercial service facility – New Orleans International – also is vulnerable to a 5.5-m (18-ft) surge. Vulnerability of the region's infrastructure to storm surge is summarized in table 4.4.

Table 4.4. Storm Surge Impacts on Gulf Coast Transportation Modes: Percentage of Facilities Vulnerable

Storm Surge Height	Interstate Highways	Ports (Freight and Nonfreight)	Rail Lines	Airports
5.5 m (18 ft)	56%	98%	33%	22 airports
7.0 m (23 ft)	64%	99%	41%	29 airports

The effects of storms on the transportation network go beyond the impacts of storm surge. Severe winds and rainfall events throughout the study region can cause damage and flooding, disrupting system performance. Wind damage risk contours were not mapped as part of this project. Experience shows that the highest hurricane velocities are experienced along the coasts, diminishing as storms move inland, but that severe damaging winds can be sustained well inland. Hurricanes also spawn tornados, which can have substantially higher velocities over much smaller areas. The entire study area is within 100 mi of the Gulf of Mexico shoreline, and all of it could be considered potentially vulnerable to significant wind damage. As noted in chapter 3.0, while historical and projected increases in summer minimum temperatures for the study area suggest an increase in the probability of severe convective weather (Dessens, 1995; Groisman et al., 2004), GCMs currently lack the capacity for simulating small-scale phenomena such as thunderstorms, tornadoes, hail, and lightning.

One factor that complicates the effects of both storm surge and relative sea level rise is the condition of the barrier islands. As noted in section 3.5.1, wave heights in coastal bays will tend to increase due to the combined erosional effects of sea level rise and storms on coastal barrier islands and wetlands. As the barrier islands erode, their role in shielding Gulf Coast waterways and infrastructure from the effects of waves will diminish, which means their ability to protect coastal infrastructure from waves at current sea levels and future sea levels, as well as from storm surge, will likely diminish.

Any facility subject to flooding may incur structural damage or be rendered inoperable due to debris or other obstructions. Restoring facility and system performance necessitates considerable time and investment on the part of facility owners. The secondary economic costs to both businesses and communities who rely on these transportation networks could be considerable as well, depending on the time required to restore system performance.

This report does not attempt to estimate the total costs of protecting, maintaining, and replacing Gulf Coast transportation infrastructure due to damage caused by climate change. It does, however, include a case study on Hurricane Katrina in section 4.3.1 that provides examples of the efforts associated with addressing the impacts of the hurricane.

4.1.5. Climate Impacts on Freight Transport

The private sector has made massive investments in transportation infrastructure in the Gulf Coast study area, a large portion of which revolves around moving freight. Almost all of the roads and major airports are publicly owned, but the vehicles that operate over them, and the commercial and freight services that they accommodate, are private. Many of the ports are private, and the vessels and commercial services using them are private. Almost all of the Nation's rail infrastructure is privately owned and operated.

Disruption of privately owned infrastructure can have huge costs for the owners and users of these facilities. Repair costs for the more than 65-km (40-mi) CSX railroad segment damaged in Hurricane Katrina, $250 million, could be dwarfed by the costs of moving the line if the company chose to relocate the line further inland; Congressional proposals have considered authorizing $700 million in Federal funding to help relocate the damaged portion of the CSX segment. This is just a small share of the 1,915 km (1,190 mi) of rail line in the study area that are vulnerable to sea level rise and storm surge. Critical transportation-dependent industries – petroleum, chemical, agricultural production and transportation, etc. – are heavily concentrated in the study area. The private sector, therefore, has a significant interest in the impacts of climate change on transportation infrastructure, because it potentially affects hundreds of billions of dollars annually in commercial activity over study area roads, railroads, airports, seaports, and pipelines.

One of the key issues that draws the private sector into the discussion of climate impacts on transportation is the movement of freight. The private sector has proven adept at using intermodal freight systems – involving ports, highways, rail, and aviation – to transport goods as inexpensively as possible. However, this lean and efficient system is vulnerable: a disruption that seemingly affects a limited area or a single mode can have a ripple effect throughout the supply chain.

The loss of direct freight transportation service or connectivity in the Gulf Coast would likely have a substantial impact beyond the transportation provider and the local economy. The interruption of freight transportation service in the Gulf Coast could impact the distribution of goods nationally and, therefore, impact the national economy. Costs of raw materials or products that have to be rerouted or transported by an alternate mode would likely increase to absorb higher transportation costs. Further, most businesses and industries that once held large inventories of products have shifted to low inventory, justin-time delivery business models, managing much of their inventories in transit. Therefore, they have lower tolerance for delays in shipment and receipt of goods and now demand greater reliability and visibility from their freight carriers. This system is very cost effective, but it leaves shippers with little cushion when the freight transportation system fails. A large failure such as that caused by a hurricane can quickly disrupt thousands of supply chains, undermining the operations and profitability of many shippers, carriers, and customers. For example, after Hurricane Katrina, CSX rerouted trains and experienced an increase in operating costs of the railroad through increased fuel usage, crew costs, equipment delays, and a loss of overall system capacity. Other freight transportation impacts included the disruption in the distribution of petroleum by pipelines and the failure of ships being able to make port along the Gulf Coast. An increase in transportation costs such as these is likely to increase the price of the final product and could jeopardize the national and global competitiveness of affected businesses.

4.2. Climate Impacts on Transportation Modes

This section begins with an in depth examination of the impacts of climate change on each individual mode. It continues by looking at how these impacts could affect emergency management and evacuation and closes with a look at key cross-modal issues.

4.2.1. Highways

As in most parts of the nation, roads are the backbone of the transportation network in the Gulf Coast. Highways[3] are the chief mode for transporting people across the region, and together with rail, highways are essential for moving freight throughout the region and to other parts of the United States. Thus, impacts to the highway network could serve as choke points to both passenger and freight traffic that emanates in or flows through the region. While temperature and precipitation changes have some implications for highway design and maintenance, the key impacts to the highway network result from relative sea level rise and storm surge.

Temperature

Impacts related to projected changes in average temperatures appear to have moderate implications for highways, while increases in extreme heat may be significant. Maintenance and construction costs for roads and bridges are likely to increase as temperatures increase. Further, higher temperatures cause some pavement materials to degrade faster, requiring earlier replacement. Such costs will likely grow as the number of days above 32 °C (90 °F) – projected to grow from the current average of 77 days to a range of 99 to 131 days over the next century – increases and as the projected maximum record temperatures increase in the region.

While maintenance and construction costs are expected to rise as the number of very hot days increase, the incremental costs have not been calculated as part of this analysis. These additional, excessive temperature-related costs are incorporated into the total maintenance and construction costs for all pavements and bridges. Changes in materials used may help reduce future temperature-induced maintenance costs. For example, Louisiana Department of Transportation and Development (DOTD) has begun to use asphalts with a higher polymer content, which helps pavement better handle higher temperatures, though at a higher initial cost than standard asphalt.

There are measures that could be taken to mitigate the loss in productivity associated with maintenance and construction, such as evening work hours, but these measures also would increase costs. In subsequent phases of this study, the implications on construction, maintenance, and operation budgets in specific sublocations should be examined.

The designs of steel and concrete bridges and of pavements in the study area typically are based on a maximum design temperature of 46 °C (115 °F) to 53 °C (125 °F). The increase in maximum record temperatures implied by the climate model projections are less than these values, although under the climate scenarios they would approach those values over the next century. It may be prudent for future designers of highway facilities to ensure that joints in steel and concrete bridge superstructures and concrete road surfaces can adequately accommodate thermal expansion resulting from these temperatures. The State DOT design manuals generally establish the maximum design temperature at a value near 53 °C (125 °F), well above the current maximum recorded temperatures in the study area, but as temperatures increase there may well be more failures of aging infrastructure. Consideration should be given to designing for higher maximum temperatures in replacement or new construction.

Precipitation

As previously noted, the analysis generally indicates little change in mean annual precipitation (152 cm or 60 inches per year) through either 2050 or 2100, but the range of possible futures includes both reductions and increases in seasonal precipitation. In either case, the analysis points to potential reductions in soil moisture and runoff as temperatures and the number of days between rainfall events increase. The research team analyzed average annual precipitation separately from potential changes in intensity of rain events.

Under a scenario of insignificant change or a reduction in average precipitation, coupled with drier soils and less runoff, there would be decreases in soil moisture, which may result in a decline of slides in slopes adjacent to highways. It also would mean less settling under pavements, with a decrease in cracking and undermining of pavement base courses. While uniform decreases in runoff could reduce scouring of bridge piers in rivers and streams, greater frequency of high-intensity events could result in more scour. Stresses on animal and plant populations brought about by higher temperatures and changes in rainfall patterns could make it more difficult and expensive to mitigate the impacts of highway development on the natural environment.

Pavement settling, bridge scour, and ecosystem impacts may not be significantly impacted by modest increases in average annual rainfall because of the effects of increasing temperature on evaporation rates. However, while potential changes in average annual precipitation are likely to have minor impacts, an increase in the intensity of individual rainfall events may have significant implications for highways. An increase in the frequency of extreme precipitation events – as discussed in chapter 3.0 – would increase accident rates, result in more frequent short-term flooding and bridge scour, as well as more culvert washouts, and exceed the capacity of stormwater management infrastructure. More instances of intense rainfall also may contribute to more frequent slides, requiring increased maintenance. However, some states, such as Louisiana, already address precipitation through pavement grooving and sloping and thus may have adequate capacity to handle some increase in precipitation.

Relative Sea Level Rise

As discussed above, the effects of 61- and 122-cm (2- and 4-ft) RSLRs were analyzed to assess their implications on highways. The presence or absence of protective structures was not considered in this baseline analysis but would be an important factor in subsequent sublocation assessments.

As shown in figure 4.1, the majority of the highways at risk from a 61-cm (2-ft) increase in relative sea level are located in the Mississippi River delta near New Orleans. The most notable highways at risk are I-10 and U.S. 90, with 220 km (137 miles) and 235 km (146 miles), respectively, passing through areas that will be below sea level if sea levels rise by 61 cm (2 ft). Overall, 20 percent of the arterial miles and 19 percent of the interstate miles in the study area are at elevations below 61 cm (2 ft) and thus are at risk from sea level rise unless elevated or protected by levees (table 4.5).

The majority of the highways at risk from a 122-cm (4-ft) increase in relative sea level are similarly located in the Mississippi River Delta near New Orleans (figure 4.2). The most notable highways at risk remain I-10 and U.S. Highway 90, with the number of miles increasing to 684 km (425 mi) and 628 km (390 mi) passing through areas below sea level, respectively. Overall, 28 percent of the arterial miles and 24 percent of the interstate miles are at elevations below 122 cm (4 ft). Currently, about 130 mi (209 km) or about 1 percent of major highways (interstates and arterials) in the study region are located on land that is at or below sea level.

As shown in figure 4.3, many of the National Highway System (NHS) intermodal (IM) connectors pass through low-lying areas concentrated in the Mississippi River Delta, where sea level rise is expected to have the most pervasive impact. Intermodal connectors are primarily necessary to provide highway access for various transportation facilities, such as rail, ports, and airports, some of which will be below sea level with a relative sea level rise of 61 to 122 cm (2 to 4 ft). Of the 1,041 km (647 mi) of IM connectors, 238 km (148 mi), or 23 percent, are at risk to a 61-cm (2-ft) increase in relative sea levels; and a total of 444 km (276 mi), or 43 percent, are at risk to a 122-cm (4-ft) increase. In addition to the terminals at risk under the 61-cm (2 ft) RSLR scenario (the New Orleans International Airport, Port Fourchon, most rail terminals in New Orleans, ferry terminals in New Orleans, and ferry terminals outside of the Mississippi River Delta in Galveston and Houston), additional terminals at risk under the 122 cm (4 ft) RSLR scenario include port facilities in Lake Charles, Galveston, Pascagoula, and Gulfport.

The cost of various adaptation options – including relocating, elevating, or protecting highways and IM connectors – is not addressed by this study. Additionally, the costs of right-of-way and environmental mitigation for relocating or elevating such facilities are unknown at this time. The adaptation and investment plans for specific facilities will be determined by local and regional decision makers.

As discussed in section 4.2.1, the available elevation data for the study area is sufficient to make first order conclusions about roads that are at risk of flooding; it does not indicate the elevation of specific highways. However, it is worth noting that the loss of use of a small individual segment of a given highway may make significant portions of that road network impassable. Further, even if a particular interstate or arterial is passable, if the feeder roads are flooded, then the larger road becomes less usable.

Storm Activity

As discussed in chapter 3.0, the intensity of hurricanes making landfall or striking in the Gulf Coast study area can be expected to increase. About half of the region's arterial miles and about three-quarters of the IM connectors are vulnerable to a storm surge of 5.5 m (18 ft), and these proportions are even higher for a 7-m (23-ft) storm surge.

Surge Wave Crests and Effects on Bridges

The wave energy during storm surge events is greatest at the crest of the wave. The facilities most at risk are bridge decks and supports that are constructed at the wave heights reached during a storm. The impact of the 2005 hurricanes vividly illustrated some of the factors involved in infrastructure vulnerability (see section 4.3.1.) While only a small percentage of the study area's bridges are located at the shore and have bridge decks or structures at these heights, when storm waves meet those bridges the effect is devastating; spans weighing 300 tons were dislodged during Hurricane Katrina. Although these bridges are few in number compared to the over 8,000 bridges on the functionally classified system, over two dozen bridges were hit by wave surges resulting from Hurricane Katrina and experienced serious damage.

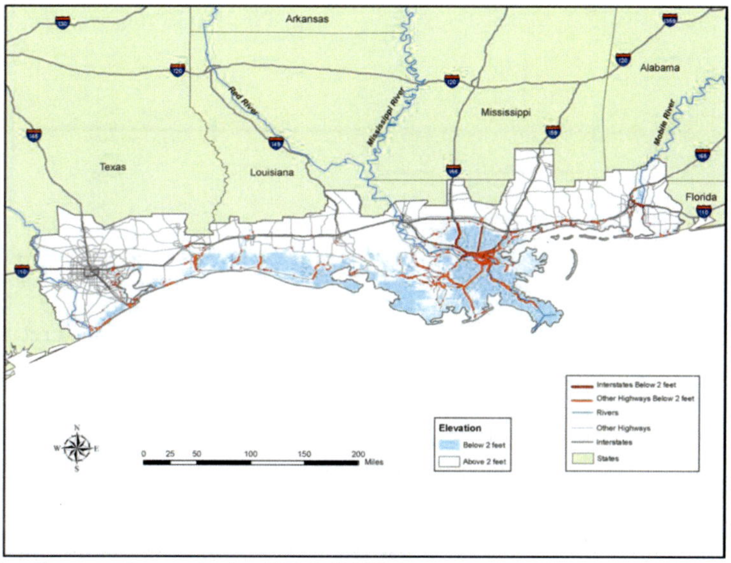

Source: Cambridge Systematics analysis of U.S. DOT data.

Figure 4.1. Highways at risk from a relative sea level rise of 61 cm (2 ft).

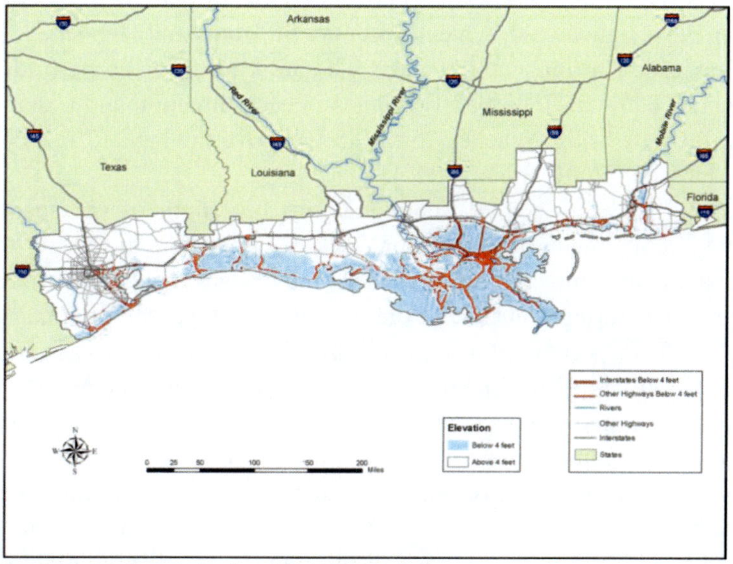

Source: Cambridge Systematics analysis of U.S. DOT data.

Figure 4.2. Highways at risk from a relative sea level rise of 122 cm (4 ft)

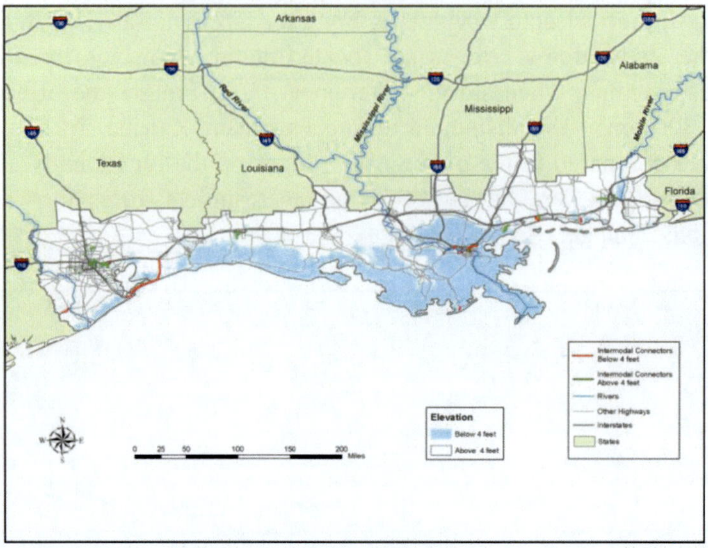

Source: Cambridg e Systematics analysis of U.S. DOT data.

Figure 4.3. National Highway System (NHS) intermodal connectors at risk from a relative sea level rise of 122 cm (4 ft)

Table 4.5. Relative Sea Level Rise Impacts on Highways: Percentage of Facilities Vulnerable

Relative Sea Level Rise	Arterials	Interstates	Intermodal Connectors
61 cm (2 ft)	**20%**	**19%**	**23%**
122 cm (4 ft)	**28%**	**24%**	**43%**

An example is shown in figure 4.4. In perhaps the most spectacular example, the Bay St. Louis Bridge on U.S. Highway 90, which links Bay St. Louis and Henderson Point, MS, was destroyed by Hurricane Katrina's storm surge. The 3.2-km- (2-mi-) long bridge was recently replaced at a cost of $267 million, with two lanes in each direction and a shared-use path. At it highest point, the new bridge reaches 26 m (85 ft) above the bay, 17 m (55 ft) higher than its predecessor (Nossiter, 2007; Sloan, 2007).

Design features such as lack of venting along the length of the span, solid railings (preventing water from flowing through), and lack of connectors anchoring the spans to the pilings or corrosion in existing connectors made some bridges more susceptible than others to the force of the water during Katrina. In the absence of standard American Association of State Highway and Transportation Officials (AASHTO) design factors for storm surge, both the Louisiana DOTD and Mississippi DOT have developed their own approaches to designing for future storms. For instance, Louisiana DOTD is developing standards calling for new bridges to be elevated beyond a 500-year event for the main span (9.1-11.6 m, or 30-3 8 ft) and a 100-year event for transition spans close to shore. In addition, new bridges will be designed with open railings to reduce the impact of pounding water (Paul, 2007). Mississippi also has adopted more stringent design standards and is rebuilding the Biloxi Bay and St. Louis Bay bridges as high-rise structures, to keep the bridge decks above future storm surges.

Source: NASA Remote Sensing Tutorial.

Figure 4.4. Hurricane Katrina damage to U.S. Highway 90 at Bay St. Louis, MS

As the sea level rises, the coastline will change. Bridges that were not previously at risk may be exposed in the future. Additionally, bridges with decks at an elevation below the likely crest of storm surges, based on experience from previous storms, will be below water during the storm event and not subject to wave damage. Only data regarding the height of bridges above navigable channels was available to this study – a small portion of all bridges in the region. Therefore, a full analysis of the possible impacts of wave crests on bridges was not feasible.

Surge Inundation

Figures 4.5 and 4.6 show areas potentially vulnerable to surge inundation at the 5.5- and 7-m (18- and 23-ft) levels and identifies interstate and arterial highways that pass through these risk areas. As illustrated, a substantial portion of the highway system across the study area is vulnerable to surge inundation: 51 percent of all arterials and 56 percent of the interstates are in the 5.5-m (1 8-ft) surge risk areas. At the 7-m (23-ft) level, these percentages increase only slightly: 57 percent of all arterials and 64 percent of the interstates are in 7-m (23-ft) surge risk areas (table 4.6).

The risk from surge inundation for NHS IM connectors is even greater than that for all highways. Seventy-three percent of IM connector miles are located in areas that would be inundated by a 5.5-m (18-ft) surge, and the proportion of IM connectors that is vulnerable at the 7-m (23-ft) level is only slightly higher (see figure 4.7).

As noted above, the elevation data is sufficient to make only general conclusions about roads that are at risk of inundation. Local conditions for specific segments and facilities may be important, and individual roads that may be vulnerable should be studied in detail.

While inundation from storm surges is a temporary event, during each period of inundation the highway is not passable, and after the surge dissipates, highways must be cleared of debris before they can function properly. Of particular concern is that a substantial portion of all of the major east-west highways in the study area, particularly I-10/I-12, are at risk to storm surge inundation in some areas, and during storm events and the recovery from these events, all long-distance highway travel through the study area is likely to be disrupted.

The expense of these poststorm cleanups can be considerable and is often not included in State DOT budgets. For instance, the Louisiana DOTD spent $74 million on debris removal alone following Hurricanes Katrina and Rita (Paul, 2007). In the 14 months following the hurricanes, the Mississippi DOT spent $672 million on debris removal, highway and bridge repair, and rebuilding the Biloxi and Bay St. Louis bridges (Mississippi DOT, 2007). See section 4.3.1 for a fuller discussion of poststorm cleanup costs.

Moreover, data from the Louisiana DOTD suggests that prolonged inundation can lead to long-term weakening of roadways. A study of pavements submerged longer than three days during Katrina (some were submerged several weeks) found that asphalt concrete pavements and subgrades suffered a strength loss equivalent to two inches of pavement. Portland concrete cement pavements suffered little damage, while composite pavements showed weakening primarily in the subgrade (equivalent to one inch of asphalt concrete). The study estimated a $50 million price tag for rehabilitating the 320 km (200 mi) of submerged state highway pavements and noted that an additional 2,900 km (1,800 mi) of nonstate roads were submerged in the New Orleans area. The data was collected several months after the waters had receded; there has not been a subsequent analysis to test whether any strength was restored over time (Gaspard et al., 2007).

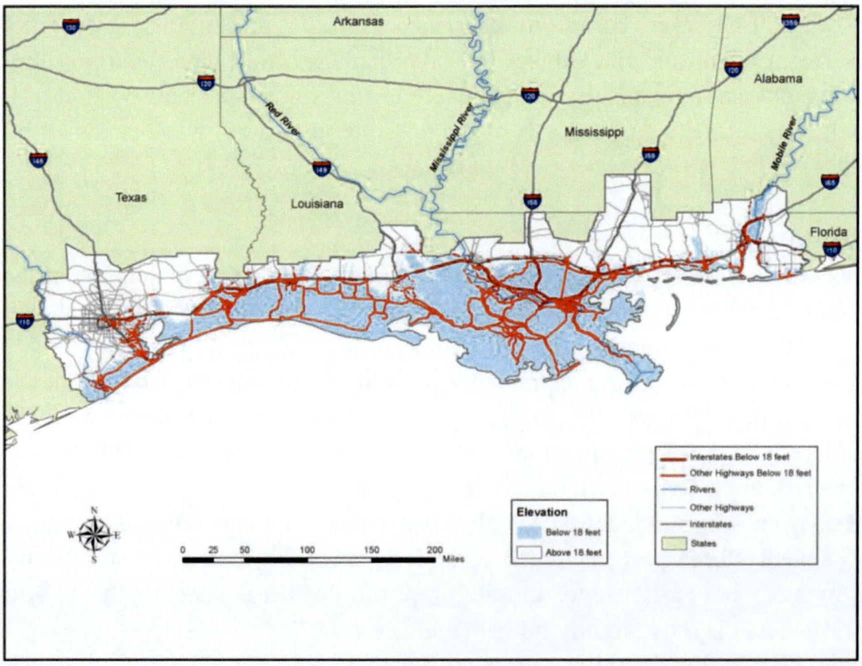

Source: Cambridge Systematics analysis of U.S. DOT data.

Figure 4.5. Highways at risk from storm surge at elevations currently below 5.5 m (18 ft)

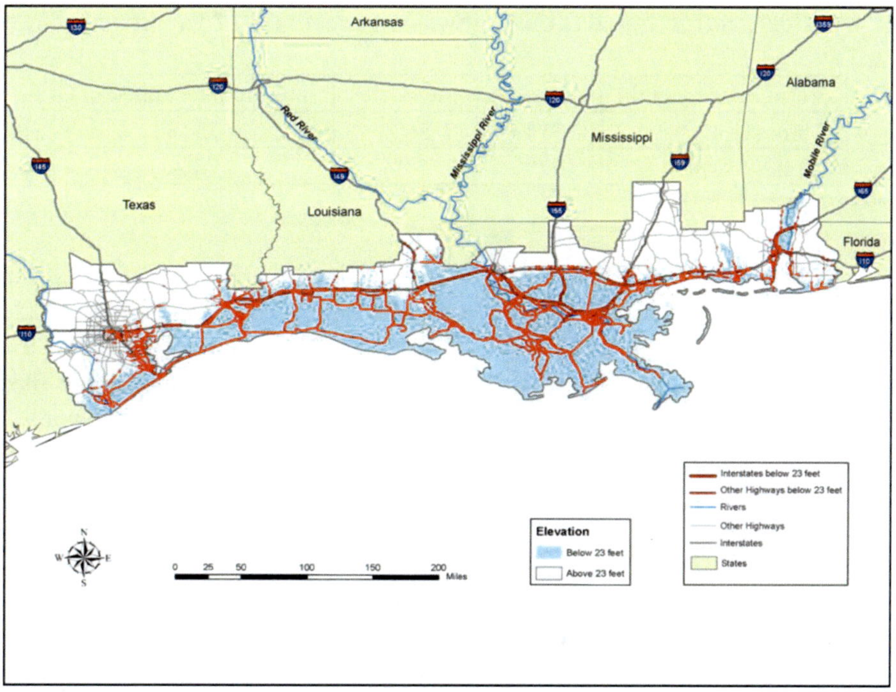

Source: Cambridge Systematics analysis of U.S. DOT data.

Figure 4.6. Highways currently at risk from storm surge at elevations currently below 7.0 m (23 ft)

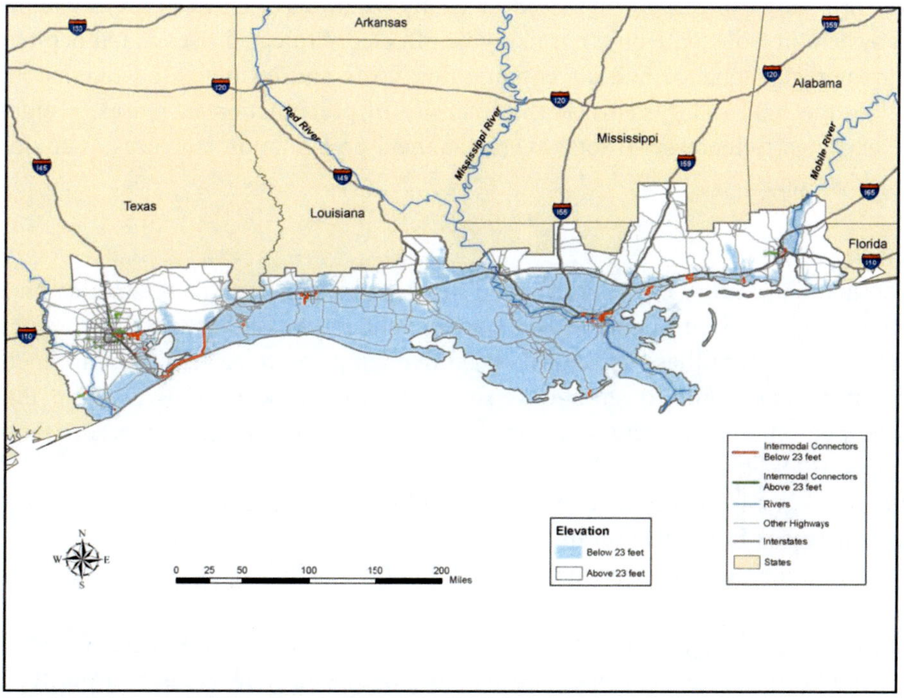

Source: Cambridge Systematics analysis of U.S. DOT data.

Figure 4.7. National Highway System (NHS) intermodal connectors at risk from storm surge at elevations currently below 7.0 m (23 ft)

Table 4.6. Storm Surge Impacts on Highways: Percentage of Facilities Vulnerable

Storm Surge Height	Arterials	Interstates	Intermodal Connectors
5.5 m (18 ft)	**51%**	**56%**	**73%**
7.0 m (23 ft)	**57%**	**64%**	**73%**

Wind

Wind from storms may impact the highway signs, traffic signals, and luminaries throughout the study area. The wind design speed for signs and supports in the study area is typically 160 to 200 km/h (100 to 125 mi/h). These designs should accommodate all but the most severe storm events. More significant safety and operational impacts are likely from debris blown onto roadways and from crashes precipitated by debris or severe winds.

4.2.2. Transit

Transit in the region consists of bus systems as well as light rail in New Orleans, Houston, and Galveston. While bus routes could be affected by relative sea level rise, transit operators can presumably adjust their routes as needed, particularly since the location of transit users and routes also might change. Storm surge could be a more serious, if temporary, issue. For the light rail systems in New Orleans and Galveston, an increase in relative sea level of 61 or 122 cm (2 or 4 ft) would affect at least some of the routes, especially in New Orleans; storm surge of 5.5 or 7.0 m (18 or 23 ft) would have an even greater impact. The light rail system in Houston would not likely be affected. Projected rises in temperature could lead to greater maintenance and air conditioning costs and an increased likelihood of rail buckling for the light rail systems. If the intensity of precipitation increases, accident rates could be expected to increase. If total average annual precipitation increases, it could lead to higher accident rates.

Temperature

Given the temperature projections noted in chapter 3.0, temperature stresses on engines and air conditioning systems could possibly affect vehicle availability rates, disrupting overall scheduled service. Since these additional, excessive temperature-related costs are included in the total maintenance and construction costs of transit agencies, it is possible that those amounts will at a minimum increase by an amount proportional to the increase in the number of days above 32 °C (90 °F).

Furthermore, temperature increases, especially increases in extremely high temperatures, will cause increases in the use of air conditioning on buses to maintain passenger comfort. This will exacerbate the issue of vehicle availability rates and raise costs due to increased fuel consumption.

Increases in (record maximum) temperatures are likely to only impact fixed guideway rail networks and have little or no impact on bus or paratransit systems, aside from the vehicle maintenance issues noted above. As discussed in greater detail in section 4.2.3, rail networks are subject to "sun kinks" (the buckling of sections of rail) at higher temperatures; sun kinks are likely to occur more frequently as (record maximum) temperatures increase. The possibility of rail buckling can lead to speed restrictions to avoid derailments. The track used by the trolley systems in Galveston and New Orleans have expansion joints that generally are not significantly affected by sun kinks, while Houston's METRORail uses continuously welded rail (CWR) track. CWR track lacks expansion joints and thus is more prone to sun kinks.

Precipitation

The climate model results point to potential increases or decreases in average annual precipitation. If precipitation increases, it very likely would lead to an increase in accidents involving buses, as well as increased costs and disruptions associated with such accidents. The same also is likely if the intensity of precipitation increases. Even an increase in roadway accidents not involving buses will lead to congestion that could disrupt bus schedules.

Relative Sea Level Rise

If relative sea level increases to an extent that transit service would pass through areas under water in the future, either the connectivity provided by that transit would be lost or corrective actions to reroute the transit would be needed. Since the vast majority of transit service is provided by buses, schedules and routes can be modified easily, though the same is not true for terminals and maintenance facilities. Therefore, minimal impact on bus systems is expected from RSLR. For light rail systems in the region, however, RSLR could potentially be a much more serious issue. Moving tracks and permanent facilities is a major undertaking; tracks would need to be protected or moved to higher ground.

With the exception of the RTA and St. Bernard buses in New Orleans and a small portion of the routes traveled in Galveston, bus and paratransit service is not expected to be affected by either a 61- or 122-cm (2- or 4-ft) increase in relative sea levels. If bus routes are not affected, ancillary facilities such as terminals and maintenance facilities may not be affected either. Figure 4.8 shows the effect of a 122-cm (4-ft) rise in relative sea level on fixed bus routes in New Orleans. This clearly illustrates the vulnerability of the transit network in New Orleans without levees or other protection.

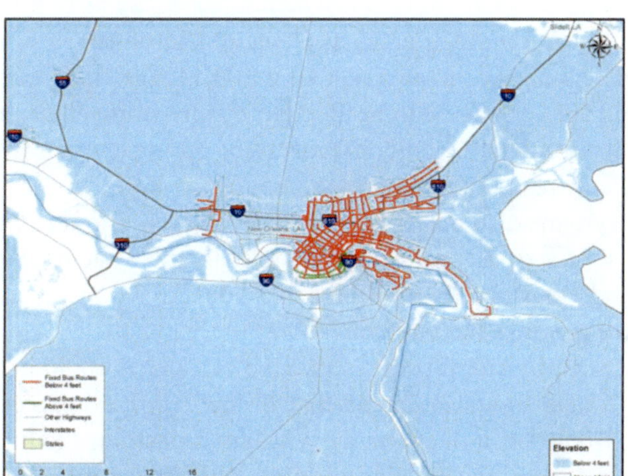

Source: Cambridge Systematics analysis of U.S. DOT data.

Figure 4.8. Fixed bus routes at risk from a relative sea level rise of 122 cm (4 ft), New Orleans, LA

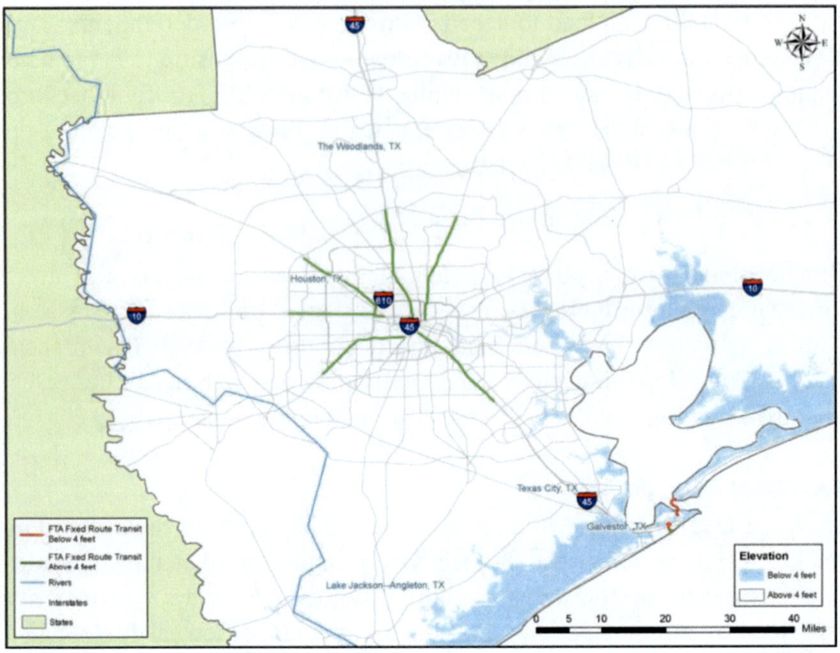

Source: Cambridge Systematics analysis of U.S. DOT data.

Figure 4.9. Fixed transit guideways at risk from a relative sea level rise of 122 cm (4 ft), Houston and Galveston, TX

The New Orleans streetcars system operated by the RTA and some small portions of the streetcar system operated by Island Transit in Galveston are similarly at risk of inundation at either the 61- or 122-cm (2- or 4-ft) sea level rise scenarios. Like the city itself, portions of many of the streetcar routes in New Orleans currently are below sea level, and it is only the levee system that maintains the ability of these streetcars to function. In contrast, the fixed transit system in Houston is not at risk at these levels, as show in figure 4.9.

Storm Activity

Transit facilities passing through areas at elevations at or below 5.5 and 7.0 m (18 and 23 ft) were identified. As shown in figures 4.10 and 4.11, the fixed transit systems in New Orleans and Galveston are very likely to be affected by any storms that generate surges of 5.5 m (18 ft) or more. This inundation would affect service during and immediately after a storm, though it would not likely result in long-term disruptions.

Fixed bus route systems also are at risk to storm surges. The bus route systems that are vulnerable to storm surges of 5.5 m (18 ft) include all the systems except those in Baton Rouge, Beaumont, and Houston (figure 4.12 and 4.13). At 7.0 m (23 ft), the risk of storm surge inundation also extends to the fixed bus routes in Beaumont.

The risk of inundation by storm surge is that the bus routes could not operate on flooded or obstructed roads. It also should be noted that in low surge events, even if the buses can operate, their utility would be influenced by whether pedestrian facilities are passable and riders can walk to bus stops. Consideration should be given to developing contingency plans for alternative routes during storms.

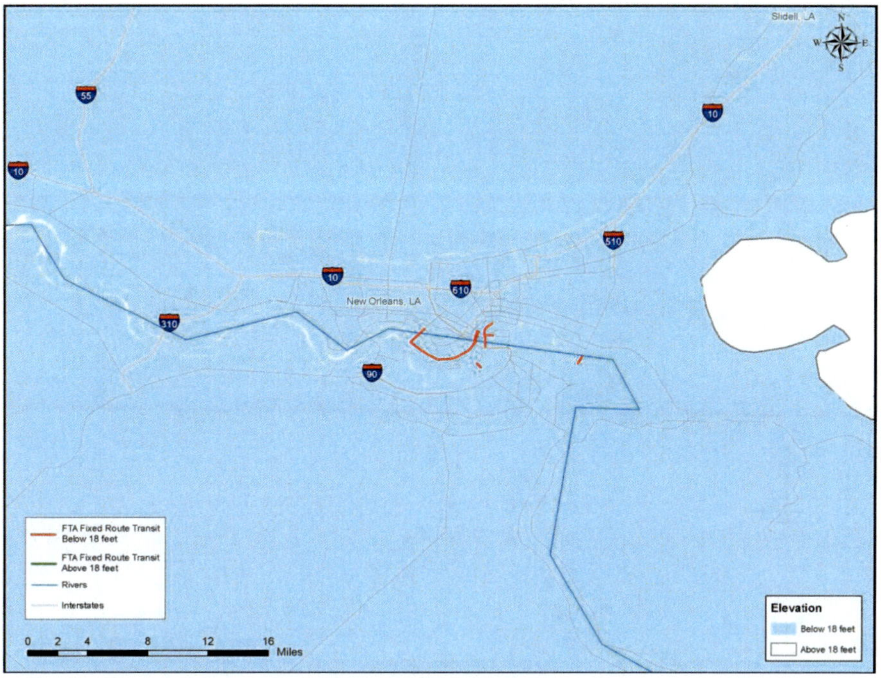

Source: Cambridge Systematics analysis of U.S. DOT data.

Figure 4.10. Fixed transit guideways at risk from storm surge at elevations currently below 5.5 m (18 ft), New Orleans, LA

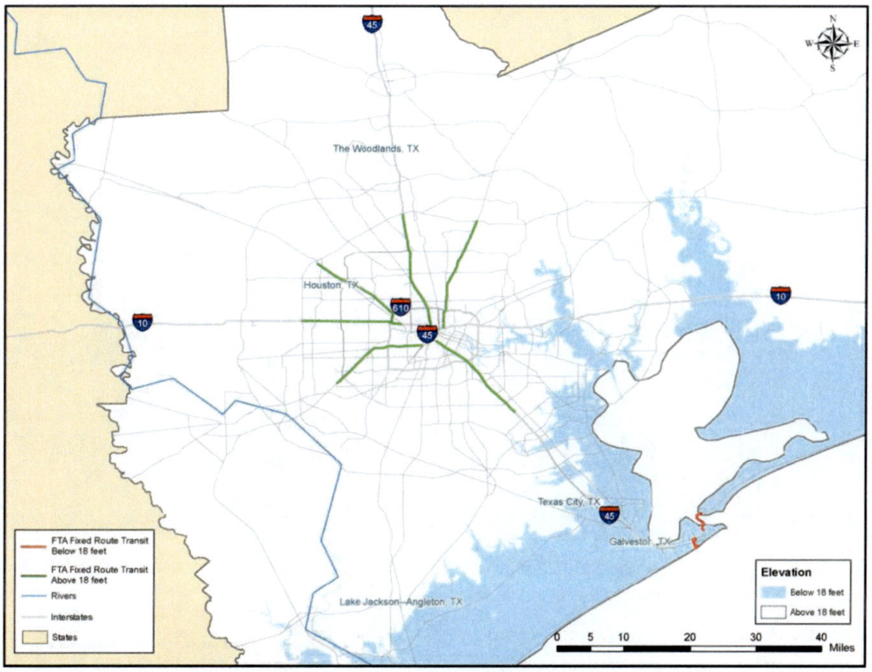

Source: Cambridge Systematics analysis of U.S. DOT data.

Figure 4.11. Fixed transit guideways at risk from storm surge at elevations currently below 5.5 m (18 ft), Houston and Galveston, TX

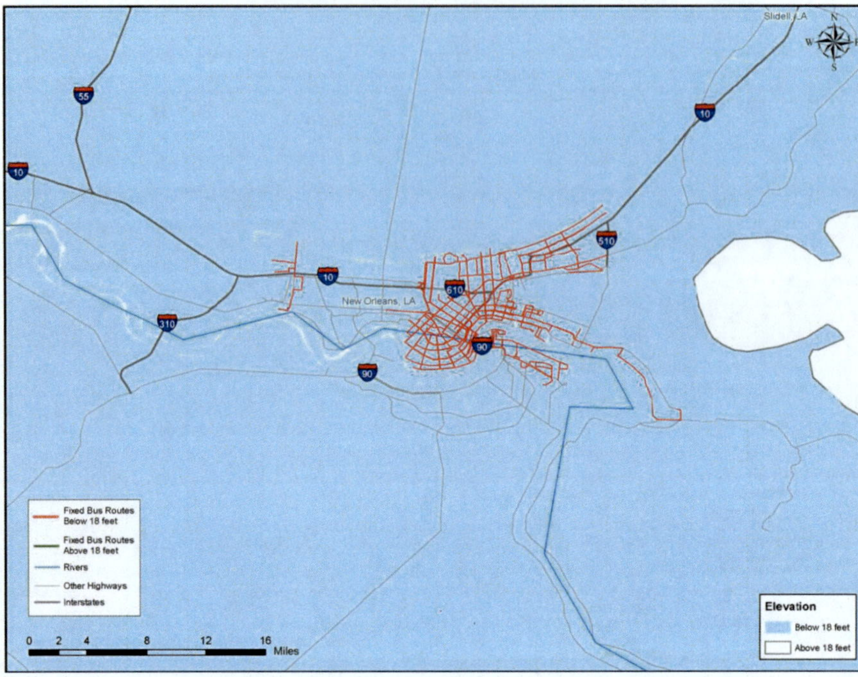

Source: Cambridge Systematics analysis of U.S. DOT data.

Figure 4.12. Fixed bus routes at risk from storm surge at elevations currently below 5.5 m (18 ft), New Orleans, LA

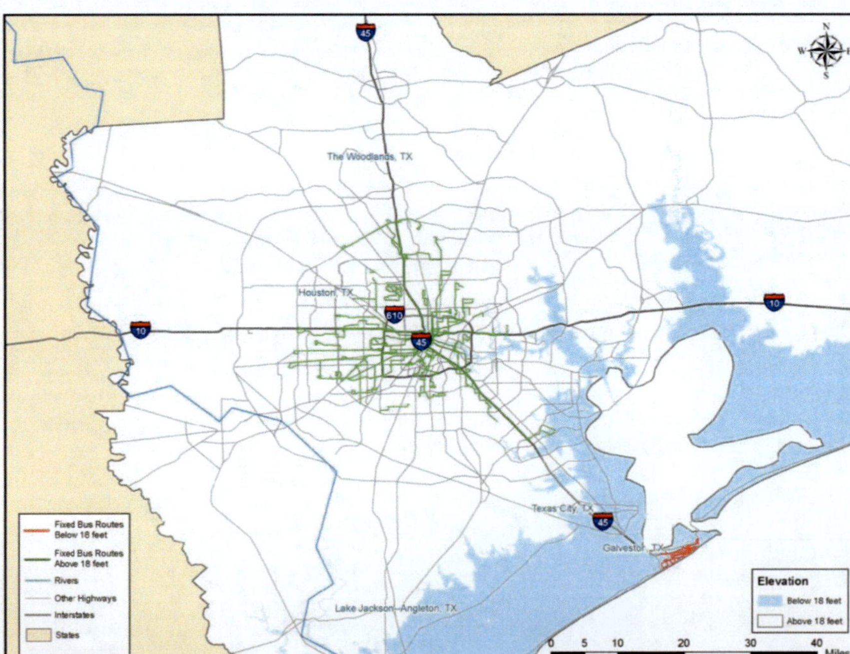

Source: Cambridge Systematics analysis of U.S. DOT data.

Figure 4.13. Fixed bus routes at risk from storm surge at elevations currently below 5.5 m (18 ft), Houston and Galveston, TX

Storm Winds

The transit infrastructure that is most vulnerable to impacts by the winds associated with increases in the number of intense storms are the overhead catenary lines that power street cars in New Orleans and Houston. Transit signs and control devices also are subject to wind damage.

However, rather than wind damage to transit facilities, the most widespread impact may be from fallen trees and property debris blocking the streets on which transit routes operate. This impact would occur during and immediately after storm events and should be addressed by highway clean up operations.

Storm Waves

With the exception of light rail and Bus Raid Transit (BRT) systems, transit equipment can be moved away from areas subject to wave impacts, and therefore, storm wave impacts during surge events are not expected to impact most transit systems. Even in the case of fixed guideways, storm waves will mostly affect areas immediately on the shoreline, which is not where fixed guideway facilities in the New Orleans and Houston systems are located. However, the trolley tracks in Galveston are at risk to these impacts.

4.2.3. Freight and Passenger Rail

Rail lines in the region play a key role in transporting freight and a minor role in intercity passenger traffic. Much of the traffic on class I rail lines in the region is for transshipments as opposed to freight originating or terminating in the region (figure 2.12). Rail connectivity and service also is vital to the functioning of many, if not most, of the marine freight facilities in the study area.

Of the four main climate drivers examined in this study, storm surge could be the most significant for rail. One-third of the rail lines in the study region are vulnerable to a storm surge of 5.5 m (18 ft), and 41 percent are vulnerable to a storm surge of 7.0 m (23 ft). Fifty-one freight facilities and 12 passenger facilities are vulnerable to storm surges of 7.0 m (23 ft). Sea level rise is of less concern for rail; a 122-cm (4-ft) RSLR would affect less than 10 percent of rail miles, as well as 19 freight facilities and no rail passenger facilities. Temperature increases could raise the danger of rail buckling, but would be unlikely to necessitate design changes. Projected precipitation patterns do not indicate that design changes are warranted to prevent increased erosion or moisture damage to railroad track.

Temperature

The level of average temperature increases discussed in chapter 3.0 is unlikely to require immediate design changes to track or other rail infrastructure, as these ranges generally fall within the current standards for existing rail track and facilities. However, the increase in temperature extremes – very hot days – could increase the incidence of buckling or "sun kinks" on all the rail tracks in the study area. This occurs when compressive forces in the rail, due to restrained expansion during hot weather, exceed the lateral stiffness of the track, causing the track to become displaced laterally. The amplitude of track buckles can reach 75 cm (30 inches) or more.

Track buckling occurs predominately on continuously welded track, though it also can occur on older jointed track when the ends of the track become frozen in place. Track buckling is most prevalent on an isolated hot day in the springtime or early summer, rather than mid to late summer when temperatures are more uniformly hot. Buckling also is more likely to occur in alternating sun/shade regions and in curves.

The most serious problem associated with track buckling is derailments. A derailment can occur when a buckled section of track is not observed in time for the train to safely stop. One way to overcome this is through blanket slow orders. In hot weather (more than 35 °C, or 95 °F), railroads issue blanket slow orders (generally to reduce all train speeds by 16 km/h or 10 mi/h) to help prevent derailments caused by buckling. This has several negative consequences, such as longer transit times, higher operating costs, shipment delays, reduced track capacity, and increased equipment cycle time leading to larger fleet sizes and costs. Reduced train speeds similarly affect passenger rail schedules, causing delays in travel schedules.

Research into improved track design and installation has greatly reduced the derailments attributable to buckling. For example, concrete cros sties with improved fasteners can withstand greater track stress than wooden ties with spikes. During installation, the rail is prestressed to a target neutral temperature. Since the track is more stable when the rail is in tension at temperatures below the neutral temperature, the target neutral temperature is generally 75 percent of the expected maximum temperature of the region. In the Gulf Coast region, the neutral temperature is typically 38 °C (100 °F), while 32 °C (90 °F) is used in more northern climates. Prestressing can occur either thermally (by actually heating the steel during installation) or mechanically by stretching the steel to introduce the desired stress prior to fastening it to the crossties.

A temperature change of 1.5 °C (2.7 °F) over the next 50 years may slightly raise the neutral temperature used for installation but would have little impact on track design otherwise. A temperature increase in this range would not necessitate replacing existing track. It would most likely be replaced as part of normal maintenance, upgrades to handle increased traffic volumes, or replacement due to storm surge or other catastrophic events. The typical cost to upgrade track can vary greatly depending upon the type of upgrade, the slope and curvature, and the number of bridges and tunnels. Costs to replace track range from $0.3 million to $1.9 million per kilometer ($0.5 million to $3 million per mile), excluding any additional right-of-way expenses.

If incidences of buckling rise it will be increasingly important to develop improved methods of detection. It is relatively easy to detect a broken rail by running a light electric current through track, but manual observation remains the best method for identifying track buckling. Research is underway to develop improved methods that measure temperature and stress of the track.[4]

The projected increases in average temperature and number of hot days, coupled with possible increases in humidity, would create serious safety concerns for workers in rail yards and other rail facilities and would require investments to protect rail workers. This might include increases in crew size to allow for more frequent recovery breaks or greater use of climate-controlled facilities for loading and unloading the railcars. Regardless of the solution, providing the necessary relief for workers will lead to increased operating or capital expenses, which will be reflected in higher transportation costs.

Precipitation

The primary impacts on rail infrastructure from precipitation are erosion of the track subgrade and rotting of wooden crossties. Erosion of the subgrade can wash away ballast and weaken the foundation, making the track unstable for passage of heavy locomotives and railcars. Ballast is typically granite or other hard stone used to provide a flat, stable bed for the track, and also to drain moisture from the track and ties. Without ballast, wooden crossties would rot at a faster rate, leading to more buckling and unstable track. As with buckling, subgrade erosion and rotting crossties are difficult to detect using methods other than visual inspection. This situation is improving, though, through remote sensing advances that detect standing water and air pockets.

The precipitation projections do not indicate that design changes are warranted to prevent increased erosion or moisture damage to railroad track, even with a potential change of 13 percent in precipitation levels. The runoff projections point to even fewer problems with erosion over the next century than are present today, due to possibly less precipitation and slightly higher temperatures. However, if the frequency and/or the intensity of extreme rainfall events increases, it could lead to higher rates of erosion and railroad bridge scour, as well as higher safety risks and increased maintenance requirements.

Relative Sea Level Rise

The effects on rail lines and facilities of relative sea level of 61 and 122 cm (2 and 4 ft) over the next 50 to 100 years were analyzed. The obvious impacts for both of these sea level rise scenarios are water damage or complete submersion of existing rail track and facilities. These ground elevations affect the vulnerability of rail segments to storm surge as well.[5] Table 4.7 indicates the percent of rail lines and facilities vulnerable to sea level rise at 6-1 and 122-cm (2- and 4-ft) levels. Currently, about 50 miles or about 2 percent of rail lines in the study region are located on land that is at or below sea level.

Figure 4.14 displays the rail network, used by both freight trains and Amtrak, with the RSLR elevation projections. Rail lines located in areas with a ground elevation of 0 to 61 cm (0 to 2 ft) are vulnerable to a relative sea level rise of 61 cm (2 ft) or more. Lines located in slightly higher areas, with a ground elevation of 61 to 122 cm (2 to 4 ft), are vulnerable to a relative sea level rise of 122 cm (4 ft).

Most of the rail lines in and around New Orleans would likely be impacted by RSLR. The heavily traveled CSX line between Mobile and New Orleans, which was damaged during Hurricane Katrina, also is at risk, as are several area short lines. A listing of the rail lines impacted if relative sea level rises 61 cm (2 ft) includes the following:

- Most rail lines in and around New Orleans;
- Burlington Northern Santa Fe (BNSF) line between Lafayette and New Orleans;

Table 4.7. Relative Sea Level Rise Impacts on Rail: Percentage of Facilities Vulnerable

Relative Sea Level Rise	Rail Lines (track miles)	Rail Freight Facilities (94)	Rail Passenger Stations (21)
61 cm (2 ft)	5%	12%	0
122 cm (4 ft)	9%	20%	0

- Canadian Nation (CN) line into New Orleans;
- CSX line between Mobile and New Orleans;
- CSX line north of Mobile;
- Louisiana and Delta Railroad west of New Orleans;
- Portions of the Mississippi Export (MSE) rail line in Mississippi;
- The New Orleans and Gulf Coast Railway line between New Orleans and Myrtle Grove, LA;
- Norfolk Station (NS) line into New Orleans;
- Portions of the Port Bienville Railroad;
- Segments of the Union Pacific (UP) line west of New Orleans; and
- Various segments of track around Lake Charles and Galveston.

Further degradation of these lines is very likely to occur should relative sea level increase by 122 cm (4 ft), with additional problems on the Kansas City Southern (KCS) route into New Orleans, the NS line north of Mobile, and selected track segments around Beaumont and Houston.

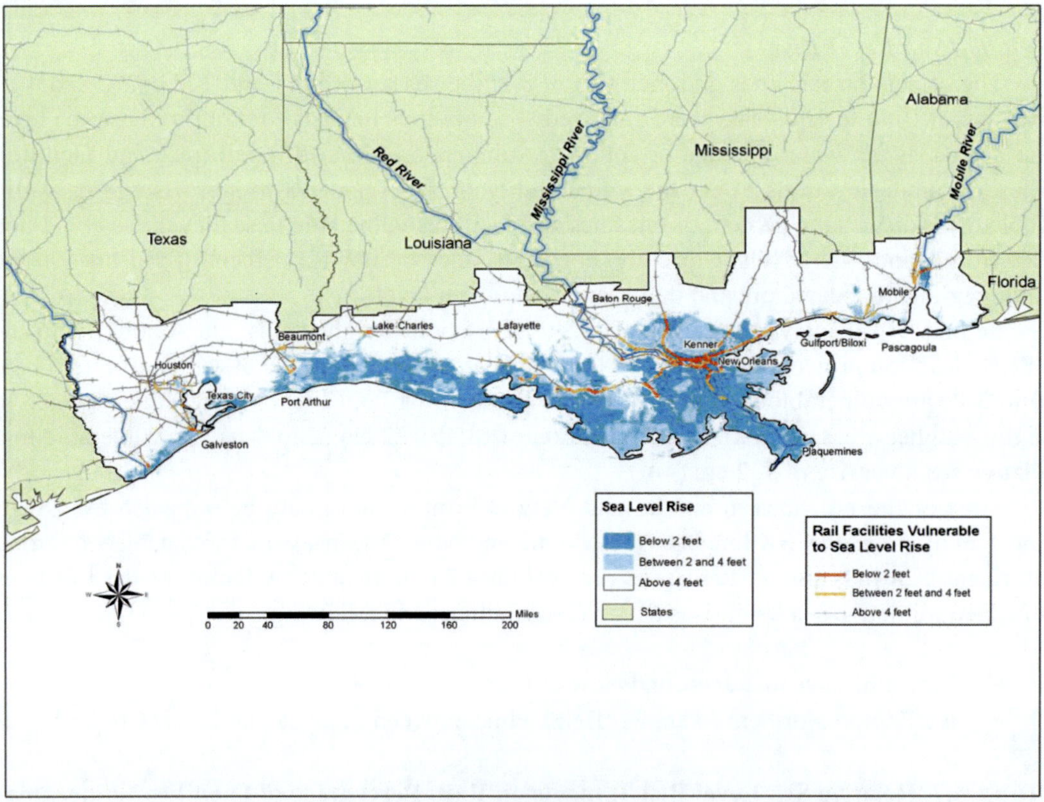

Source: Cambridge Systematics analysis of U.S. DOT data.

Figure 4.14. Rail lines at risk due to relative sea level rise of 61 and 122 cm (2 and 4 ft). Of the 4,722 km (2,934 mi) of rail lines in the region, 235 km (146 mi), or 5 percent, are at risk from a relative sea level rise of 61 cm (2 ft) or less. An additional 195 km (121 mi), for a total of nine percent, are at risk from an increase of 61 to 122 cm (2 to 4 ft).

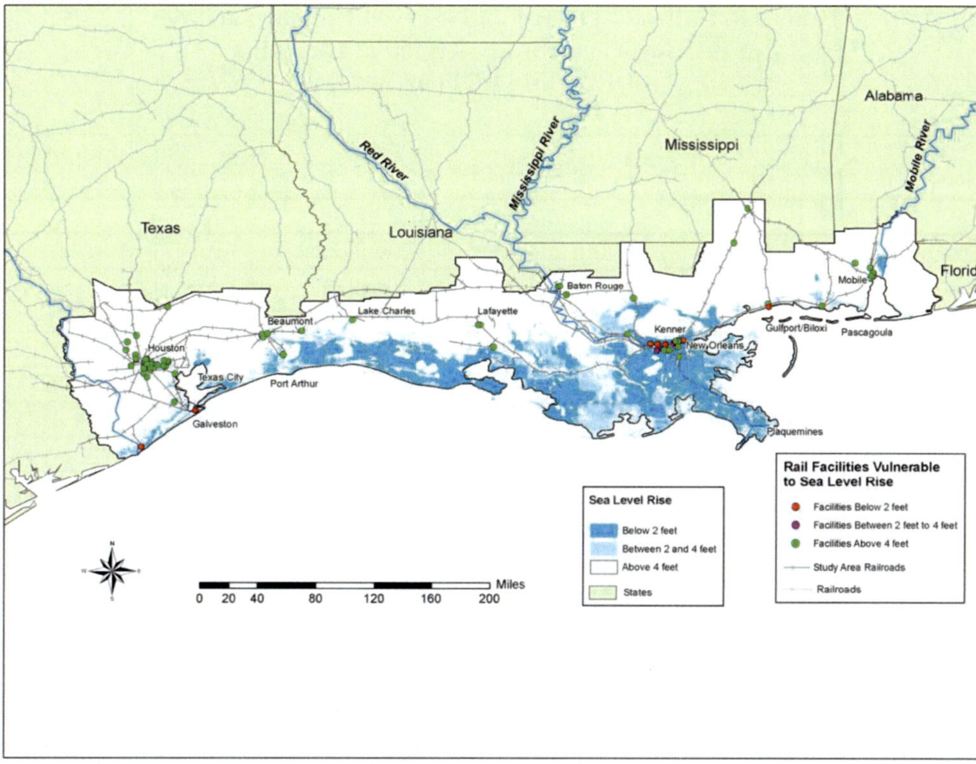

Source: Cambridge Systematics analysis of U.S. DOT data.

Figure 4.15. Railroad-owned and -served freight facilities at risk due to relative sea level rise of 61 and 122 cm (2 and 4 ft). Of the 94 facilities in the region, 11 are at risk from a 61-cm (2-ft) increase in relative sea level, and an additional 8 facilities are at risk from a 122-cm (4-ft) increase

Figure 4.15 shows the potential impacts of RSLR on railroad-owned and served facilities in the study region. Facilities located at less than 61 cm (2 ft) of elevation are very likely to be affected by a rise in relative sea level of 61 cm (2 ft). These include the KCS, NS, and UP rail yards in the New Orleans area. Facilities between 61 and 122 cm (2 and 4 ft) of elevation are very likely to be affected by a rise in relative sea level of 122 cm (4 ft). A listing of facilities with elevation 122 cm (4 ft) or less is contained in table 4.8. A listing of all freight rail facilities in the Gulf Coast study region, along with their elevation grid codes, is provided in appendix C.

A related issue is how railroad customers will respond to these rising relative sea levels and storm surge, and how these decisions will affect the demand for rail services. For example, to what extent will customers choose to relocate or modify their shipping and production patterns? Some industries, most notably the ports, need to remain at or near the water's edge to send and receive shipments. There will be a continued need for rail service into these locations. Other rail customers, however, may begin to relocate to higher ground or to different regions entirely. This will in turn affect the type and scale of rail network needed to meet the demand for inbound and outbound freight shipments. While it is difficult to predict the future choices of rail customers, it seems likely that climate change will negatively impact growth in goods movement at the lower elevations, and thus could lead to significantly reduced, and costlier, rail service in the region.

Table 4.8. Railroad-Owned and -Served Freight Facilities in the Gulf Coast Study Region at Elevation of 122 CM (4 FT) or Less

Name	Modal Access	City	State	Elevation cm (ft)
KCS	Rail and truck	Metairie	LA	< 0
Larsen Intermodal, Inc.	Rail and truck	Metairie	LA	< 0
New Orleans Cold Storage and Warehouse, Ltd.	Rail and truck	Metairie	LA	< 0
Port of Gulfport	Truck, port, rail	Gulfport	MS	< 0
Port of Galveston	Truck, port, rail	Galveston	TX	< 0
NS – New Orleans, Louisiana	Rail and truck	New Orleans	LA	0-30 (0-1)
UP Intermodal Facility	Rail and truck	Avondale	LA	0-30 (0-1)
Port of Freeport	Truck, port, rail	Freeport	TX	0-30 (0-1)
Dry Storage Corporation of Louisiana	Rail and truck	Kenner	LA	30-61 (1-2)
DSC Logistics	Rail and truck	Kenner	LA	30-61 (1-2)
Yellow Terminal	Rail and truck	New Orleans	LA	30-61 (1-2)
BNSF – New Orleans, Louisiana	Rail and truck	Westwego	LA	61-91 (2-3)
BNSF 539 Bridge	Rail and truck	Westwego	LA	61-91 (2-3)
BNSF Intermodal Facility	Rail and truck	New Orleans	LA	61-91 (2-3)
Intermodal Cartage Company	Truck, port, rail	New Orleans	LA	61-91 (2-3)
Transflo	Rail and truck	New Orleans	LA	61-91 (2-3)
BNSF 101 Avonda	Rail and truck	Avondale	LA	91-122 (3-4)
Downtown Transfer, Inc.	Rail and truck	Avondale	LA	91-122 (3-4)
Port of New Orleans	Truck, port, rail	New Orleans	LA	91-122 (3-4)

Turning to passenger rail service, none of the Amtrak passenger rail stations are at a high risk of impact due to a 122-cm (4-ft) increase in relative sea level. However, the rail lines used by Amtrak are at risk. These include the Sunset Limited routes between Mobile and New Orleans on the CSX-owned track and between New Orleans and Houston on the UP- owned track.

Table 4.9 summarizes the impacts of RSLR and storm surge on the freight and passenger rail lines and facilities in the region. These calculations are based on ground-level elevations of the rail facilities. All facilities and lines at low elevations are included, even though some of them are surrounded by higher land that may block the rising sea levels. The actual inland flow of the water is due to higher relative sea levels that was not availableat this time for this study.

One final factor, not directly addressed by the maps and tables discussed in this section, is the extent to which rising relative sea levels create a higher water table that leads to additional flooding during periods of normal precipitation. As the water table rises, the ground is less able to absorb normal rainfall. This could cause frequent flooding of rail track and facilities beyond the levels identified in the maps and tables.

Table 4.9. Vulnerability from Sea Level Rise and Storm Surge by rail Distance and Number of Facilities

		Cumulative		
Elevation Risk Gridcode	**Ground Elevation m (Ft)**	**Length of Railway Segments Vulnerable km (mi)**	**Freight Facilities Vulnerable**	**Passenger Facilities Vulnerable**
0 and 1	<0.3 (<1)	26 (86)	8	0
2	0.3-0.6 (1-2)	45 (146)	11	0
3	0.6-0.9 (2-3)	58 (191)	16	0
4	0.9-1.2 (3-4)	81 (267)	19	0
5	1.2-1.5 (4-5)	126 (412)	22	0
6	1.5-5.5 (5-18)	294 (966)	40	9
7	5.5-7.0 (18-23)	363 (1,190)	51	12
8	>7.0 (>23)	894 (2,934)	94	21

Storm Activity

Hurricane Katrina provided a vivid example of the devastating impacts of severe storm events to the rail system in the Gulf Coast study area. Making landfall on August 29, 2005, Katrina caused damage to all of the major railroads in the region. BNSF, CN, KCS, and UP all suffered damage, mostly to yards in and around New Orleans. CSX track and bridges also were damaged. NS had nearly 8 km (5 mi) of track washed away from the 9.3-km- (5.8-mi-) long Lake Pontchartrain Bridge. By September 13, 2005, most of these railroads had resumed operations into New Orleans, at least on a partial basis. There were still yards that had not fully opened, though this was due to a mixture of storm damage to the yard and customers not being fully operational. By October 8, 2005 most rail service on these carriers had been restored, except CSX (Association of American Railroads, 2005). (See section 4.3.1 for more on the impacts of the 2005 hurricanes.)

Figure 4.16 illustrates the rail lines most at risk from storm surge at the 5.5- and 7.0-meter (18- and 23-ft) marks. One-third of the rail lines in the study region are vulnerable to a storm surge of 5.5 m (18 ft), and 41 percent are vulnerable to a storm surge of 7.0 m (23 ft) (table 4.10). This includes the heavily traveled CSX line from New Orleans to Mobile and the UP and BNSF lines from New Orleans to Houston. Cities at risk include Mobile, Gulfport, Biloxi, New Orleans, Baton Rouge, Lafayette, Lake Charles, Beaumont, Port Arthur, and Galveston.

Similarly, figure 4.17 shows the potential impacts of storm surge on railroad-owned and served facilities in the study region. Facilities at less than 5.5 m (18 ft) of elevation have the highest risk of 5.5-m (18-ft) storm surge impacts. These include 43 percent of the rail facilities in the study region. An additional 11 facilities are between 5.5 and 7.0 m (18 and 23 ft) of elevation and are very likely to be affected by a 7.0-m (23-ft) storm surge. A listing of all freight rail facilities in the Gulf Coast study region, along with their elevation grid codes, is provided in appendix C.

Figure 4.18 shows the risks for Amtrak passenger rail stations due to storm surge at 5.5 and 7.0 m (18 and 23 ft). The data indicates that there is low risk overall to Amtrak stations from storm surge, but the nine stations listed in table 4.11 are very likely to be affected by a storm surge of 5.5 m (18 ft). Two of the stations, Galveston and La Marque, TX, do not have direct passenger rail service but are connected to the Amtrak services by bus. At the 7.0-m (23-ft) storm surge level, an additional three stations are likely to be affected: New Iberia, LA, and Bay St. Louis and Biloxi, MS. A listing of all Amtrak stations in the Gulf Coast study region, along with their elevation grid codes, is provided in appendix C.

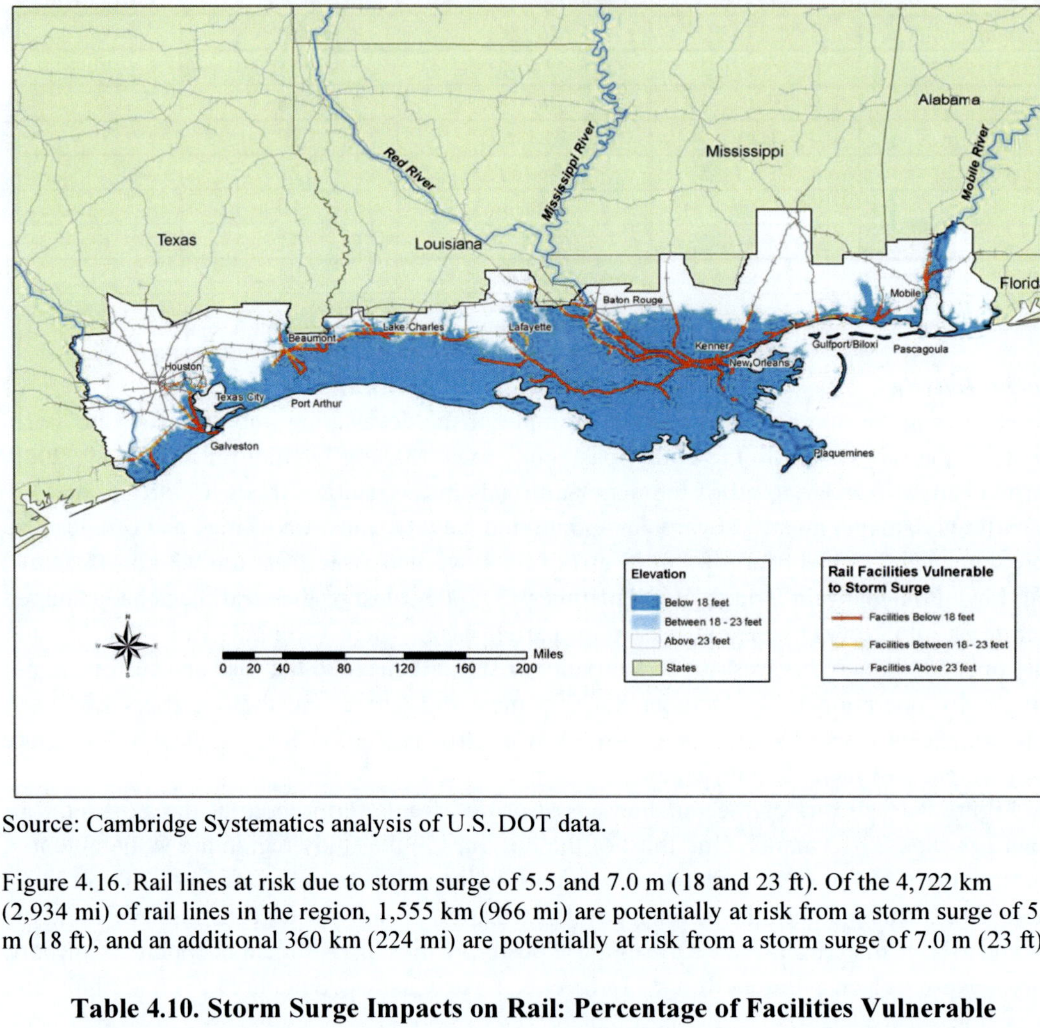

Source: Cambridge Systematics analysis of U.S. DOT data.

Figure 4.16. Rail lines at risk due to storm surge of 5.5 and 7.0 m (18 and 23 ft). Of the 4,722 km (2,934 mi) of rail lines in the region, 1,555 km (966 mi) are potentially at risk from a storm surge of 5.5 m (18 ft), and an additional 360 km (224 mi) are potentially at risk from a storm surge of 7.0 m (23 ft)

Table 4.10. Storm Surge Impacts on Rail: Percentage of Facilities Vulnerable

Storm Surge Height	Rail Lines (Track Miles)	Rail Freight Facilities (total of 94)	Rail Passenger Stations (total of 21)
5.5 m (18 ft)	33%	43%	43%
7.0 m (23 ft)	41%	54%	57%

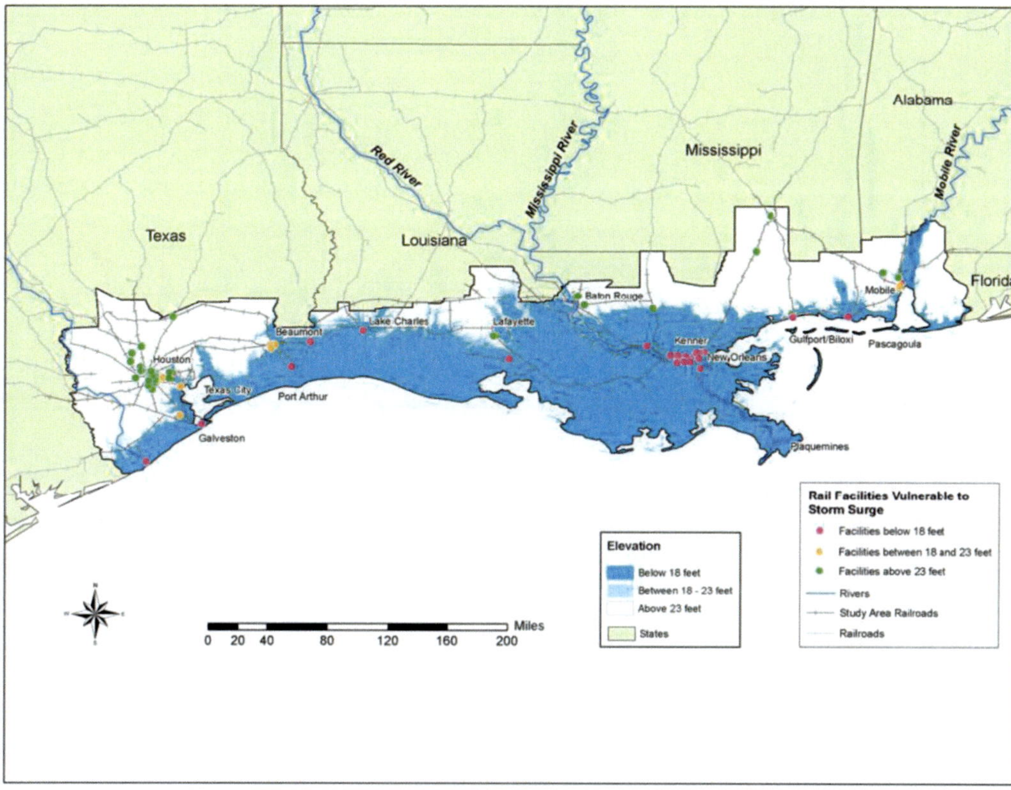

Source: Cambridge Systematics analysis of U.S. DOT data.

Figure 4.17. Railroad-owned and -served freight facilities at risk due to storm surge of 5.5 and 7.0 m (18 and 23 ft). Of the 94 facilities in the region, 40 are at risk from a storm surge of 5.5 m (18 ft) or less, and an additional 11 facilities are at risk from storm surge of 5.5 to 7.0 m (18 to 23 ft)

Table 4.11. Amtrak Stations Projected to be Impacted by Storm Surge of 5.5 and 7.0 m (18 and 23 FT)

Station	State	Amtrak Services
5.5-m (18-ft) Storm Surge		
Mobile	AL	Sunset Limited[1]
Pascagoula	MS	Sunset Limited[1]
Lake Charles	LA	Sunset Limited
New Orleans	LA	City of New Orleans, Crescent, Sunset Limited
Schriever	LA	Sunset Limited
Slidell	LA	Crescent
Beaumont	TX	Sunset Limited
Galveston	TX	Service by bus
La Marque	TX	Service by bus
7.0-m (23-ft) Storm Surge		
New Iberia	LA	Sunset Limited
Bay St. Louis	MS	Sunset Limited[1]
Biloxi	MS	Sunset Limited[1]

[1] Stations are currently inactive due to Hurricane Katrina.

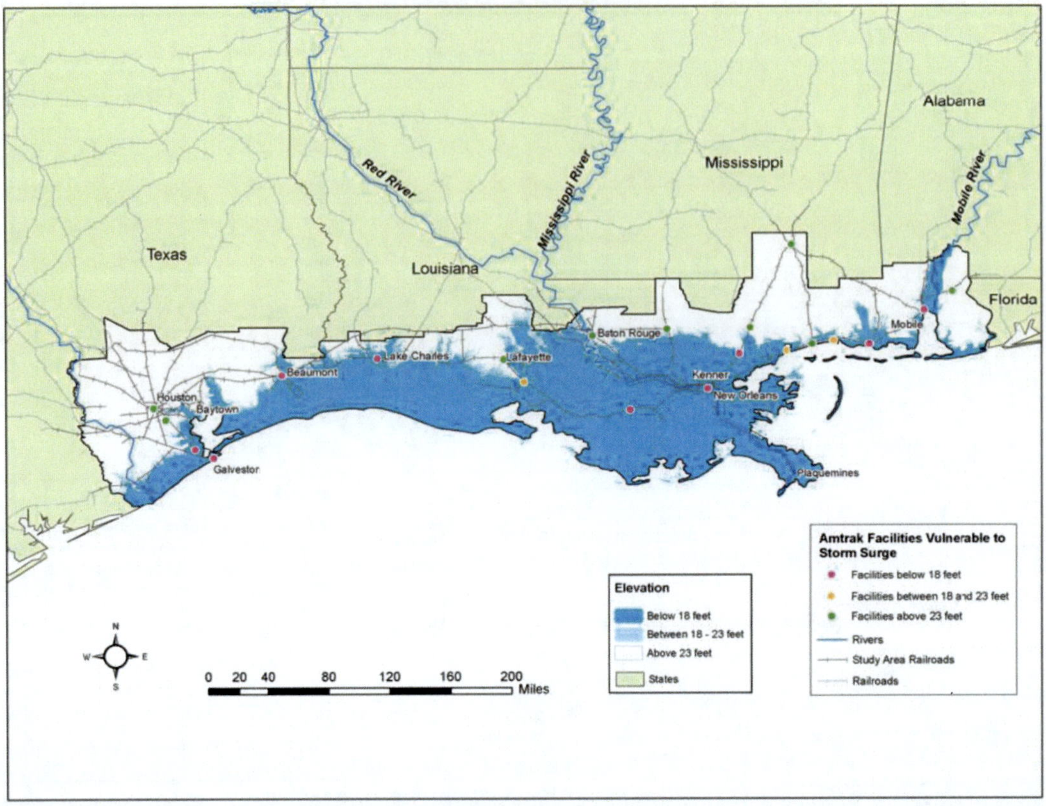

Source: Cambridge Systematics analysis of U.S. DOT data.

Figure 4.18. Amtrak facilities at risk due to storm surge of 5.5 and 7.0 m (18 and 23 ft). Of the 21 Amtrak facilities in the region, 9 are at risk from a storm surge of 7.0 m (18 ft) or less, and an additional 3 facilities are at risk from storm surge of 5.5 to 7.0 m (18 to 23 ft)

Railroad Response to Hurricane Damage

In the immediate aftermath of a hurricane, one of the largest problems facing railroad operators who are trying to restore service is safety issues at road-rail, at-grade crossings. Without power to operate the crossing gates, the railroads either need to manually flag each crossing or not run the trains. The larger railroads purchase electric generators that can be deployed after a hurricane to operate the gates, thus allowing trains to offer emergency response services and resume economic activity. For prolonged outages, as was the case with Hurricane Katrina, the railroads need to reeducate the public on the dangers of at-grade crossings once train service resumes.

Other short-term responses are directed at protecting revenues and controlling costs. Business customers within a region impacted by a hurricane are likely facing the same difficulties as the railroads and may not be fully operational. Once a company is fully operational, though, a railroad needs to be ready to offer service or risk losing business to other railroads, trucks, or barges. Delays in rail service availability can lead to a long-term loss of revenue. The other issue is continued long-haul service to businesses outside of the impacted area. After Hurricane Katrina, CSX rerouted trains that previously passed through the New Orleans gateway to junctions at St. Louis and Memphis. This extra routing increases the operating costs of the railroad through increased fuel usage, crew costs, equipment delays, and a loss of overall system capacity. There is a strong financial incentive to return to normal operations as soon as possible after a catastrophic event.

The long-term response of the railroads to increased storm intensity currently is being evaluated. The railroads are participating with both public and private groups to identify the best ways to serve the Gulf Coast region in the future. CSX Chief Operating Officer Tony Ingram stated, "We are open to ideas that are in the best interests of CSX, its customers, and its communities." Mr. Ingram further stated, "Our recent rebuild of the Gulf Coast line restores vital service and underscores our commitment, but does not foreclose other long-term alternatives for the rail line." (CSX, 2006a).

One obvious response is to begin relocating rail track and facilities further away from coastal areas and making expanded use of intermodal shipping. For example, CSX recently announced a new 1,250-acre integrated logistics center (ILC) in Winter Haven, FL, to serve the Tampa and Orlando markets. This ILC will include truck, rail, and warehousing for the storage and transfer of consumer goods to these two urban markets (CSX, 2006b). Although this ILC location was driven by proximity to the expanding Tampa and Orlando markets and the availability of affordable land – rather than as a risk reduction strategy – it does provide an interesting model for redesigned approaches to long-haul shipping by using inland locations and trucks to serve sensitive coastal markets.

Other proposals have included the relocation of CSX rail lines in Mississippi. As proposed, the rail relocation would occur in the Gulfport area and would bypass the Bay St. Louis Bridge that was damaged by Hurricane Katrina. However, much of the rail line on this CSX route might remain in storm surge danger, as illustrated in figure 4.16.

Another issue related to moving rail lines further away from coastal areas is that it will, in most cases, move passenger rail service further from population centers. The highest density populations tend to occur along coastal regions, making it the most desirable location for passenger rail stations. If the rail track is moved further inland to areas with lower population density, it would have a negative impact on intercity service and the potential of any future commuter passenger rail service that might be warranted by population growth along the coast. On the other hand, this effect could be obviated if rail facilities and passenger centers migrate inland in tandem, but coordinated responses cannot be assumed, in part because the entities involved – private rail companies, citizens, and governments – face different decisions related to the impacts of climate change, and their decision making processes are also necessarily different.

The temperature and precipitation changes projected under the climate scenarios and models used in this study likely would not necessitate any rebuilding of rail facilities or any significant design changes in the Gulf Coast study area rail network. The larger issue is damage due to RSLR, storm surge, and hurricanes. Rail lines totaling 1,915 km (1,190 mi) and 40 rail facilities are at risk from storm surge as examined above. (See figures 4.16 and 4.17.) Railroads may begin slowly relocating track and facilities further away from coastal areas, though this will be largely driven by customer location and needs. Increased use of rail-truck transloading from ILCs further from the coast might be an alternative. Any effort to move rail lines from the higher density coastal areas will have a negative impact on intercity passenger rail ridership and the potential utility of the line for commuter rail service as the population along the coast increases.

4.2.4. Marine Facilities and Waterways

Due to their location, marine facilities are most vulnerable to storm surge and relative sea level rise. Marine facilities include both freight and nonfreight facilities: ports, marinas, and industry-support facilities. Virtually all of the region's port facilities, or 98 percent, have the potential to be inundated by a storm surge of 5.5 m (18 ft), and 99 percent would be affected by a surge of 7.0 m (23 ft). A RSLR of 61 cm (2 ft) has the potential to affect 64 percent of the region's port facilities, while a 122-cm (4-ft) rise in relative sea level would affect nearly three-quarters of the port facilities. Impacts related to increased temperatures and changes in precipitation are expected to include increased costs related to maintenance as rising temperatures place greater stress on facilities, higher energy costs for refrigeration, and changes in the quantity and type of products shipped through the region as production and consumption patterns change both in and outside the region due to climate change.

Marine facilities and waterways are vital to the region and to the Nation as a whole. As noted in chapter 2.0, the study area is one of the Nation's leading centers of marine activity. Much of the region's economy is directly linked to waterborne commerce. and in turn, this waterborne commerce supports a substantial portion of the U.S. economy.

While some of these functions could be considered "replaceable" by facilities and waterways elsewhere, many of them – by virtue of geography, connections to particular industries and markets, historical investments, or other factors – represent unique and essentially irreplaceable assets. It might be possible to provide capacity equivalent to the Gulf Intracoastal Waterway or the Mississippi River on land, via highway and/or rail. It might even be possible to provide landside connections to, and sufficient capacity at, alternative international seaports, but the capital costs to provide such "replacement capacity" would undoubtedly be huge, and the costs to system users would be dramatically higher, if not prohibitively higher.

Higher Temperatures

Higher temperatures may affect port facilities in three key ways. First, higher temperatures will increase costs of terminal construction and maintenance, particularly of any paved surfaces that will deteriorate more quickly if the frequency of high temperatures increases. Many terminals – especially container and automobile handling terminals – have very large and open paved surfaces for storing cargo that in some cases can range up to hundreds of paved acres, while most others have at least some open paved area for storage. Nearly all provide on-terminal circulation space for trucks and wheeled terminal equipment. All such areas would be vulnerable to higher temperatures. Second, higher temperatures will lead to higher energy consumption and costs for refrigerated warehouses or "reefer slots" (electrical plug-ins for containers with on-board cooling units). Third, higher temperatures would likely lead to increased stress on temperature-sensitive structures. Container handling cranes, warehouses, and other marine terminal assets are made of metals. With increasing record temperatures and days over 32 °C (90 °F), it may be necessary to design for higher maximum temperatures in replacement or new construction. On the other hand, most dock and wharf facilities are made of concrete and lumber, which are generally less sensitive to temperature fluctuations. It is possible that lock and dam structures could be affected, although this will require further investigation. While this analysis examines existing facilities, it should be noted that development of new types of surfaces and structures that can better tolerate high temperatures, for example, would counteract some adverse impacts.

Temperature changes in other parts of the country may prompt some changes in consumption and production patterns in the United States that in turn would affect shipping patterns in the study region. Compared to the freight movement patterns of today, increases in temperature in the southeast or other regions could possibly lead to increases in shipments of coal or other energy supplies that pass through the region's ports. (This assumes that the current mix of power plants and fuels remains the same; however, changes in energy consumption patterns and improvements in energy efficiency are certainly possible, which could lead to changes in demand for fossil fuels.) Additionally, temperature changes in other regions could possibly lead to changes in the quantity and location of grain production, thus changing shipping patterns involving Gulf Coast ports; such changes could have economic ramifications for the Nation as a whole as well as for regional ports.

Precipitation

As noted previously, projections of future annual average rainfall suggest a slight increase or decrease in average annual precipitation depending on choice of GCM and emissions scenario. The prospect of more intense precipitation events, as indicated in chapter 3.0, could require the capacity of some stormwater retention and treatment facilities to be increased. The handling of stormwater can be a significant expense for container terminals, auto terminals, and other terminals with large areas of impervious surface. Increasing environmental regulatory requirements also may add to costs of adapting stormwater handling infrastructure.

Relative Sea Level Rise

Typically, the highest portion of the marine terminal is the wharf or pier structure, where a vessel actually berths. Structures and open storage areas behind the wharf or pier may be at the same level or may be lower. The highway and rail connections serving the terminal will be at land level, unless they are on bridge structures. Depending on their design, different terminals will have different areas of particular vulnerability with respect to RSLR.

It is important to note that many existing facilities at low elevations are protected by levees and other physical structures, which should provide resistance to gradual changes in sea levels. The specific effects of existing protections have not been considered in this study. For facilities that are not appropriately protected, either by elevation or by structures, rising water levels pose an increased risk of chronic flooding, leading in the worst case to permanent inundation of marine terminal facilities, either completely or in part, and rendering them inoperable.

Of freight facilities in the study area, about 72 percent are vulnerable to a 122-cm (4-ft) rise in relative sea level. Of the 994 freight facilities in the United States Army Corps of Engineers (USACE) database, 638 (64 percent) are in areas with elevations between 0 and 61 cm (2 ft) above sea level, and another 80 (8 percent) are in areas with elevations between 61 and 122 cm (2 and 4 ft). More than 75 percent of facilities are potentially vulnerable in Beaumont, Chocolate Bayou, Freeport, Galveston, New Orleans, Pascagoula, Plaquemines, Port Arthur, Port Bienville, and Texas City; between 50 percent and 75 percent of facilities are potentially vulnerable in Gulfport, Houston, Lake Charles, Mobile, South Louisiana, and the Tenn-Tom. Only Baton Rouge, with 6 percent of facilities potentially at risk, appears to be well-positioned to avoid impacts of sea level rise (see figure 4.19).

A similar situation faces nonfreight facilities. Seventy-three percent of study area marine, nonfreight facilities in the study area are potentially vulnerable to a 122-cm (4-ft) increase in relative sea level. Of the 810 nonfreight facilities in the USACE database, 547 (68 percent) are in areas with elevations between 0 and 61 cm (2 ft) above sea level, and another 47 (6 percent) are in areas with elevations between 61 and 122 cm (2 and 4 ft). More than 75 percent of facilities are potentially vulnerable in Beaumont, Chocolate Bayou, Freeport, Galveston, New Orleans, Pascagoula, Plaquemines, Port Arthur, the Tenn-Tom, and Texas City; between 50 percent and 75 percent of facilities are potentially vulnerable in Houston, Lake Charles, Mobile, and South Louisiana. Twenty-seven percent of Gulfport facilities and no Baton Rouge facilities are potentially at risk (see table 4.10).

Navigable depths are likely to increase in many harbors and navigation channels as a result of rising sea levels. This could lead to reduced dredging costs, but higher costs where rising water levels require changes to terminals. The functionality and/or protections of lock and dam structures controlling the inland waterway system also may be impacted by relative sea level rise.

Various indirect impacts could potentially affect operations and need for ports. As discussed in earlier sections, impacts on highways and rail connections could affect the ability to utilize and transport goods to and from affected ports. Rail connections to the Ports of New Orleans, Mobile, Pascagoula, and Gulfport/Biloxi are at greatest risk.

Production and consumption patterns within the study area are likely to be significantly affected by changes in sea level, which could lead to increased demand for certain types of shipments and reduced demand for others. As residential populations relocate from affected areas, demand for transported goods would decline. Similarly, as commercial activities relocate, transportation services would shift with them. Further, shifts in population could cause labor shortages for transportation and commercial facilities.

Storm Activity: Water and Wind Damage

While the actual facilities that would be flooded depend on the particulars of a given storm – the landfall location, direction, tidal conditions, etc. – fully 99 percent of all study area facilities are vulnerable to temporary and permanent impacts resulting from a 7.0-m (23-ft) storm surge, while almost 98 percent are vulnerable to temporary and permanent impacts resulting from an 5.5-m (18-ft) storm surge (figure 4.20 and table 4.13). All facilities are vulnerable to wind impacts. Similar to sea level rise, storm surge impacts on highway and rail connections could affect the ability to utilize ports for transport of goods to and from affected ports.

Table 4.12. Relative Sea Level Rise Impacts on Ports: Percentage of Facilities Vulnerable

Relative Sea Level Rise	Ports	
	Freight	Nonfreight
61 cm (2 ft)	64%	68%
122 cm (4 ft)	72%	73%

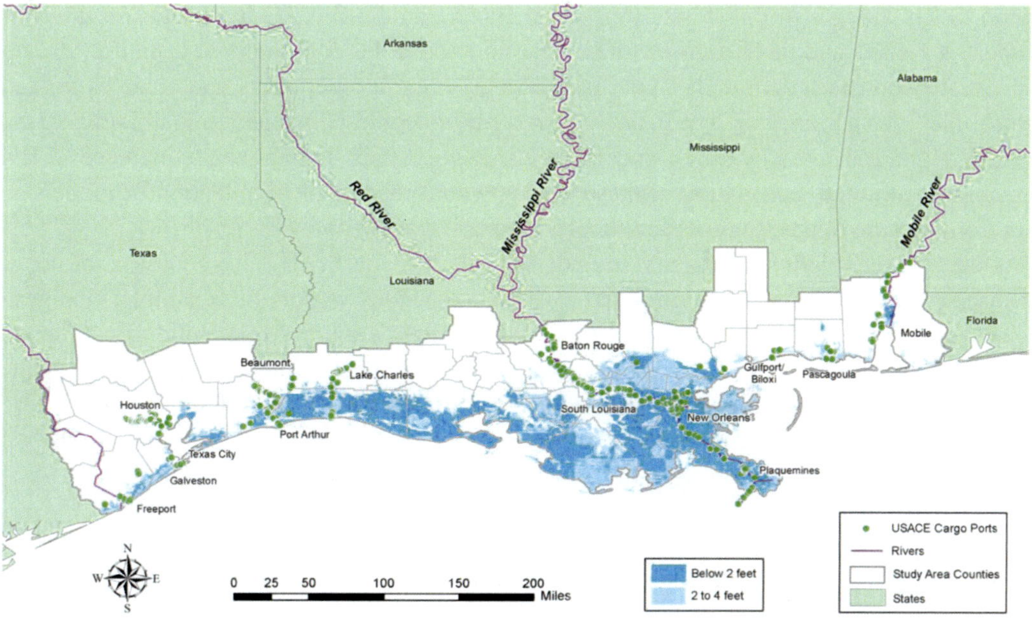

Source: Cambridge Systematics analysis of U.S. Army Corps of Engineers data.

Figure 4.19. Freight handling port facilities at risk from relative sea level rise of 61 and 122 cm (2 and 4 ft)

As evidenced by Katrina, fast moving water can be incredibly damaging to marine facilities. Water can physically dislodge containers and other cargo from open storage areas, knock down terminal buildings, damage or destroy specialized terminal equipment, damage wharf and pier structures, temporarily inundate and submerge large areas, and undermine or damage pavement and foundations. Wind has its most damaging effects on unreinforced terminal structures, such as metal warehouses that feature large surface areas and relatively light construction. **Much of Katrina's damage to the Port of New Orleans** – which mostly escaped water damage – was due to wind tearing off warehouse roofs and doors.

Wind and water can result in navigation channels becoming inoperable due to blockages and/or loss of markers. One of the first recovery tasks following Katrina was locating and clearing the channel in the Mississippi River, allowing it to reopen to barge and vessel traffic. Wind and water also can affect the location and protection afforded by the barrier islands that help define the Gulf Intracoastal Waterway.

Further, as mentioned earlier, highway and rail connectivity is vital to the functioning of nearly all port facilities in the study area. The road and rail facilities that are potentially at risk of surge at 5.5 and 7.0 m (18 and 23 ft) are shown in figures 4.5, 4.6, 4.7, 4.16, and 4.17. While the actual highways that would be flooded depends on the particulars of a given storm, a substantial portion of the highway system is at risk of surge inundation, including roads in all four states in the study area. The resulting potential loss of access to ports is obviously a critical vulnerability to reliable intermodal operations.

Secondary Impacts

Water levels in navigable rivers, and thus the ability to move freight, would be affected by higher or lower levels of precipitation, evapotranspiration, and runoff occurring outside the region. Such changes in the Mississippi River Basin could affect the ability to use the upper Mississippi River and its tributaries to export grain and other commodities from the Midwest and Great Plains States through Gulf Coast ports. Dredging operations and changes in water control facilities and marine terminals at up-river ports could be needed to maintain access to them. Freight transport by truck and rail outside the study region could increase if river transport is curtailed. Estimation of these effects would require the application of models and data from outside of the study area to incorporate up-river hydrology.

Demand for freight services that include use of Gulf Coast ports also could be influenced by changes in precipitation and temperature outside the study region. For example, changes in the amount and frequency of precipitation as well as temperature levels could affect demand for U.S. grain products overseas, just as changes in the same climate drivers in the United States could affect the ability of U.S. grain producers to supply export markets and domestic consumers. Such changes could have implications for Gulf Coast ports in particular, as well as for national highway and rail systems.

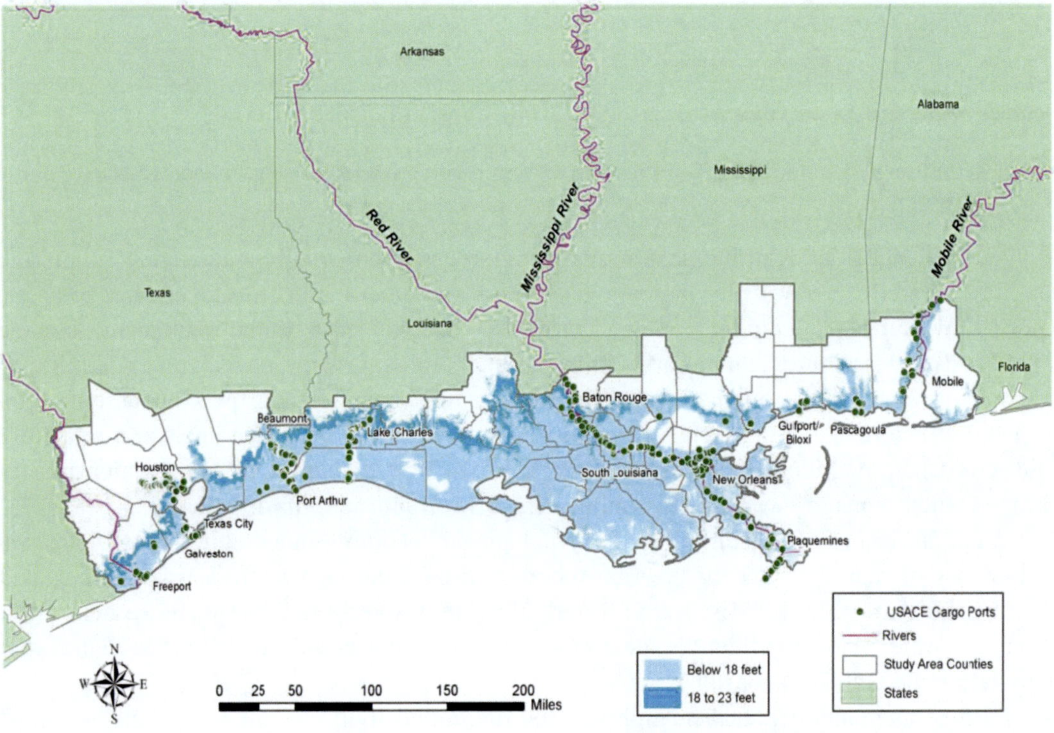

Source: Cambridge Systematics analysis of U.S. Army Corps of Engineers data.

Figure 4.20. Freight handling port facilities at risk from storm surge of 5.5 and 7.0 m (18 and 23 ft)

Table 4.13. Storm Surge Impacts on Ports: Percentage of Facilities Vulnerable

Storm Surge Height	Ports (Freight and Nonfreight)
5.5 m (18 ft)	98%
7.0 m (23 ft)	99%

Similarly, transport of energy supplies through Gulf Coast ports could be influenced by changes in temperature across the globe. Increases in temperature in the United States could affect the demand for energy products transported through Gulf Coast ports; demand for natural gas and coal to power electricity plants in the southeast, for example, could lead to greater production and/or importation of natural gas and liquefied natural gas (LNG) LNG through the ports and could put downward pressure on coal exports through the Gulf in favor of domestic consumption. On the other hand, coal exports through Gulf Coast ports could increase as export demand increased. Of course, climate mitigation policies could lead to significant shifts in preferred energy resources, leading to changes in energy transport demand. Such changes would have implications for pipelines (natural gas, petroleum), as well as rail (coal) and ports (coal). These secondary effects may prove to be important in the future, and such changes need to be monitored closely to track and adapt to changing demand levels.

4.2.5. Aviation

It is possible that existing patterns and intensity of severe weather events could be adversely affected by climate change, and such events could have the greatest impacts on aviation. These changes in severe weather may be widespread geographically such that they could profoundly affect the operational aspects of aviation and overall air traffic and air space management. If the climate becomes wetter, more general aviation pilots would need to learn to fly by instruments or avoid flying during inclement weather. Increased precipitation also could affect commercial service operations, particularly by raising the potential for delays. However, it should be noted that predicting how severe weather patterns would change as a result of climate change is extremely difficult and uncertain. Ultimately, the impact on the operational aspects of aviation could potentially supersede the overall magnitude of combined effects on aviation due to other factors discussed below.

A total of 29 airports could be vulnerable to a storm surge of 7.0 m (23 ft). The analysis suggests that 3 airports may be vulnerable to an increase in RLSR of 1.2 m (4 ft). Temperature increases considered by this report would indicate a small increase in baseline runway length requirements, assuming other relevant factors are held constant; however, the changes will very likely not be sufficient, especially accounting for ongoing technological change in commercial aircraft, to have any substantial impact on runway length requirements. Nevertheless, aircraft manufacturers may want to determine whether the generic hot day temperatures used in their specifications for civilian aviation aircraft are sufficiently high.

Temperature

Runway Design and Utilization

Required runway length is a function of many variables, including airport elevation, air temperature, wing design, aircraft takeoff weight and engine performance, runway gradient, and runway surface conditions.[6] Runways are designed to accommodate the most stringent conditions aircraft can experience. Climate model simulations as discussed in chapter 3.0

have conclusively noted that future change in climate will be accompanied by increases in temperature. Generally speaking, the higher the temperature the longer the runway that is required. In fact, initial runway construction planning takes into account, as a matter of course, a range of temperatures that can very well capture the extent of the increase in mean maximum temperature derived from the model results. If increases in temperature exceed the range initially expected, then considerations for additional adjustment in runway length may be necessary, depending on other relevant considerations such as payload and elevation. However, this is considered unlikely.

With rising temperatures, it is possible that there could be an impact on aircraft performance that would warrant aircraft manufacturers considering field length requirements in their design specifications. However, current trends in aircraft design point to shorter takeoff distances as airframes become lighter and engines become more powerful. Thus, due to technological innovation, runway length requirements may actually decrease even if temperatures increase.

Forecasting aircraft manufacturer's product offerings beyond 20 to 30 years is speculative, but trends toward increased fuel efficiency, more powerful engines, and lighter weight aircraft are anticipated to continue, which could offset the need for longer runway length as temperatures rise. Analysis of passenger jet aircraft performance indicates that newer aircraft entering the market over the last 50 years use less runway length per pound of aircraft. A comparison of two similar Boeing aircraft illustrates this point: the Boeing 737-200 aircraft entered commercial service in 1968 with an engine thrust of 6,580 kg (14,500 lb) and a per passenger seat thrust ratio of 53 kg (117 lb). In 2008, the company's first 7 87-800 "Dreamliner," made of up to 50 percent light weight composite products, will enter service. Compared to its predecessor, the 737-200, the GE Aircraft Engines on the 787 will provide more than four times as much thrust and twice as much engine thrust per passenger seat. This design, paired with more fuel efficient engines, translates into increased fuel efficiency, producing fuel savings up to 20 percent versus similar sized aircraft as well as shorter takeoff distances.

In order to better understand how changes in temperature could affect the current generation of aircraft, we looked at both general aviation and civil aviation applications. Generally, assessments of required runway length are conducted along two tracks for general aviation and civil aviation airports, and our analysis below reflects this difference:[7]

- Using the procedures outlined in the Federal Aviation Administration (FAA) Advisory Circulars (AC) (for general aviation aircraft); and
- Using the manufacturer's performance curves, published by aircraft manufacturers[8] (primarily large commercial service aircraft).

General Aviation

While planning for runway design generally accounts for a range of temperatures, this analysis of general aviation airports looks solely at how changes in assumptions about temperature would affect the baseline analysis of runway length requirements for a hypothetical general aviation airport by using the FAA's Airport Design for Microcomputers software.[9] The software allows for four variable inputs: airport elevation; runway slope measured in difference in elevations at each end of the runway; mean maximum temperature for the hottest day of the month; and runway conditions. Aircraft performance during takeoff varies significantly based on runway elevation, although generally speaking, there is only moderate difference in runway length needed between an airport at sea level and one at 91 m (300 ft) above sea level. Runways located in mountainous areas, however, have significantly longer runways than those at sea level. Mean maximum temperature is used by airport planners to identify the average hottest temperature during the hottest month of the year. Generally speaking, longer runways are required at hotter temperatures. Requirements for wet runways, which have less friction for braking or slowing the aircraft, are set out in regulation.

Table 4.14 lists the FAA design standards for a hypothetical general aviation airport and shows that all small airplanes (defined as having a maximum takeoff weight of less than 5,670 kg or 12,500 lb) could operate in the study area with a 1,308-m (4,290-ft) runway on days as hot as 33 °C (91.5 °F). On cooler days, less runway length is required. Large aircraft with maximum takeoff weights greater than 5,670 kg (12,500 lb) require longer runways. As noted in table 4.14, 1,637 m (5,370 ft) of runway is recommended to accommodate 75 percent of large airplanes up to 27,200 kg (60,000 lb) at up to a 60 percent useful load when runway surfaces are wet. Wet runway conditions require more length, and these conditions are typically used when calculating runway length.

Table 4.14. FAA Recommended Runway Lengths for Hypothetical General Aviation Airport

Airport Data	
Airport Elevation	30
Maximum Difference in Runway Centerline Elevation (feet)	1
Temperature (°F)	91.5
Runway Condition	Wet
Small Airplanes	
Small Airplanes with Approach Speeds of Less than 30 Knots	330
Small Airplanes with Approach Speeds of Less than 50 Knots	870
Small Airplanes with Less than 10 Passenger Seats	
75 Percent of these Small Airplanes	2,530
95 Percent of these Small Airplanes	3,100
100 Percent of these Small Airplanes	3,660
Small Airplanes with 10 or More Passenger Seats	4,290
Large Airplanes	
Large Airplanes of 60,000 Pounds[1] or Less	
75 Percent of these Large Airplanes at 60 Percent Useful Load	5,370
75 Percent of these Large Airplanes at 90 Percent Useful Load	7,000
100 Percent of these Large Airplanes at 60 Percent Useful Load	5,500
100 Percent of these Large Airplanes at 90 Percent Useful Load	8,520

[1] Maximum takeoff weight.

Source: U.S. DOT Federal Aviation Administration, Airport Design Version 4.2D.

While planning for airport construction generally accounts for a range of temperatures, this analysis looks solely at how changes in assumptions about temperature would affect the baseline results generated using the FAA's Airport Design for Microcomputers. The research team analyzed the effect of changes in mean maximum temperature for the hottest month of the year on runway length requirements as indicated by the climate scenarios reviewed in chapter 3.0. Mean maximum temperature was the only variable changed; airport elevation, centerline elevation, and runway surface conditions (wet) were held constant.[10] The 5th , 50th , and 95th percentile temperature increases demonstrated in scenarios A1B, B1, and A2 were applied to the FAA design standards for the hypothetical airport presented. The increases in runway length based on the increase in temperature associated with each scenario are discussed below. Mean maximum monthly temperature is derived by averaging the daily high temperature for the month with the highest average maximum temperature, which for the Gulf Coast is August. The projected temperature increases used were then added to the base year's mean maximum monthly temperature. The current average mean maximum temperature is estimated to be 33 °C (91.4 °F), based on 1972-2002 data from 12 research stations from the Carbon Dioxide Information Analysis Center (CDIAC) located in the region. For example, for scenario A1B the 50th percentile temperature increase of 2.5 °C (4.5 °F) was added to the 33 °C (91.4 °F) base year mean maximum temperature, indicating that in 2050 the mean maximum temperature is projected to be 35.5 °C (95.9 °F).

Below is a brief discussion of the results of this analysis that indicates the range of potential changes in baseline runway length requirements under the climate scenarios, conveying the full range of results based on the models and scenarios. For 2100, we point out the lowest and highest results. These results indicate the change in baseline runway length requirements for this hypothetical airport by using the FAA's airport design software, given a specific change in mean max temperature.

The analysis confirms that generally speaking, the possible increases are quite small. Given the long lead times and ongoing changes in aircraft technology, this means that possible temperature increases most probably will have little effect on runway length for commercial aircraft.

The potential temperature increases for the month of August are summarized in chapter 3.0, table 3.11. Over the longer term (to 2100), the analysis indicates an increase of between 1.8 °C (3.2 °F) (B1, 5th percentile) and 6 °C (10.8 °F) (A2, 95th percentile). An increase at the lower end would indicate a potential need to increase runway length by 9 m (30 ft) for small aircraft and by 12 to 15 m (40 to 50 ft) for large general aviation aircraft. At the 95th percentile, an increase of 6 °C (10.8 °F) could require lengthening the runway by 30 to 46 m (100 to 150 ft) for small airplanes and by 40 to 219 m (130 to 720 ft) for large aircraft.

Generally speaking, the possible increases in baseline runway length requirements are very low, especially for small aircraft (see table 4.15). The scale of these runway length requirement increases range from 8 to 16 percent for corporate jets to 2 to 3 percent for light general aviation aircraft. While these limited analyses are illustrative of the potential influence of temperature increase on runway length based on existing aircraft technology, whether more detailed analyses would need to be conducted would be decided by airport managers on a case by case basis in order to determine possible investment considerations.

Table 4.15. Summary of Impacts of Temperature Change to Runway Length (General Aviation) under Three Climate Scenarios (SRES Scenarios A2, B1, and A1B)

Analysis Category	Base Year	50th Percentile					
		2050 Climate Scenarios			2100 Climate Scenarios		
		A2	B1	A1B	A2	B1	A1B
		35.2	34.8	35.5	37.7	35.7	36.9
Possible Mean Maximum Temperature of Hottest Month, °C (°F)	**33.0 (91.4)**	**(95.5)**	**(94.6)**	**(95.9)**	**(99.9)**	**(96.3)**	**(98.4)**
Runway Length Analysis by Aircraft Type	**Runway Length m (Ft)**	**Runway Length Percent Increase**					
Small Airplanes with Less than 10 Passenger Seats							
75 Percent of these Small Airplanes	**771 (2,530)**	1.6%	1.2%	1.6%	3.2%	1.6%	2.8%
95 Percent of these Small Airplanes	**945 (3,100)**	1.3%	1.0%	1.6%	2.9%	1.6%	2.6%
100 Percent of these Small Airplanes	**1,116 (3,660)**	1.6%	1.1%	1.6%	3.3%	1.6%	2.7%
Small Airplanes with 10 or More Passenger Seats	**1,308 (4,290)**	1.6%	1.2%	1.9%	3.3%	1.9%	2.8%
Large Airplanes of 60,000 Pounds or Less							
75 Percent of these Large Airplanes at 60 Percent Useful Load	**1,637 (5,370)**	0.9%	0.7%	1.1%	2.4%	1.1%	2.0%
75 Percent of these Large Airplanes at 90 Percent Useful Load	**2,134 (7,000)**	2.1%	0.9%	2.7%	7.9%	2.7%	6.0%
100 Percent of these Large Airplanes at 60 Percent Useful Load	**1,676 (5,500)**	2.5%	1.6%	3.3%	8.0%	3.3%	6.2%
100 Percent of these Large Airplanes at 90 Percent Useful Load	**2,597 (8,520)**	6.8%	4.9%	7.9%	16.3%	7.9%	13.1%

Source: U.S. DOT Federal Aviation Administration (FAA) Airport Design Version 4.2D.

Commercial Service Airports

Commercial service, military airfields, and industrial airport master plans determine the size of "critical" aircraft anticipated to operate at an airport in the future, then design the runway system to accommodate the critical aircraft. Runways at commercial airports are designed by using aircraft manufacturer's specifications. Figure 4.21 is a table showing runway lengths for airport design issued by Boeing for the 757-200 aircraft. These specifications provide length of runway required for aircraft based on payload, temperature, and elevation. In general, the higher the temperature, elevation, and payload weight, the longer the runway needs to be to accommodate the aircraft (figure 4.21).

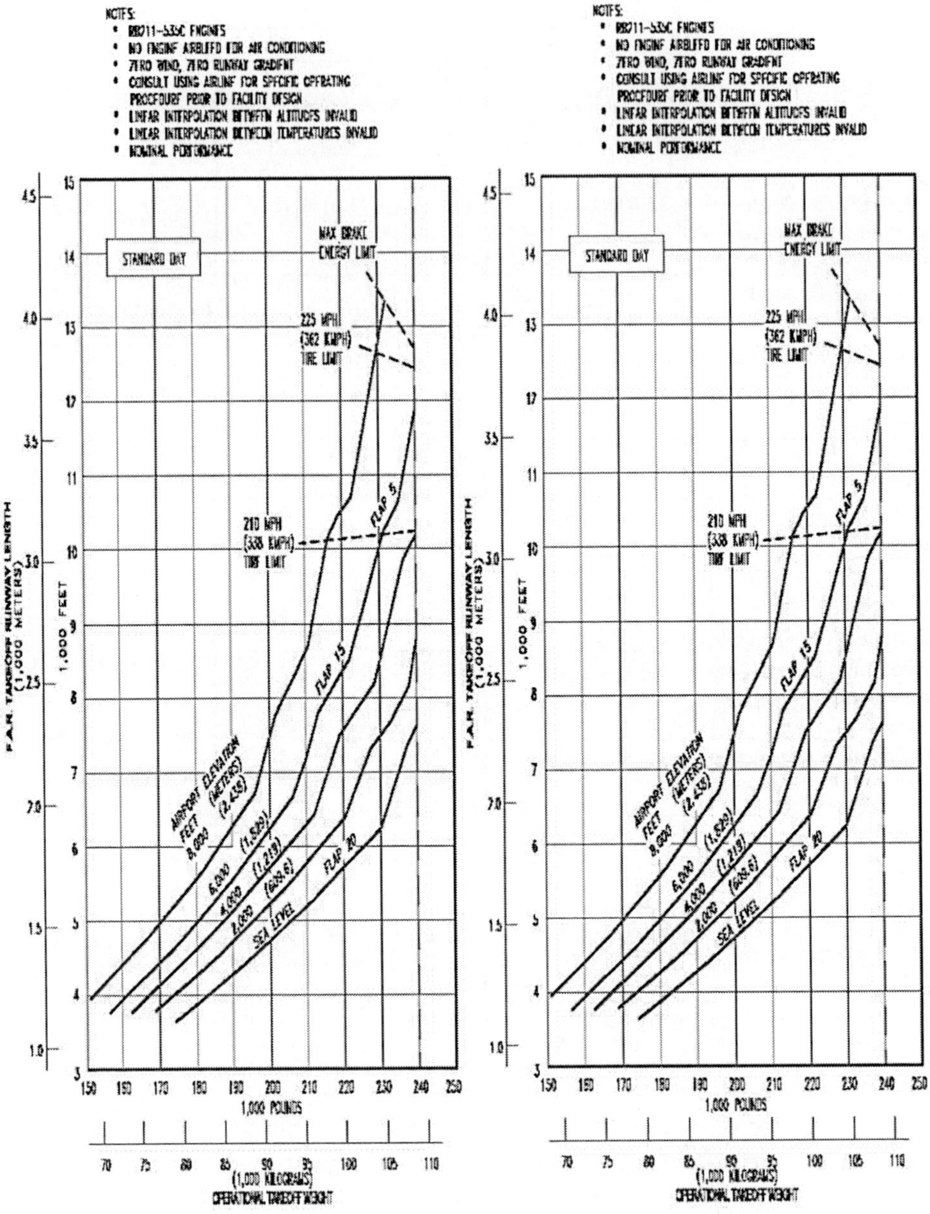

Source: The Boeing Company, 2002.

Figure 4.21. Boeing 757-200 takeoff runway requirements for design purposes

Commercial airliners offer versatility in their ability to operate at a wide assortment of airports throughout the world. Large wide-body aircraft such as the Boeing 747 are designed to seat over 300 passengers and operate at international gateway airports such as Houston, whereas narrow-body aircraft designed for medium-sized markets seat 100 to 200 passengers and serve markets such as Tallahassee, FL, and Baton Rouge, LA. Regional jets seat 34 to 70 passengers and serve markets such as Lake Charles Regional Airport in Louisiana.

Airport master plans determine the size of "critical" aircraft anticipated to operate at an airport in the future, then design the runway system to accommodate these critical aircraft. Unlike general aviation airports that rely on the FAA airport design software to calculate runway length requirements, runways at commercial airports are designed by using aircraft manufacturer's specifications. Once airports go into service, it is the pilot's responsibility to calculate aircraft performance on a given day prior to takeoff based on the following: ambient temperature, aircraft gross takeoff weight (GTW), airfield elevation, wind velocity and direction, and runway surface slope and drag. Thus, on hot days the pilot can make adjustments in cargo or passenger loads in order to takeoff on a runway, given its length. On days when the temperature is higher than the aircraft specs contemplate, the airliner would need to lower its weight to accommodate the higher temperatures.

Table 4.16 lists the required runway lengths for three groups of aircraft, fully loaded, for a generic hot day (a standard day temperature of 15 °C (59 °F) plus 15 °C (27 °F), for a total of 30 °C (86 °F)) and compares the manufacturer's specifications with the primary runway lengths of the 11 commercial service airports in the study area. Shortfalls in runway length for specific aircraft are presented in italics. Houston Bush Intercontinental (IAH) is the fourth largest market in the United States and is the only international gateway airport in the study area. Other airports in the study area do not require the same runway lengths since wide-body aircraft do not operate at these airports on a scheduled basis. On the opposite end of the spectrum, regional jets typically operate at Lake Charles Regional (LCH), Hattiesburg (HBG), and Beaumont/Port Arthur (BPT). These airports are designed to accommodate regional jets and turboprop aircraft and have shorter runway lengths. The other commercial airports in the study area are designed to accommodate medium-haul, narrow-body jets.

As shown in the discussion above, the maximum temperature contemplated by this study is 39 °C (102.2 °F), which is 33 °C (91.4 °F) plus 6 °C (10.8 °F), based on scenario A2 for the year 2100. This maximum temperature is 9 °C (16.2 °F) higher than the generic hot day. Therefore, aircraft manufacturers may want to consider the extent to which the use of a standard day temperature of 15°C (59°F) plus 15°C (27°F) as a measure of a typical hot day will continue to be applicable for aircraft design or whether to increase this temperature based on any projected temperature increase associated with a change in climate.

Temperature Conclusions

As is the case today, pilots will need to address how temperature increases may affect aircraft takeoff performance capabilities and payload requirements, and airports will need to address any such increases in the context of current runway utilization and future runway design. Given past trends, it is likely that future aircraft will be able to operate on shorter runways. Airports serving large commercial aircraft in the future, however, are anticipated to continue to utilize aircraft manufacturer's specifications to determine runway lengths.

Precipitation

In general, airlines, airports, and aircraft operate more efficiently in dry weather conditions than wet. Weather is a critical influence on aircraft performance and the outcome of the flight operations while taking off, landing, and while aloft. Precipitation affects aircraft and airports in several ways such as decreasing visibility, slowing air traffic by requiring greater separation between aircraft, and decreasing braking effectiveness. On the ground, effects include creating turbulence, increasing the risk of icing of wings, and affecting engine thrust.

The climate scenarios for the years 2050 and 2100 developed as part of this research generally indicate that the Gulf Coast study area could become a warmer but drier climate. However, the models do indicate the possibility that the climate could be warmer with increased annual precipitation. In either scenario, the increased intensity of individual rainfall events is likely.

Implications of a drier climate to airport and aircraft operations may include positive and negative effects. Less precipitation would most likely reduce aircraft and air traffic delays; reduce periods of wet surfaces on runways, taxiways, and aprons; and in the winter months, reduce the risk of wing icing. A drier climate also may increase the number of days of visual flight rules[11] (VFR) operations. A warmer climate with less precipitation may, however, increase convective weather (turbulence), as well as increase the number and severity of thunderstorms. In addition, increased water vapor in the atmosphere, particularly during the summer months, may increase haze and reduce pilot visibility, thereby reducing the number of VFR days.

A wetter climate would reduce the number of VFR-operating time periods and would impact the general aviation sector. General aviation pilots would either learn to fly in instrument flight rules (IFR) conditions by becoming "instrument rated" or not fly during periods of reduced visibility and precipitation. In order for pilots to fly in IFR conditions, aircraft flight decks must be equipped with complex navigation instruments, which is a significant investment for aircraft owners.

Increased extreme precipitation events also would impact commercial service aircraft operations. During severe thunderstorm activity it is not unusual for an airline to cancel flights or at a minimum experience delays in operations. Navigation in heavy precipitation is possible and currently occurs on a daily basis in the national air system. However, precipitation almost always creates delays, particularly at the most congested airports.

If the Gulf Coast study area climate proves to have more intense precipitation events, airport planners and engineers would need to consider the implications of periods of increased heavy rainfall in airport design and engineering. This is particularly true of airports located on floodplains in the study area since they are more susceptible to flash flood events. Eight of the 61 airports in the study area are located on 100-year floodplains. These airports are identified in table 4.17.

Relative Sea Level Rise

As indicated in chapter 3.0, RSLR scenarios developed as part of this research indicate that coastal zones in the Gulf Coast study area are very likely to be inundated by rising sea level combined with geologic subsidence. As a result, some airport infrastructure would most likely be susceptible to erosion and flooding.

Table 4.16. Commercial Aircraft Runway Length Takeoff Requirements

Aircraft Group	Aircraft Type[1]	Required Runway Length[2]	Commercial Service Airport Primary Runway Lengths (m)										
			EFD 2,744	IAH 3,658	HOU 2,317	BPT 2,057	MSY 3,080	LFT 2,332	BTR 2,135	LCH 1,981	MOB 2,597	GPT 2,744	HBG 1,859
A. Measured in Meters													
Wide-Body	747-400	3,170	*-426*	488	*-853*	*-1,113*	*-90*	*-838*	*-1,035*	*-1,189*	*-573*	*-426*	*-1,311*
	MD 11	3,597	*-853*	61	*-1,280*	*-1,539*	*-517*	*-1,265*	*-1,462*	*-1,615*	*-999*	*-853*	*-1,738*
	777-200LR	3,505	*-762*	153	*-1,188*	*-1,448*	*-426*	*-1,173*	*-1,370*	*-1,524*	*-908*	*-761*	*-1,646*
Medium-Haul[3] Narrow Body	737-900	2,652	92	1,006	*-335*	*-594*	428	*-320*	*-517*	*-671*	*-55*	92	*-793*
	DC-9-15	2,499	244	1,159	*-182*	*-442*	580	*-167*	*-365*	*-518*	98	244	*-640*
	737-800	2,225	518	1,433	92	*-168*	855	107	*-90*	*-244*	372	519	*-366*
	MD-80	2,195	549	1,463	123	*-137*	885	137	*-60*	*-213*	403	549	*-336*
	737-300	2,012	732	1,646	305	46	1,068	320	123	*-30*	586	732	*-153*
	A300-600	1,981	762	1,677	336	76	1,098	351	154	0	616	763	*-122*
	737-500	1,920	823	1,738	397	137	1,159	412	215	61	677	824	*-61*
	A319	1,859	884	1,799	458	198	1,220	473	276	122	738	885	*-0.3*
	757-200	1,829	915	1,829	488	229	1,251	503	306	152	768	915	30
	737-600	1,768	976	1,890	549	290	1,312	564	367	213	829	976	91
Regional Jets and Turboprops	ERJ 145	1,951	793	1,707	366	107	1,129	381	184	30	646	793	*-92*
	ERJ 135	1,951	793	1,707	366	107	1,129	381	184	30	646	793	*-92*
	CRJ	1,829	915	1,829	488	229	1,251	503	306	152	768	915	30
	DASH8-300	1,554	1,189	2,103	763	503	1,525	778	580	427	1,043	1,189	304
Aircraft Group	**Aircraft Type[1]**	**Required Runway Length[2]**	**Commercial Service Airport Primary Runway Lengths (Ft)**										
			EFD 9,001	**IAH 12,001**	**HOU 7,602**	**BPT 6,750**	**MSY 10,104**	**LFT 7,651**	**BTR 7,004**	**LCH 6,500**	**MOB 8,521**	**GPT 9,002**	**HBG 6,099**
B. Measured in Feet													
Wide-Body	747-400	10,400	*-1,399*	1,601	*-2,798*	*-3,650*	*-296*	*-2,749*	*-3,396*	*-3,900*	*-1,879*	*-1,398*	*-4,301*
	MD 11	11,800	*-2,799*	201	*-4,198*	*-5,050*	*-1,696*	*-4,149*	*-4,796*	*-5,300*	*-3,279*	*-2,798*	*-5,701*
	777-200LR	11,500	*-2,499*	501	*-3,898*	*-4,750*	*-1,396*	*-3,849*	*-4,496*	*-5,000*	*-2,979*	*-2,498*	*-5,401*
Medium-Haul[3] Narrow Body	737-900	8,700	302	3,301	*-1,098*	*-1,950*	1,404	*-1,049*	*-1,696*	*-2,200*	*-179*	302	*-2,601*
	DC-9-15	8,200	801	3,801	*-598*	*-1,450*	1,904	*-549*	*-1,196*	*-1,700*	321	802	*-2,101*
	737-800	7,300	1,701	4,701	302	*-550*	2,804	351	*-296*	*-800*	1,221	1,702	*-1,201*
	MD-80	7,200	1,801	4,801	402	*-450*	2,904	451	*-196*	*-700*	1,321	1,802	*-1,101*

Table 4.16. (Continued)

Aircraft Group	Aircraft Type[1]	Required Runway Length[2]	Commercial Service Airport Primary Runway Lengths (Ft)										
			EFD 9,001	IAH 12,001	HOU 7,602	BPT 6,750	MSY 10,104	LFT 7,651	BTR 7,004	LCH 6,500	MOB 8,521	GPT 9,002	HBG 6,099
	737-300	6,600	2,401	5,401	1,002	150	3,504	1,051	404	*-100*	1,921	2,402	*-501*
	A300-600	6,500	2,501	5,501	1,102	250	3,604	1,151	504	0	2,021	2,502	*-401*
	737-500	6,300	2,701	5,701	1,302	450	3,804	1,351	704	200	2,221	2,702	*-201*
	A319	6,100	2,901	5,901	1,502	650	4,004	1,551	904	400	2,421	2,902	*-1*
	757-200	6,000	3,001	6,001	1,602	750	4,104	1,651	1,004	500	2,521	3,002	99
	737-600	5,800	3,201	6,201	1,802	950	4,304	1,851	1,204	700	2,721	3,202	299
Regional Jets and Turboprops	ERJ 145	6,400	2,601	5,601	1,202	350	3,704	1,251	604	100	2,121	2,602	*-301*
	ERJ 135	6,400	2,601	5,601	1,202	350	3,704	1,251	604	100	2,121	2,602	*-301*
	CRJ	6,000	3,001	6,001	1,602	750	4,104	1,651	1,004	500	2,521	3,002	99
	DASH8-300	5,100	3,901	6,901	2,502	1,650	5,004	2,551	1,904	1,400	3,421	3,902	999

[1] MD 11 aircraft runway length based on standard day +18°C (33°F). All other aircraft based on standard day +15°C (27°F).

[2] Assumes all elevations at sea level.

[3] Medium-Haul are aircraft weights for 800 miles of fuel on-board.

EFD	Houston Ellington Field	MSY	New Orleans International	MOB	Mobile Regional
IAH	Houston Intercontinental	LFT	Lafayette Regional	GPT	Gulfport Biloxi
HOU	Houston Hobby	BTR	Baton Rouge Metropolitan	HBG	Hattiesburg Regional
BPT	Beaumont/Port Arthur Regional	LCH	Lake Charles Regional		

Table 4.17. Airports Located on 100-Year Flood Plains

Associated City	State	Airport Name
Gonzales	LA	Louisiana Regional
Sulphur	LA	Southland Field
Galliano	LA	South Lafourche
New Orleans	LA	Lakefront
Reserve	LA	St. John The Baptist Parish
Thibodaux	LA	Thibodaux Municipal
Winnie/Stowell	TX	Chambers County-Winnie Stowell
Galveston	TX	Scholes International at Galveston

Sources: Wilbur SMITH Ssociates; USGS.

Table 4.18. Gulf Coast Study Area Airports Vulnerable to Submersion by Relative Sea Level Rise of 61 to 122 CM (2 to 4 ft)

State	Associated City	Airport Name	Airport Type	Elevation in Feet
Louisiana	Galliano	South LaFourche	GA	1
Louisiana	New Orleans	New Orleans NAS JRB	MIL	3
Louisiana	New Orleans	Louis Armstrong-New Orleans International	CS	4

Geographic information system (GIS) analysis indicates three airports in the study area would be below mean sea level (MSL) if relative sea level increases by 122 cm (4 ft). Each of these airports currently is protected by preventive infrastructure such as dikes and levees, which will need to be maintained. If feeder roads in the area are inundated, however, access to these airports may be disrupted. Table 4.18 lists these airports and their elevations. All three airports are located in Louisiana and range from New Orleans International (122 cm or 4 ft elevation), one of the study area's large commercial service airports, to South LaFourche (30 cm or 1 ft), a very small general aviation facility. The third is a military airport, New Orleans Naval Air Station Joint Reserve Base (NASJRB New Orleans) (91 cm or 3 ft).

Storm Activity

Both storm surge and hurricane force winds can damage airport facilities. As indicated in chapter 3.0, the study team analyzed the vulnerability of facilities to storm surge heights of 5.5 and 7.0 m (18 and 23 ft). At these elevations a variety of airports in the region would be vulnerable to the impacts of storm surge, though this depends on the specific characteristics of each individual storm event, including landfall location, wind speed, wind direction, tidal conditions, etc.

Figure 4.22 depicts airports within the study that are vulnerable to storm surges of 5.5 or 7 m (18 or 23 ft). Table 4.19 lists these airports by location, type, and elevation. There are 22 airports in the 0- to 5.5-m (18-ft) MSL category and seven airports in the 5.8- to 7-m (19- to 23-ft) MSL category. This list includes some major airports in the region, such as New Orleans International. Also, the commercial service airport in Lake Charles, LA, would be vulnerable. See section 4.3.1 for a discussion of the wind impacts of the 2005 hurricanes on airport facilities.

4.2.6. Pipelines

There is a combined total of 42,520 km (26,427 mi) of onshore liquid (oil and petroleum product) transmission and natural gas transmission pipelines in the Gulf Coast area of study, as shown in figure 4.23.[12] This includes 22,913 km (14,241 mi) of onshore natural gas transmission pipelines and 19,607 km or 12,186 mi of onshore hazardous liquid pipelines (Pipeline and Hazardous Materials Safety Administration [PHMSA], 2007). This region is essential to the distribution of the Nation's energy supply through pipeline transportation, and historically the landside pipelines have been relatively secure from disruption by increased storm activity and intensity. A number of risks and vulnerabilities to climate-related impacts have been revealed, however, particularly for submerged or very low elevation pipelines. PHMSA of the U.S. DOT has jurisdiction over onshore pipeline facilities and some offshore pipeline facilities. PHMSA has jurisdiction over offshore pipeline facilities that are exposed or are hazards to navigation when the offshore pipeline facilities are between the mean watermark and the point where the subsurface is under 4.6 m (15 ft) of water as measured from mean low water. The U.S. Department of the Interior Minerals Management Service (MMS) has jurisdiction over about 36,000 miles of offshore pipelines in the Gulf of Mexico.

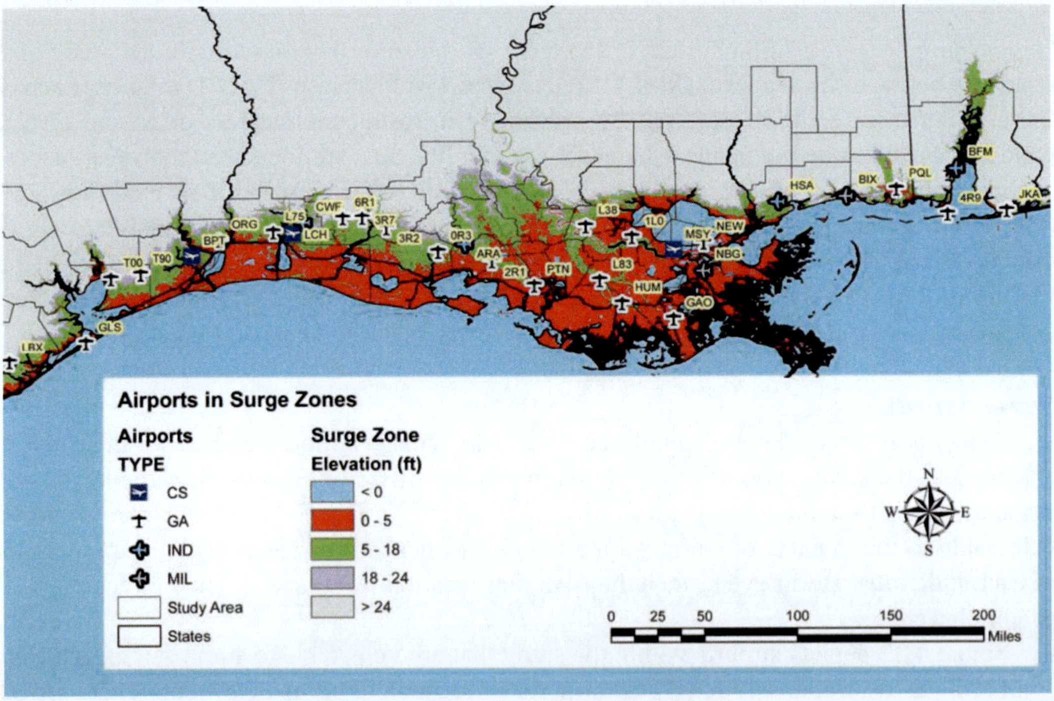

Source: Cambridge Systematics analysis of U.S. DOT and USGS data.

Figure 4.22. Gulf Coast study area airports at risk from storm surge

Table 4.19. Gulf Coast Study Area Airports Vulnerable to Storm Surge

State	Associated City	Airport Name	Airport Type	Elevation m (ft)
Airports 0 to 5.5 m (18 ft) Elevation				
Alabama	Gulf Shores	Jack Edwards	General Aviation	4.9 (16)
Alabama	Mobile	Dauphin Island Airport	General Aviation	1.5 (5)
Louisiana	Abbeville	Abbeville Chris Crusta Memorial	General Aviation	4.6 (15)
Louisiana	Crowley	Le Gros Memorial	General Aviation	5.2 (17)
Louisiana	Galliano	South LaFourche	General Aviation	0.3 (1)
Louisiana	Gonzales	Louisiana Regional	General Aviation	4.6 (15)
Louisiana	Houma	Houma-Terrebonne	General Aviation	3.0 (10)
Louisiana	Jeanerette	Le Maire Memorial	General Aviation	4.3 (14)
Louisiana	Lake Charles	Lake Charles Regional	Commercial Services	4.6 (15)
Louisiana	Lake Charles	Chennault International	Industrial	5.2 (17)
Louisiana	New Orleans	New Orleans NAS JRB	Military	0.9 (3)
Louisiana	New Orleans	Louis Armstrong-New Orleans International	Commercial Services	1.2 (4)
Louisiana	New Orleans	Lakefront	General Aviation	2.4 (8)
Louisiana	Patterson	Harry P. Williams Memorial	General Aviation	2.7 (9)
Louisiana	Reserve	St. John The Baptist Parish	General Aviation	2.1 (7)
Louisiana	Sulphur	Southland Field	General Aviation	3.4 (11)
Louisiana	Thibodaux	Thibodaux Municipal	General Aviation	2.7 (9)
Louisiana	Welsh	Welsh	General Aviation	5.5 (18)
Mississippi	Pascagoula	Trent Lott International	General Aviation	5.2 (17)
Texas	Beaumont/Port Arthur	Southeast Texas Regional	General Aviation	4.6 (15)
Texas	Galveston	Scholes International at Galveston	General Aviation	1.8 (6)
Texas	Orange	Orange County	General Aviation	4.0 (13)
Airports 5.6 – 7.0 m (19 to 23 ft) Elevation				
Alabama	Mobile	Mobile Downtown	Industrial	5.8 (19)
Louisiana	Iberia	Acadiana Regional	Industrial	6.1 (20)
Louisiana	Jefferson Davis	Jennings	General Aviation	6.1 (20)
Mississippi	Hancock	Stennis International	Industrial	7.0 (23)
Mississippi	Harrison	Keesler AFB	Military	6.1 (20)
Texas	Brazoria	Brazoria County	General Aviation	6.7 (22)
Texas	Chambers	Chambers County-Winnie Stowell	General Aviation	6.4 (21)

Sources: U.S. DOT FAA Records, April 2006; FEMA Storm Inundation Data.

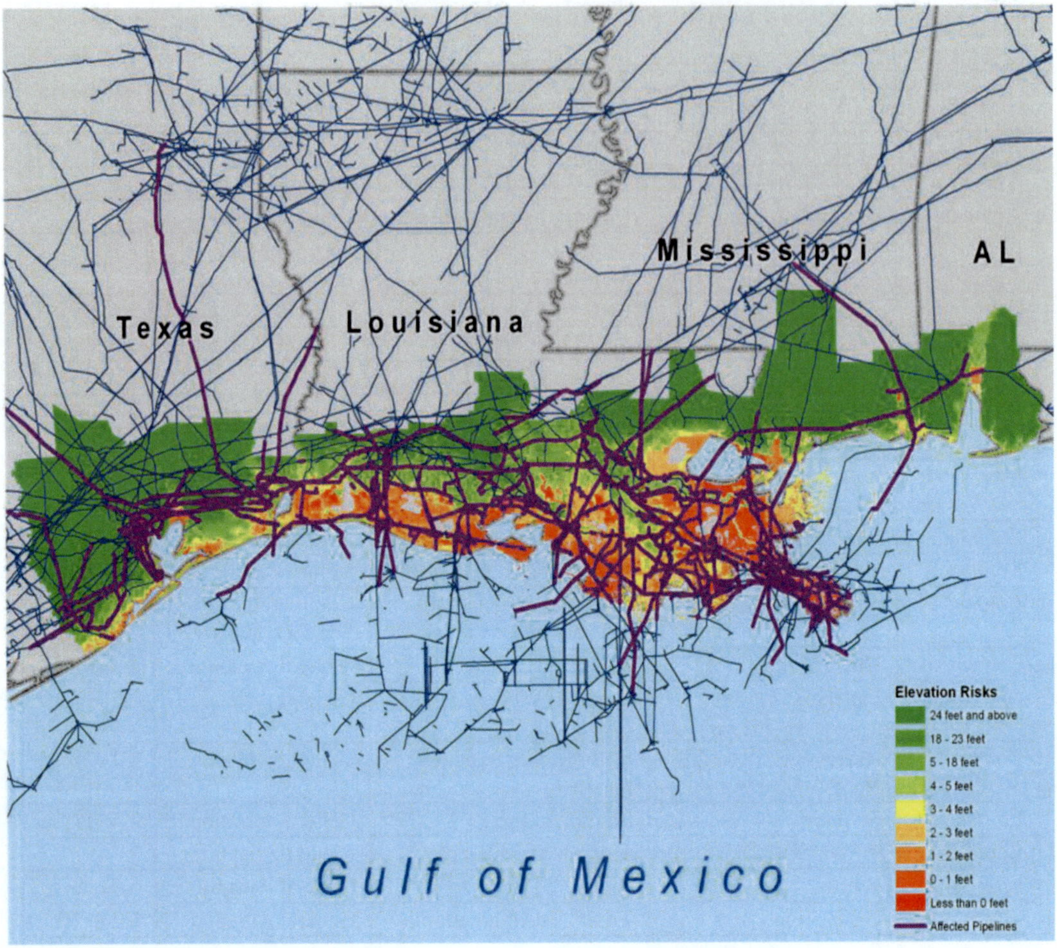

Source: Texas Transportation Institute.

Figure 4.23. Landside pipelines having at least one GIS link located in an area of elevation 0 to 91 cm (3 ft) above sea level in the study area

Some historical weather events have resulted in only minor impacts on pipelines, with the notable exceptions of Hurricanes Andrew, Ivan, Katrina and Rita, which caused fairly extensive damage to underwater pipelines, and flooded distribution lines in areas where houses were destroyed. Storm surge and high winds historically have not had much impact on pipelines – either onshore transmission lines or offshore pipelines – since they are strong structures, well-stabilized, and/or buried underground. Yet offshore pipelines have been damaged in relatively large numbers on occasion, as during Hurricanes Andrew and Ivan. Temperature shifts resulting from climate scenario projections are not expected to have much direct or indirect impact on pipelines. Increases or decreases in precipitation – either long-term or in the frequency or extent of droughts or inundation – could impact soil structure. Sea level rise would likely have little direct effect but could affect water tables, soil stability, and the vulnerability of pipelines to normal wave action as well as sea surge.

Changes in soil structure, stability, and subsidence – whether undersea, landside, or in wetlands or transition elevations – could play an important role in pipeline-related risks. However, there is little information on this topic outside of earthquake risks. There has recently been concern about how wave action could affect the seabed, either by liquefying/destabilizing the sand or silt surface above a buried pipeline or by gradually eroding away seabed that had been covering the pipeline. It is unclear at present whether a changing climate might lead to conditions that exacerbate these effects and cause additional damage.

The possible effects on pipelines from climate change – storm surge and extreme winds, temperature shifts, precipitation changes, and sea level rise – were considered in this analysis. Both pipeline companies and governmental agencies have considered pipeline risks, vulnerability, and safety and have well-developed inspection, maintenance, and response plans. However, these plans do not appear to address a number of risks that may be arising. This study did not examine the adequacy of those plans. While some issues regarding impacts have been addressed here, there is still significant uncertainty about the overall risk to pipelines from climate change.

Importance of Pipeline Operations in the Study Area

Onshore natural gas transmission pipelines are primarily located in Louisiana. Approximately 49 percent of natural gas wellhead production either occurs near the Henry Hub, which is the centralized point for natural gas futures trading in the United States or passes close to the Henry Hub as it moves to downstream consumption markets. The Henry Hub – located near the town of Erath in Vermillion Parish, north central Louisiana – interconnects nine interstate and four intrastate pipelines, including: Acadian, Columbia Gulf, Dow, Equitable (Jefferson Island), Koch Gateway, Louisiana Resources Company (LRC), Natural Gas Pipe Line, Sea Robin, Southern Natural, Texas Gas, Transco, Trunkline, and Sabine's mainline.

Temperature

Pipelines are more protected than other types of infrastructure from the effects of temperature changes projected for the region due to the moderating and insulating effects of soil and water. The great majority of the transmission pipeline system is buried under at least 91 cm (3 ft) of soil cover, both onshore and offshore. Federal regulations require that all pipelines in navigable waters be buried. Moreover, pipelines are designed to carry product at significant temperature variations (natural gas is pressurized in their system, while petroleum products are heated considerably above ambient temperatures). There is extensive experience with pipelines in much more extreme ambient temperature conditions (Alaska, Saudi Arabia, West Africa) than would be expected in the Gulf States region. Sea temperatures will vary even less than land temperatures. Thus, there is not expected to be any significant effect on pipelines due to direct effects from increased (or decreased) temperatures.

Precipitation Changes

Sustained periods of increases or decreases in precipitation, whether over months or the cumulative effect across years, can cause substantial soil changes due to drought or saturation. Changes in water tables may occur both from local climate changes as well as from global effects such as sea level rise. An increase in water table level or increased surface water runoff can cause erosion or slumping (collapse) of the soil surface, thereby leading to potential for pipeline exposure.

In the lowland and marsh areas particularly associated with the coastal regions of Louisiana, the soil is being washed away due to storm activity. With the disappearance of the soil, the pipelines in these regions are losing cover.

Detailed analysis of geology and pipeline-specific conditions are required to draw more precise conclusions regarding the potential for serious disruption of the transmission pipeline system from climate-related soil changes. Nonetheless, this is an area of concern, because a considerable and unpredictable portion of the pipeline system could be vulnerable to these climate change and sea level induced impacts.

Another vulnerability is from expected short-term changes (such as torrents and floods), where significant change in water flow rate and water flow energy are a result of increased precipitation. Risk analysis of the impacts of extreme events is required to determine appropriate adaptation or mitigation actions.

Storm Impact Preparation, Mitigation, and Response

Wave action during storms may impact pipelines. For offshore pipelines, in instances where significant subsidence occurs and the pipeline segment is exposed, that section is exposed to wave action. High-energy waves may subject a pipeline to stress levels it was not designed to withstand, causing a fracture. An exposed offshore pipeline also could be vulnerable to lateral and vertical displacement, exposure to vessel traffic and fishing trawls, or rupture by currents, which may be very important in this context.

Pipeline operating companies are required to have an emergency plan in place to cover all known or expected situations that may require response to repair the pipeline system due to damage, including, storms, excavation, and even sabotage or terrorist attack. Pipeline systems are segregated by sections between valves in order to allow isolation and shutdown of segments for routine maintenance, malfunctions, or response to emergency incidents. During shutdowns, pipelines are pressurized at the value given in the emergency plan design, and all valves are closed to prevent flow problems during the rest of the event and to facilitate repair. The operating portion of the emergency plan generally presumes that operations will be interrupted for a period of only up to 10 days.

As soon as a storm or other event dissipates in the pipeline area of concern, the pipeline response team initiates their response plan. An inspection begins as a visual flyover the pipeline in the affected region to examine it for exposure or other obvious indications of damage. Some damage also can be detected through sensors measuring pressure in different pipeline segments or through other physical indicators, although these approaches may miss some damage (e.g., structural damage not yet causing a leak) or not be able to isolate damage location more precisely. If damage is apparent, then a thorough closeup inspection will take place, including divers if necessary. After damage has been identified, a repair team initiates repairs.

The damages caused to pipelines by Hurricanes Andrew and Ivan were severe and fairly widespread throughout the storm front region, as documented by the MMS study discussed below (Skinner, 2006). After Ivan, oil refineries had ample products to supply, but the pipelines could not deliver due to damages. In contrast, damage to pipelines from Katrina and Rita was relatively minor; most pipelines were ready to take product but were hampered by the lack of available product due to refinery damage and/or power shortages.

One representative of a hazardous liquid pipeline company stated that, prior to Ivan, obtaining pipeline maintenance and repair contract commitments was relatively easy, "a foregone conclusion of commitment," but after Katrina and Rita, it has become increasingly difficult to obtain solid commitments from suppliers to respond to emergency calls. While suppliers are still offering contracts, the commitment is now only an offering to put the customer on a response list for a front-end fee. There is no longer a guarantee that the supplier will respond to an emergency call within a fixed time period or otherwise provide service, because all their assets and personnel may be engaged in a prior commitment.

Response capabilities and reliability have thus declined, even while the acknowledged storm threat has increased due to Ivan's illustration of a previously unknown level of damage. Furthermore, while there were emergency operating plans before Ivan that matched the committed response time, we now know that responses may take longer, and operating plans will need to be adapted to meet these eventualities.

Hurricane Damage Studies

One of the more substantial studies of hurricane damage to pipelines in the Gulf of Mexico was done by Det Norske Veritas (DNV) Technology Services upon a request from the MMS (Skinner, 2006). This was an assessment of damage to the Gulf of Mexico offshore pipelines resulting from the passage of Hurricane Ivan in September 2004. The DNV study also summarized the impacts of Hurricanes Andrew, Lili, Katrina, and Rita.

Hurricane Ivan reached Category 5 strength three times and was a Category 3 hurricane when it made landfall. Hurricane Ivan resulted in approximately 168 pipeline damage reports, although the vast majority of Gulf of Mexico offshore pipelines performed well during its passage. According to the MMS DNV report, the impact on the environment from pipeline spills was minimal. The majority of pipeline damage occurred at or near platform interfaces, in areas of mudflows, or as a result of an indirect hurricane impact, such as platform failure or anchor dragging. Localized failures at pipeline crossings and excessive movements in shallow water depths indicate that more hurricane-resistant design considerations might be needed on a site-specific basis but do not warrant industry-wide design or construction code revisions. The report suggests that design assumptions used for shallow water pipelines need to be evaluated in areas dominated by silty soils, particularly where self-burial is intended as the method of installation.

Hurricane Andrew passed through the Gulf of Mexico in August 1992 as a Category 4 storm. It damaged more than 480 pipelines and flow lines. Prior to Hurricane Andrew, minimal damage to pipelines had been experienced as a result of passing hurricanes, with combined pipeline failures from hurricanes for the period of 1971 through 1988 resulting in about 100 damage reports. Most of the pipeline failures were in depths less than 30 m (100 ft) of water.

Hurricane Lili was a Category 4 storm offshore in the Gulf of Mexico and was downgraded to a Category 2 hurricane at landfall in October 2002. There were 120 pipeline damage reports to the MMS following Hurricane Lili. The majority of the pipeline failures in Lili occurred in small diameter pipelines with no apparent correlation for age, which also was true for damages reported from Andrew.

According to the MMS, there were 457 offshore oil and gas pipelines that were damaged as a result of Hurricanes Katrina and Rita (MMS Press Release, 2006). Most of the damage was relatively minor. Disruptions also occurred due to power outages, and pipeline operators procured portable electric power generators necessary to resume operations. The closure of major pipelines originating in the Gulf of Mexico region in the wake of both storms served to exacerbate the petroleum product supply situation (EIA, 2005).

Storm Activity: Erosion

The above information and an interview with a hazardous liquids (mostly petroleum products) pipeline company revealed that damage from erosion and soil stability due to storm wave action has focused new interest on this phenomenon. The results of Hurricane Ivan, when erosion occurred in waters up to 76 m (250 ft) in depth, demonstrated that this effect can occur at depths previously considered impervious. The problem and solution is still being investigated in joint industry programs, along with Office of Pipeline Safety and MMS.

Erosion typically has been found to occur in what the industry has termed "ultra shallow waters." This phenomenon was prevalent from Hurricane Ivan (Skinner, 2006) but almost completely lacking from Hurricanes Katrina and Rita. This indicates that risk is not only due to storm intensity but also may be based on more complex meteorological and fluid dynamics factors, making the risk less predictable than assumed.

In ultra shallow waters where erosion occurs, the general concurrence of industry specialists is that the seabed is "liquefying" (when the sand or silt shifts from a wet solid to a suspended state) in certain wave-action conditions. Pipeline design incorporates negative buoyancy (a present regulatory requirement and previously considered good design practice), but, if the sands are liquefying, the negative buoyancy may become positive and the pipeline ends up on the seabed surface. Documentation shows that the seabed level has not changed in these occurrences, but the pipeline has changed its elevation from 0.9 m (3 ft) below the seabed surface to resting on the surface. While possible solutions are being developed, impacts from more frequent or severe storms currently can be considered a vulnerability of the pipeline system (EIA, 2005).

Storm Activity: Increased Storm Severity

In the Gulf Coast area of study, transmission pipelines have been designed to maintain their integrity for a (historical) 100-year storm event. Interviews with natural gas transmission pipeline company representatives indicate that the potential of pipeline damage due to increased storm activity or increased severity of storms is considered to be of marginal concern. They framed the issue as: to what extent can increased weather damage be effectively planned for, and what level of risk exposure should be assumed, beyond regulatory requirements? While there is an extensive regular inspection process that may identify weaknesses that could be expanded by a storm or by more gradual soil structure changes, it only partially prepares for and mitigates potential storm damage.

Discussions regarding the potential for transmission pipeline damage consistently centered on the issue that nearly all the transmission lines are buried with 0.9 m (3 ft) of top cover, more in urban and populated areas, and they are regularly inspected for integrity. There is a need to better understand issues regarding damage to exposed pipeline portions (which may be the most vital), such as valves, pumping stations, etc., and damage to underground portions from previously unconsidered factors, such as changing water tables and soil subsidence due to sea level changes..

Researchers interviewed MMS regulatory officials in the New Orleans, LA, office regarding the effects they see concerning climate change. The offshore pipelines are regulated by MMS regarding design, construction, operations, and maintenance requirements. MMS representatives indicated that they do not anticipate increased storm severity and frequency will appreciably affect the pipelines under their regulatory authority in the Gulf of Mexico. (Note that MMS' authority ends at the State/Federal boundary offshore.[13]) It is unclear whether their comments took into account the changing soil structure and shoreline in the region. They based their comments on the fact that the subject pipelines are at substantial depth, and the pipelines are buried 0.9 m (3 ft) below sea floor level or anchored to piers designed to prevent pipeline movement on the sea floor. It is not certain how this accounts for the results of Hurricane Ivan and the findings of the DNV study.

Secondary Impacts

The level of oil and natural gas products moved via pipelines could be influenced by changes in temperature across the globe. Increases in temperature in the United States could affect the demand for energy products transported through the Gulf Coast; demand for natural gas (and coal) to power electricity plants in the southeast, for example, could lead to greater production and/or importation of natural gas through study region pipelines. Furthermore, climate mitigation policies designed to reduce carbon emissions could favor natural gas over other fossil fuels, thus promoting greater exploration and production of natural gas and importation of LNG, with clear implications for pipelines.

Further study is necessary before firm conclusions can be drawn regarding the vulnerability of onshore and offshore pipelines. Relatively significant damage has occasionally occurred, yet other storms have produced only minor damage. Recent investigations have raised concerns about seabed conditions under which pipelines exhibit some vulnerability. It is a matter of further research whether climate change will exacerbate those conditions or whether the interface between onshore and offshore pipelines might be affected.

4.2.7. Implications for Transportation Emergency Management

Without proactive planning, climate change could complicate evacuation efforts in the region. As noted above, some highways, the chief mode for evacuation, are very likely to be inundated permanently as relative sea level rises and periodically when areas are flooded by storms. Further, higher temperatures could make evacuations more problematic, particularly in situations where there is severe congestion; higher temperatures lead to greater air conditioning usage, making it more likely that vehicles will run out of fuel and block traffic. Large-scale emergency management is further challenged by the changing demographics of the region: an increasing percentage of residents are older and/or have special needs. Also, recent experience with evacuations suggests that congestion on key evacuation routes poses serious challenges to evacuating residents quickly. The need for interoperable communications systems across the region, currently lacking, will be heightened as the number of emergencies increases with climate change.

A robust emergency management system is highly dependent on the viability of the region's transportation infrastructure. Ensuring the capability to both evacuate residents, and move emergency responders and services into affected areas will require purposeful adaptation and, thus, focused investment in the transportation system. This section examines the implications for transportation emergency management of the potential impacts to highways and to transit and passenger rail that were presented earlier in this report. Many of these routes are expected to become increasingly vulnerable to higher sea levels and storm surge.

This section also highlights some of the lessons learned from recent hurricane evacuation experiences and examines some of the issues related to the varied – and often incompatible – communications system found across the region.

Further analysis and development of institutional consensus is necessary to more fully understand the implications of climate change on transportation emergency management. However, the preliminary vulnerability issues raised here are illustrative of the kinds of interactions that climate change and variability may cause for emergency management planning and operations. These issues are compounded by the changing demographics in the region.

There are two key types of emergency management/climate change scenarios. The first involves complications for emergency response activities given climate impacts. For example, unusable roads caused by higher sea levels could disrupt road connectivity, increasing the time needed for emergency response vehicles to reach fires, medical emergencies, etc. The second involves situations where the climate impact itself causes the emergency; e.g., where hurricane induced flooding or a sudden rise in relative sea levels forces people to evacuate a particular area.

Temperature

As discussed in chapter 3.0, both mean and extreme temperatures are very likely to increase in the Gulf Coast region over the next 50 to 100 years. The increase in temperatures could cause more air conditioning usage during some evacuations and could further diminish mobility. Vehicles using air conditioning during storm evacuations, particularly on congested roads, would contribute to roadside blockages as fuel is depleted and vehicles are abandoned. Furthermore, an increase in temperatures, especially maximum temperatures, coupled with a growing number of special needs residents in the Gulf Coast study area means that more lives could be vulnerable in the absence of electrical power and air conditioning in the aftermath of a storm.

Relative Sea Level Rise

As noted above, interstates and arterials tend to serve as the major evacuation routes for emergencies in the Gulf Coast study area. This substantial reliance on a single mode of transportation may endanger many people if the highway infrastructure is damaged or made inaccessible because of relative sea level rise. If the relative sea level increases such that portions of evacuation routes are under water then the essential connectivity and evacuation provided by those highways would be lost. This will be particularly important for large-scale evacuations dependent on east-west routes. Of course, as sea levels rise over time population centers may shift to higher elevations; the segments of evacuation routes that will be most critical are likely to change with these shifts in community locations. Furthermore, if the increase in relative sea level is gradual, infrastructure development would likely follow the movement of population centers.

As discussed in section 4.2.1, the majority of the highways vulnerable to a 61- and 122-cm (2- and 4-ft) rise in relative sea level are located in the Mississippi River Delta region near New Orleans. The most prominent vulnerable highways are I-10, with 220 km (137 miles) and U.S. Highway 90 with 235 km (146 mi) passing through areas likely to be below sea level with a 61-cm (2-ft) rise in relative sea level. Overall, 19 percent of the interstate miles and 20 percent of the arterial miles are at elevations below 61 cm (2 ft). With a 122 cm (4-ft) rise, the miles affected increase to 684 km (425 mi) of I-10 and 628 km (390 mi) of U.S. Highway 90. Overall, 24 percent of the interstate miles and 28 percent of the arterial miles currently are at elevations below 122 cm (4 ft).

Storm Activity

As noted in chapter 3.0, studies suggest that as radiative forcing (that is, greenhouse gas [GHG] concentrations) and sea surface temperatures continue to increase, hurricanes may be more likely to form in the Atlantic and Pacific and more likely to intensify in their destructive capacity. Storm surge disperses debris that blocks highways and makes many roads, including evacuation routes, impassable. In addition, storm surge may damage bridges and other structures, potentially compromising mobility for extended periods. While the actual highways that would be flooded and impacted by debris depends on the specific characteristics of any given storm, a substantial portion of the highway system is vulnerable to surge inundation, including roads in all four States in the study area. The areas that are potentially vulnerable to 5.5- and 7.0-m (18- and 23-ft) storm surge levels are shown in section 4.2.1 above. At the 5.5-m (18-ft) level, 51 percent of all arterial highways and 56 percent of the interstates in the study area are affected (figure 4.24). At 7.0 m (23 ft), these percentages rise to 57 percent of all highways and 64 percent of the interstates.

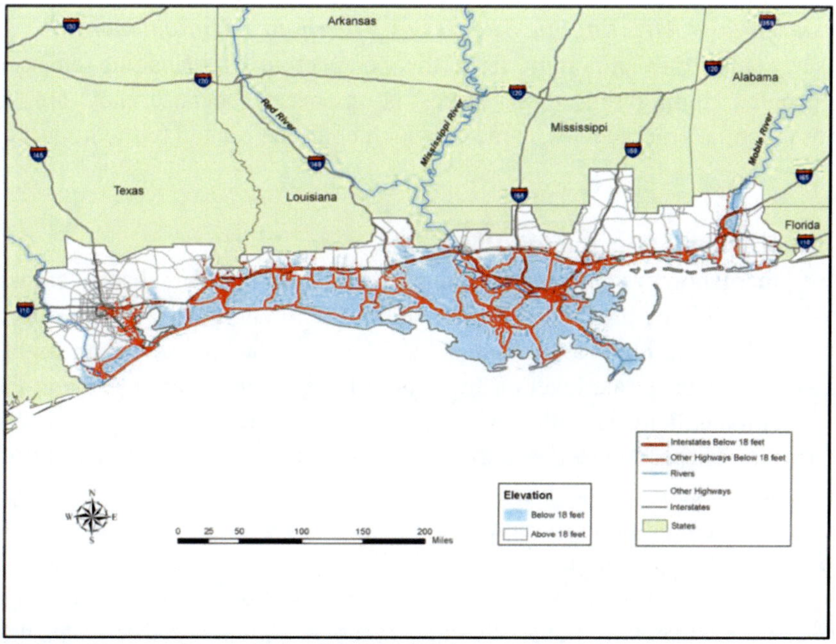

Source: Cambridge Systematics analysis of U.S. DOT data.

Figure 4.24. Evacuation route highways potentially vulnerable from storm surge of 5.5 m (18 ft)

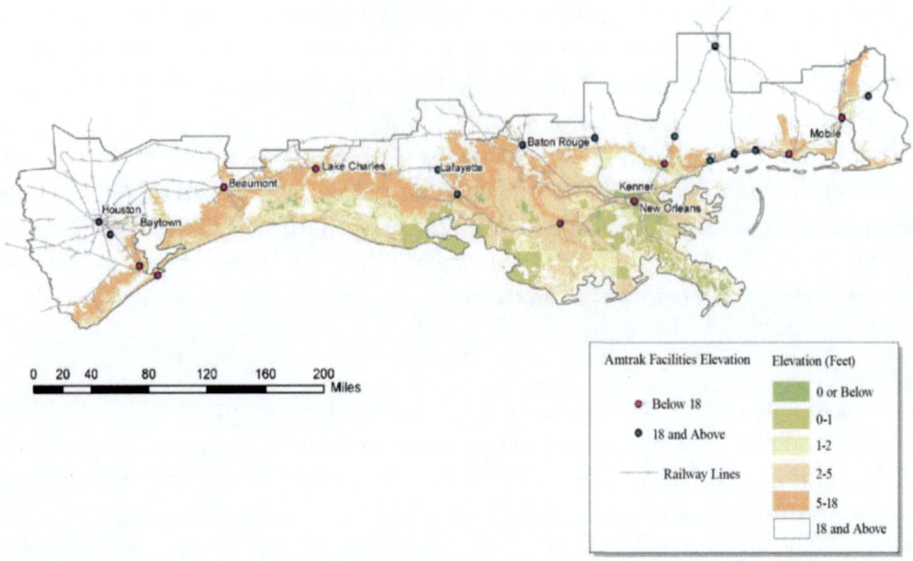

Source: Cambridge Systematics analysis of U.S. DOT data.

Figure 4.25. Risks to Amtrak facilities due to relative sea level rise and storm surge

Although not traditionally used for evacuation and emergency management purposes, railroads also could provide a transportation choice, especially for evacuees with special needs. Figure 4.25 illustrates the impacts on Amtrak facilities due to relative sea level rise and storm surge and identifies the Amtrak stations that are vulnerable to storm surge at the 5.5-m (18-ft) level.

Other Considerations Affecting the Success of Emergency Management

The issues below are important from the perspective of managing emergencies and protecting people. Highlighting these issues is important because they are relevant to preparing for potential emergencies, some of which could be related to the impacts of climate change.

Adapting Emergency Management Plans

Effective emergency evacuation plans must be living documents that incorporate current and anticipated conditions, procedures, and resources. Climate change will likely exacerbate the need to update these plans and procedures. The 2005 hurricane season highlighted the need to reassess the appropriate level of investment for emergency management planning. As discussed, the climate analysis indicates a rise in temperature and relative sea level for the Gulf Coast region. These changes – coupled with continued increases in overall population, and of particular concern, major increases in the elderly and special needs populations – translate into a difficult situation for emergency evacuations in the Gulf Coast region without thoughtful and proactive planning.

The requirement to transport those with special needs is especially challenging along the Gulf Coast, where many elderly people live in rural areas. Figure 4.26 illustrates the State and county/parish boundaries and the population over 65 that were impacted by Hurricane Katrina.

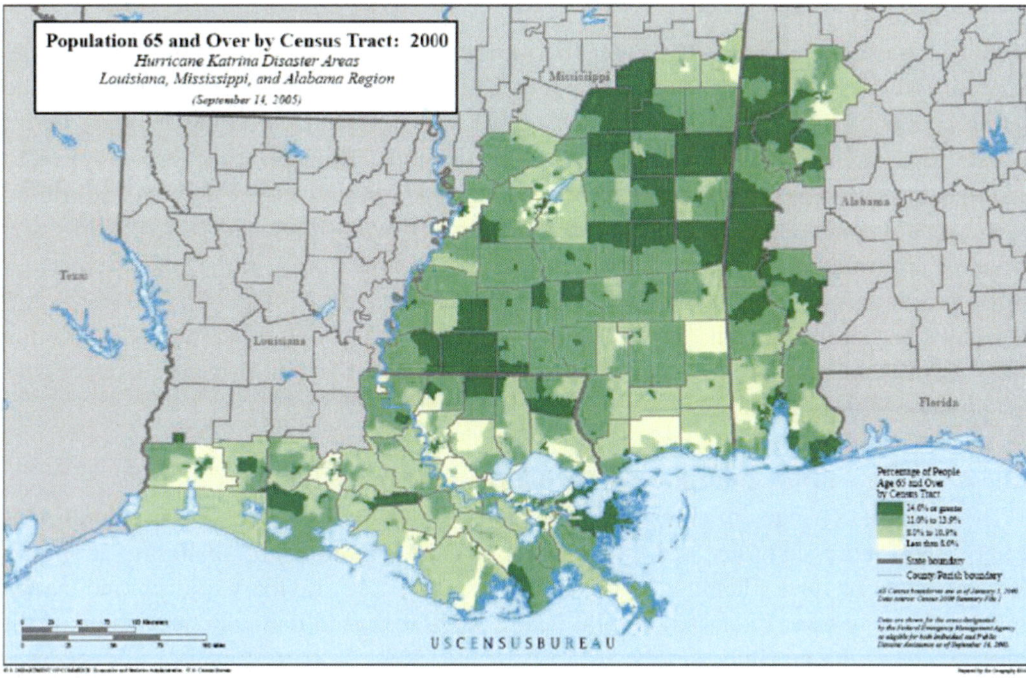

Source: U.S. Census Bureau.

Figure 4.26. Population over age 65 impacted by Hurricane Katrina

Interdependent Communications Infrastructure

Successful emergency management depends not only on the transportation infrastructure but also on interdependent communications infrastructure that allows emergency management personnel and responders to dynamically accommodate changing needs and infrastructure availability. Lessons learned from recent events indicate that significant breakdowns in communication can occur across multiple jurisdictions and agencies during major emergencies. Although not linked to or caused by changes in climate characteristics directly, cell phones and land lines quickly become unreliable both before and after major regional emergencies. Changes in climate may exacerbate this dynamic as greater penetration of storm surge and wind fields may disable the "day-to-day" communications infrastructure.

A recent study released by First Response Coalition, a public safety group, suggests that many wireless communications systems in hurricane-prone states are still unlikely to function well during major regional emergencies. Communication plans and infrastructure remain largely uncoordinated, even after concerted efforts to improve these dynamics following the 2005 hurricane season (First Response Coalition, 2006).

The use of new surveillance technologies such as unmanned aerial vehicles (UAV) may help ameliorate problems with existing communications systems that would be exacerbated by future storm events. These relatively new, but increasingly efficient and more affordable devices, could be effective new tools in the critical 72-hour period leading up to evacuations, as well as postevent recovery and response operations. These and other strategies may serve as a new means of acquiring and relaying real-time information when existing infrastructure is disabled during a storm.

Traffic Management

Traffic management related to emergency evacuations will become increasingly critical as the population in the Gulf Coast region grows. This may lead to increased instances such as that experienced during Hurricane Rita in 2005, when Galveston's long-standing plan to evacuate to Houston was complicated by the evacuation of Houston itself. Many Galveston (and other) residents tried to evacuate "through Houston," only to encounter hours of gridlock in the oversaturated transportation system that already was filled with Houston residents evacuating from the approaching storm. Other coastal communities in the region could fare similar evacuation problems in future storms. Also, as storm impacts and the resulting evacuations do not follow State lines, it is important that States not only plan for evacuations of their own residents but also account and allow for potential multistate evacuations that cross multiple State boundaries.

Critical Care Facilities and Shelters for Those with Special Needs

The predicted changes in climate over the next century will make the care of those with special needs more complex and problematic. In the instance of "sheltering in place," increased attention and planning will need to be given to auxiliary power and backup communication systems to sustain critical health services and to maintain acceptable quality of life (air conditioning, water supply, etc.).

The 2005 hurricane season also produced numerous instances of evacuees with special needs arriving at their "designated shelters," only to be turned away due to lack of capacity or the facility not even being open. Many of these shelters that denied evacuees shelter received funding from the U.S. Department of Homeland Security to support the facility infrastructure and operation. With evacuee demands only expected to increase in the future, the need to ensure reliable shelter services becomes increasingly important.

Local Development Policies

As it relates to the ability to support regional evacuations during emergencies, the potential for climate impacts – particularly storm surge and wind field during major hurricanes – should be mapped (and otherwise illustrated) to determine probable zones of risk. This information can inform local development policies and guide the location of new housing and critical care facilities to areas of lower vulnerability.

Fiscal Impacts

Revenue data collected by the State of Florida indicates that hurricane weather events reduce toll collection and increase toll system costs. As shown in table 4.20 below, the Florida 2004 hurricane season cost the State's tolled facilities $62,600,000 (Ely, 2005). These financial impacts could negatively impact the fiscal viability of toll projects that are used for evacuation routes in emergencies. The toll operating agencies in Florida recognize their toll facilities as evacuation routes and are working to suspend payment of tolls in the event of a hurricane (Warren, 2005).

Table 4.20. Hurricane impacts on toll revenue in Florida

Entity	Hurricane Season 2004 Millions		
	Estimated Revenue Loss	Estimated Damage Costs	Estimated Total Loss
Turnpike System	$32.21	$8.50	$40.71
FDOT-Owned (5)	2.48	1.33	3.81
Garcon Point	0.27	0.22	0.49
Mid-Bay	0.52	0.25	0.77
MDX	1.03	0.00	1.03
Bob Sikes	0.30	1.76	2.06
THCEA	1.44	0.00	1.44
OOCEA	9.07	1.50	10.57
Lee County	0.70	0.87	1.57
Miami-Dade County	0.11	0.00	0.11
Monroe (Card Sound)	0.04	0.00	0.04
Total	**$48.17**	**$14.43**	**$62.60**

Source: Ely, 2005.

Increased frequency and severity of hurricanes might pose a challenge to the fiscal strategies of toll facilities, may discourage the trend to finance future infrastructure with tolls, and may thereby reduce infrastructure that can be used for emergency evacuation. If too much of an area is inundated for an extended period there may be a reduction in vehicle trips below the threshold needed to support repayment of bonds. For example, beaches that had served as a destination for toll bridges could be flooded by rising sea levels and no longer support tourism. This could in turn affect toll revenues and ultimately undermine the financial viability of key segments of evacuation routes. Bridge tolls in the northwestern Florida region (Garcon Point and Mid-Bay Bridges) offer one illustration of this potential impact.

Highways provide the majority of transportation infrastructure for emergency operations. There are limited public transportation capabilities that operate on separate rights-of-way. This substantial reliance on a single mode of transportation could endanger many residents if the highway infrastructure is damaged or made inaccessible.

The prospect of climate change may require more frequent changes to emergency management plans and procedures. After the 2005 hurricane season, many public agencies are reassessing the appropriate level of investment for this activity. Recent events, as well as the climate change projections discussed in this report, highlight the need to develop action plans for worst-case scenarios. With predictions of a warmer Gulf Coast climate, more intense storms and hurricanes, and rising relative sea levels, the future design of critical infrastructure and emergency evacuation plans will need to incorporate increased challenges to our emergency management system.

4.3. IMPACTS AND ADAPTATION: CASE EXAMPLES IN THE STUDY REGION

While sections 4.1 and 4.2 analyze the potential future impacts of climate change on the region, this section focuses on the impacts associated with Hurricanes Katrina and Rita. The challenges of responding to severe weather events are all too familiar to transportation managers in the Gulf Coast. The hurricane season of 2005 was devastating for many communities in the study area. As the region rebuilds, some areas are incorporating changes to infrastructure design to help systems better withstand flooding and storm surge. The lessons learned from the costs of clean up and repair can help managers assess the implications of infrastructure damage as they consider future adaptation options. The following case examples illustrate the issues confronting managers working to ensure a safe and reliable transportation system.

4.3.1. Impacts of Hurricanes Katrina and Rita on Transportation Infrastructure

Hurricane Katrina, which made landfall on August 29, 2005, was the most destructive and costliest natural disaster in the history of the United States and the deadliest hurricane since the 1928 Okeechobee Hurricane. Over 1,800 people lost their lives during Hurricane Katrina, and the economic losses totaled more than $100 billion (Graumann et al., 2006). More than 233,000 km^2 (90,000 mi^2) were declared disaster areas. While a single storm cannot be attributed to climate change, the impacts of Hurricanes Katrina and Rita in 2005 illustrate the types of impacts that would occur more frequently if the Gulf Coast were to experience more Category 4 and 5 hurricanes in the future.

The storm had a devastating impact on much of the transportation infrastructure of coastal Mississippi, Louisiana, and Alabama, causing major damage to highways, railroads, ports, and airports. Damage was caused by flooding, pounding waves, and high winds. In addition, when the floodwaters subsided, an enormous amount of debris still had to be removed before transportation networks could function. Forty-six million yd^3 of debris were removed from Mississippi alone (from all locations, not just transportation facilities). Louisiana DOTD spent $74 million on debris removal following Hurricanes Katrina and Rita (Paul, 2007).

Through aggressive action by public and private transportation managers, many major transportation facilities were reopened relatively quickly considering the level of damage. Most of the study area highways, rail lines, pipelines, ports, and airports were back in service within weeks to a month. Limited access across the I-10 Twin Span Bridge was available within two months and nearly full access achieved within five months. The heavily damaged CSX Gulf Coast mainline and its bridges were reopened six months after being washed out by Hurricane Katrina. The worst damaged facilities were the river and bay bridges that carry U.S. Highway 90 along the edge of the Gulf Coast. Though much of the roadway and three of the six badly damaged crossings were repaired within about three months, the three remaining bridges took considerably longer to repair or replace. The last of these bridges, the Biloxi-Ocean Springs Bridge, is scheduled to reopen in November 2007, more than two years after

Katrina. In all, the price tag of cleanup and reconstruction effort will run into the billions of dollars: the Louisiana Recovery Authority estimated costs exceeding $15 billion for Louisiana alone (Louisiana Recovery Authority, 2006). Mississippi spent more than $1 billion on cleanup and bridge replacement. (Mississippi DOT, 2007).

By most accounts, the impact of Hurricanes Katrina and Rita on national-level freight flows was modest because of redundancy in the national transportation system and timing. Truck traffic was able to divert to parallel east-west interstate routes that avoided the collapsed bridges and other barriers. Railroad operators were able to reroute intermodal and carload traffic that was not bound directly for New Orleans through Memphis and other Midwest rail hubs. Most of the Mississippi river ports and the Mississippi inland waterway were back in service in time to handle the peak export demand later in the fall of 2005. Major pipelines suffered relatively little damage and were able to open within days as electrical power was restored (Grenzeback and Lukmann, 2006).

The following text outlines some of the key impacts by mode.

Roads

The most significant impacts to roads were to the numerous bay and river crossings throughout the region. While the effects were limited in some locations and damage was repaired within days, in some coastal sections prominent elements of the transportation network remained closed many months after the storm. The worst damage was focused in the area along and to the south of the I-10/I-12 corridor, including U.S. Highway 90, State Highway 1, and the Lake Pontchartrain Causeway in Louisiana, as well as I-110 in Mississippi. Three major bridge crossings along the route were destroyed and two more sustained significant damage. The damage was largely caused by the immense force of wave action on the bridge spans, many of which were not sufficiently tied down to the bridge pilings to resist movement (figure 4.4). Spans weighing 300 tons were dislodged by the hurricane.

Inundation also caused structural problems along many miles of roadway. More than 50 km (30 mi) of coastal U.S. Highway 90, which runs through the beachfront communities of Mississippi, were completely inundated by the storm. At a cost of $267 million, the 3.2- km (2-mi), four-lane Bay St. Louis Bridge (U.S. Highway 90) reopened on May 17, 2007. The total request for emergency repairs to Mississippi highways alone after Katrina is $580 million (Mississippi Gulf Coast Regional Planning Commission, 2006). Much of the paved surface between Pass Christian and Biloxi buckled or dropped into sinkholes; in places it took weeks to repair washouts and to remove many feet of sand from the road surface: More than 3,200 km (2,000 mi) of roads were submerged in Louisiana, and the Louisiana DOTD found indications that prolonged inundation can lead to long-term weakening of roadways. A study of pavements submerged longer than three days (some were submerged several weeks) found that asphalt concrete pavements and subgrades suffered a strength loss equivalent to 5 cm (2 in) of pavement (Gaspard et al., 2007). The estimate for rehabilitating a portion of these roads, 320 km (200 mi) of submerged State highway pavements, amounted to $50 million.

The expense of poststorm cleanup and repair can be considerable. The Louisiana Recovery Authority estimated that the cost of rebuilding infrastructure (defined as roads, bridges, utilities and debris removal) damaged by the hurricanes would cost $15-18 billion. Louisiana DOTD spent $74 million on debris removal; as of June 2007, Mississippi DOT had spent $672 million on debris removal, highway and bridge repair, and rebuilding the Biloxi and Bay St. Louis bridges; it expects to spend an additional $330 million in the subsequent 18 months (Mississippi DOT, 2007; Louisiana Recovery Authority, 2006). Also, debris removal is not completely benign in terms of further impact; heavy trucks removing debris in Louisiana also damaged some roadways (Paul, 2007).

Rail

The rail infrastructure in coastal Mississippi and Louisiana suffered major damage that took weeks or months to repair. The worst storm damage was focused on a 160-km (100- mi) section of CSX's Gulf Coast line between New Orleans and Pascagoula, Mississippi. CSX had to restore six major bridges and more than 65 km (40 mi) of track, much of which was washed out or undermined. Damage was so extensive on the line that CSX required more than 5 months and $250 million to complete repairs and to reopen the line. It would take many times that if the company wanted to relocate the line further inland. In addition, New Orleans is a major rail freight interchange point for east-west rail traffic, and the railroads needed to reroute intermodal and carload traffic that was not bound directly for New Orleans through other rail hubs in Memphis and St. Louis, which increased operating expenses (Grenzeback and Lukmann, 2006).

Ports

Due to their low-lying locations, the ports were susceptible to damage from all effects of Hurricane Katrina– high winds, heavy rains, and especially the storm surge. Container cranes were knocked down, storage sheds blown apart, and navigational aids lost. In Gulfport, MS, the storm surge pushed barges hundreds of feet inland and scattered 40-ft containers throughout downtown Gulfport. The storm sank nearly 175 barges near New Orleans, disrupting navigation on the river; however, almost all ports in the central Gulf Coast were able to reopen within a month of Katrina's landfall. Nonetheless, damage was costly: more than $250 million has been allocated to repair, rebuild, and expand the Port of Gulfport in the wake of Hurricane Katrina (Grenzeback and Lukmann, 2006).

Fortunately, the timing of the storm prevented a catastrophic impact on U.S. agricultural exports. Gulf Coast ports typically handle 55 percent to 65 percent of U.S. raw corn, soybean, and wheat exports. Since the bulk of U.S. corn and soybean harvest moves down the Mississippi River from October to February, the ports were generally able to restore operations in preparation for this critical season, although agriculture still faced increased shipping costs due to a shortage of barges. The severe damage to Gulfport (which specializes in importing containerized bananas and winter fruits from Central and South America) did result in a regional shortage of tropical fruits, because major fruit importers such as Dole, Chiquita, and Crowley were forced to reroute shipments to Port Everglades, FL, or Freeport, TX, at extra expense (Grenzeback and Lukmann, 2006).

Airports

A number of airports in the study area received significant damage from the strong winds, flooding rains and embedded tornadoes associated with Hurricane Katrina. Airports sustained damage to passenger terminals, maintenance facilities, and navigational devices. Power outages also took air traffic control facilities offline and darkened nighttime runway lights. As a result, some airports were closed for days and weeks while necessary repairs could be made, but relief flights were flown in before the airport facilities were fully reopened.

Louis Armstrong New Orleans International Airport, the third largest airport in the Central Gulf Coast, sustained damage to its roofs, hangars, and fencing, but had no significant airfield damage despite sitting only 122 cm (4 ft) above sea level (making it the second lowest lying international airport in the world, after Schiphol International in The Netherlands). For the first few weeks of September, the airport was open only to military aircraft and humanitarian flights, but reopened to commercial flights on September 13, 2005. On the other hand, Lakefront Airport, one of the busiest general aviation facilities in the Gulf Coast and located directly on Lake Pontchartrain to the north of the New Orleans city center, suffered extensive damage, with a number of terminals and hangars destroyed. It took 7 weeks before it could even reopen for daytime operations. Gulfport-Biloxi International, the fifth busiest commercial airport in the central Gulf Coast, was also hard hit by the storm. Located less than a mile inland, between U.S. Highway 90 and I-10 in Gulfport, the airport's terminal building, taxiways, cargo facility, general aviation facility, and rental car facility sustained an estimated $50 million to $60 million in damage. The airport reopened to commercial flights on September 8 and returned to its normal volume of traffic in February 2006 (Grenzeback and Lukmann, 2006).

Fifty-eight airports were surveyed on how they were affected by Hurricanes Katrina and Rita – the extent of damage either hurricane caused, the ability of the airports to cope with the damage, and the use of the airports for emergency management. Twenty-nine airports, or 50 percent, responded to the survey. Forty-eight percent of respondents pointed to the following as some of the main reasons for closure: electrical outage (19 percent), wind damage (16 percent), and debris on runways (12 percent) were the top three reasons identified. Civil, military, and passenger airline operations were affected by the hurricanes. Figure 4.23 identifies airports affected by Hurricane Katrina's winds. GIS analysis indicates 16 airports experienced winds exceeding 161 km/h (100 mi/h) during Hurricane Katrina, including New Orleans International, Gulfport-Biloxi, and Hattiesburg commercial service airports. These airports are located in southeastern Louisiana and south-central Mississippi. USGS data also indicates that nine airports impacted by Hurricane Rita experienced winds exceeding 161 km/h (100 mi/h), including two commercial service airports: Beaumont-Port Arthur in Texas and Lake Charles Regional in southwestern Louisiana. Survey responses indicated additional implications to aircraft operations as follows:

- Civil aircraft operations were closed at 12 airports. The average length of closure to civil aircraft operations was 209 hours, and the maximum observed closure was 1,152 hours. Lakefront Airport in New Orleans, an outlier, was closed for 48 days and skews the data. When removing this airport from the data field, the average length of time closed to civil aircraft operations is 35 hours. It is noteworthy that although many airports "opened" soon after the hurricanes passed, many were without electricity and were only open during daylight hours.
- Military aircraft operations were closed at eight airports. The average length of closure to military aircraft operations at civil airports was 33 hours, and the maximum observed closure was 96 hours.
- Two commercial service airports, Lake Charles Regional and Houston's William P. Hobby, reported passenger airline operations were suspended at their airport.

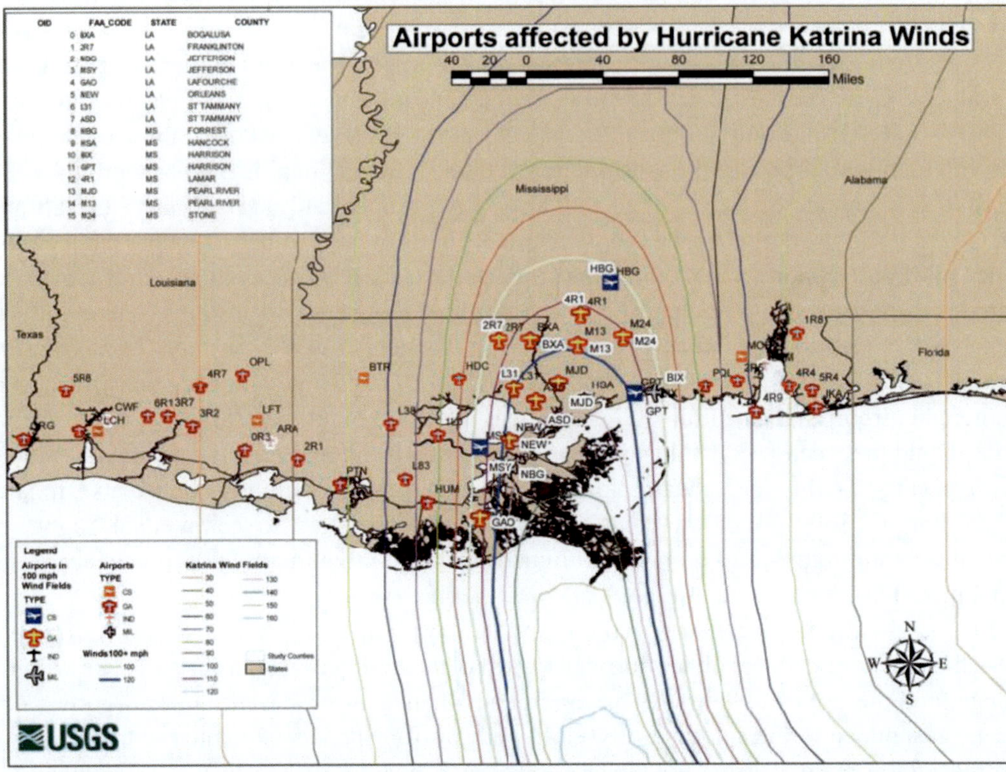

Source: USGS.

Figure 4.27. Airports affected by Hurricane Katrina winds

Hangar facilities also were damaged by Hurricanes Katrina and Rita. Thirty-eight percent of responding airports suffered damage to T-hangars, which are long rectangular structures with 12 to 20 "bays" that store single-engine and small twin-engine aircraft. Forty-five percent of responding airports experienced damage to conventional hangars, which are designed to store large aircraft and are 18 by 18 m (60 by 60 ft) to 30 by 30 m (100 by 100 ft) in size. Conventional hangars are also 6 to 9 m (20 to 30 ft) in height to accommodate large aircraft with high tails.

Pipelines

The major petroleum and petroleum-product pipelines servicing the study area received relatively little physical damage from the effects of Hurricanes Katrina and Rita, but could not operate reliably due to massive power outages in the wake of the storms and by interruptions to the supply of fresh product to transport due to refinery shutdowns, causing shortages of petroleum products in parts of the Nation. Even so, most of these systems were able to resume partial service within days of the storm and full service within a week. At the peak of the disruption caused by Hurricane Katrina, 11 petroleum refineries were shut down, representing 2.5 million barrels per day or 15 percent of U.S. refining capacity, and all major pipelines in the area were inoperable due to power outages. By September 4, 5 days after the storm, eight major petroleum refineries remained shut down (representing 1.5 million barrels per day or nine percent of U.S. refining capacity); however, all of the major crude or petroleum product pipelines had resumed operation at either full or near-full capacity (Grenzeback and Lukmann, 2006).

4.3.2. Evacuation during Hurricane Rita

Emergency evacuation is a key strategy to cope with hurricanes in the low-lying Gulf Coast study region. The evacuation of Houston/Galveston, the largest metropolitan area in the study region, prior to Hurricane Rita presents a case study of the difficulties of evacuating large urban areas and some lessons learned for future emergency planning.

Unlike New Orleans, much of Houston is high enough to be out of the storm surge zone; thus, Galveston and the low-lying eastern areas are generally supposed to evacuate first. However, Houstonians learned during Tropical Storm Allison (2001) that precipitation alone can cause massive flooding in the city from overflowing bayous and lack of drainage. With images of the devastation wrought by Hurricane Katrina fresh on their minds, up to 2.5 million people attempted to evacuate the Houston/Galveston area in the days before Rita's projected landfall (Mack, 2005) – twice as many people as the area's evacuation planning was developed for (Durham, 2006). In fact, only about half of these people lived in evacuation zones (Feldstein and Stiles, 2005).

Evacuees faced massive congestion, with 160-km (100-mi) traffic jams reported (Breckinridge et al., 2006). One fifth of the evacuees spent more than 20 hours on the road to leave the area; only half completed the trip in less than 10 hours (Mack, 2005). Worsening the congestion, households traveled in multiple cars in order to get valuable property out of harm's way: the Texas Transportation Institute (TTI) estimated that on average there were 1.2 occupants per vehicle, versus the 2.1 occupants generally assumed in evacuation planning (Durham, 2006). In an effort to ease congestion, officials improvised a last-minute contraflow system on some highways, which was not part of their original evacuation plan. Fuel shortages plagued travelers as gasoline stations on the evacuation routes were overwhelmed by demand. Tragically, 23 nursing home evacuees died when their bus caught fire on the road.

Following the storm, the Houston-Galveston Area Evacuation and Response Task Force identified several lessons learned from the experience and recommendations for the future (Durham, 2006):

- Evacuation plans should be practiced extensively prior to the hurricane season to reveal problems ahead of time.
- Plans should include a system for removing disabled vehicles – during Rita, an effective incident management service was available only within the Houston city limits. As a result, vehicle breakdowns caused significant bottlenecks along the evacuation routes.
- Contraflow plans should be developed well in advance, but operations are not a panacea. Emergency planners will need to consider the numerous drawbacks of implementing contraflow strategies: They require intensive use of law enforcement

and other personnel, disrupt day-to-day operations in areas not evacuating, and make it more difficult to move emergency vehicles and supplies back into the area.
- Thorough planning is necessary for evacuees with special needs, including ensuring an adequate supply of vehicles, identifying destination(s) capable of supporting their needs, and providing personnel sufficient training to ensure a safe trip.

The Rita evacuation also demonstrated the importance of accounting for human behavior. "Too few" people evacuated New Orleans before Katrina, but "too many" evacuated the Houston-Galveston area (Breckinridge et al., 2006). Evacuation orders are meant to reinforce the fundamental strategy of "run from the water, hide from the wind"; however, in the case of Rita it seems many evacuees ran from the wind. Similarly, the tendency of households to take as many vehicles with them as possible is a logical way to protect property but counterproductive during a mass evacuation. This illustrates the need to better understand the range of potential reactions by residents during a crisis and how best to communicate with the public to facilitate effective emergency management.

4.3.3. Elevating Louisiana Highway 1

Louisiana currently is in the process of upgrading and elevating portions of Louisiana Highway 1, a road that is very important both locally and nationally. It connects Fourchon and Port Fourchon to Leeville and Golden Meadow to the north. The project is broken into multiple phases and includes a four-lane elevated highway between Golden Meadow, Leeville, and Fourchon to be elevated above the 500-year flood level; and a bridge at Leeville with 22.3-m (73-ft) clearance over Bayou LaFourche and Boudreaux Canal. Construction has begun on both the $161 million bridge project and a segment of the road south of Leeville to Port Fourchon (Wilbur Smith, 2007).

Hurricane Katrina's impact on the energy infrastructure helped raise the profile of the dangers facing and the importance of Louisiana Highway 1. The highway floods even in low-level storms, and in addition to the effects of storm surge, the existing infrastructure also faces threats from very high rates of coastal erosion and subsidence (Smith, 2006).

The importance of this part of the Gulf Coast, and thus Highway 1, to the Nation's energy supply and infrastructure cannot be overstated. It is the only roadway linking Port Fourchon and the Louisiana Offshore Oil Port (LOOP) to the Nation. Port Fourchon supports 75 percent of deepwater oil and gas production in the Gulf of Mexico, and its role in supporting oil production in the region is increasing. The LOOP, located about 32 km (20 mi) offshore, plays a key role in U.S. petroleum importation, production, and refining as it links daily imports of 1 million barrels and production of 300,000 barrels of oil in the Gulf of Mexico to 50 percent of U.S. refining capacity. Locally, the road is the key route for transporting machinery and supplies to Port Fourchon and offshore oil workers and also for exporting seafood from the region. Perhaps most importantly, it is the evacuation route for south Lafourche and Grand Isle, as well as some 5,000 offshore oil workers (LA 1 Coalition, 2007a and b).

4.4. Conclusions

The results of this investigation shows a wide range of possible impacts on transportation infrastructure and services across the Gulf Coast study area. Given the uncertainties inherent in modeling and the complexities of the natural processes involved, this analysis does not attempt to pinpoint the precise timing of climate effects but rather provides a broad assessment of potential impacts during the coming decades. These findings provide a critical overview for transportation planners and managers of the potential implications of climate factors and indicate areas of vulnerability that warrant consideration by decision makers. Future investment decisions should be informed by the potential risks identified in this study.

Some of the most evident impacts are related to RSLR and storm surge. A 122-cm (4-ft) increase in relative sea level could inundate a substantial portion of the transportation infrastructure in the region: 28 percent of the arterials, 43 percent of the IM connectors, and 20 percent of the rail miles. Nearly three quarters of ports could be affected, as well as three airports, including Louis Armstrong International in New Orleans. Impacts associated with storm activity are more acute, although confined to the specific locations of individual storm events. Some 51 percent of arterials and 56 percent of interstates, along with almost all ports, a third of rail lines, and 22 airports are vulnerable to a storm surge of 5.5 m (18 ft), should such a surge occur. As the potential of higher-intensity storms increases and sea level rises, the vulnerability of infrastructure to storm surge becomes increasingly significant.

The direct impacts of climate factors on specific facilities can have much broader implications than implied by the percentages and maps contained in this chapter. Damage to critical links in the intermodal network can disrupt connectivity throughout the region. These disruptions can be relatively short term, as in the case of precipitation and some storm surge and weather events; moderate, as in the case of shut-downs to conduct maintenance required to repair pavement surfaces caused by higher temperatures or storm surges; or long-term interruptions of service caused by inundation and damage to entire segments of infrastructure due to storm surge or permanent sea level rise.

The safety impacts associated with climate impacts deserve further in-depth analysis beyond this effort. Storm activity and storm surge in particular have the most direct implications for safety. These include accidents caused by: debris caused by storms, washed-out roads during or after storms, or evacuations before storms. Furthermore, the other key climate drivers, including changes in precipitation patterns, temperature, and relative sea level rise, could have important safety impacts as well.

In addition to these regional impacts, the vulnerabilities of Gulf Coast transportation will have nationwide significance that merit further investigation. The resilience of Gulf Coast transportation infrastructure capabilities has implications for the country's ability to transport many key commodities into and out of the United States, including petroleum and natural gas, agricultural products, and other bulk goods.

Data and Research Opportunities

This study identified needs for additional data and research that would further advance understanding of the implications of climate change for transportation. These include information and investigation in the following areas:

- **Integration of site-specific data** – The integration of site-specific elevation and location data in a GIS-compatible format would greatly facilitate investigation of the impacts of climate change and the natural environment on transportation. This data should include information on transportation facilities as well as on protective structures such as levees and dikes

- **Additional and refined climate data and projections** – Further development of environmental trend data and climate model projections tailored to transportation decision makers is needed to facilitate integration of climate information into transportation decisions. In addition, specific data on other climate factors not fully addressed in this study would be valuable. These factors include wind speeds, isolated hot days, and fog.

- **Effects of climate change on freight transport demand** – Research is needed on the perspectives, investment considerations, relocation plans, and adaptation strategies of private sector shippers and freight transportation providers and how their requirements may evolve due to climate change and shifts in market demand.

- **Demographic response to climate change** – High-population density creates increased need for both passenger transport and movement of consumer goods. Population change will be driven by multiple factors, possibly including changing environmental conditions. Projections of population density along coastal regions and their impact on the demand for freight and passenger services need to be explored.

- **Design standards and reconstruction and adaptation costs** – Additional case information would be valuable regarding the costs of rebuilding transportation facilities following severe storms. Research is needed on how local agencies are adapting design standards during reconstruction (or construction of new facilities) to increase the resilience of their facilities, such as changes in bridge height or construction, use of new materials, and changes in design criteria. Analysis of the range of adaptation options available and the costs and benefits of specific strategies would help inform State and local transportation planners and decision makers.

- **New materials and technologies** – Research is needed to develop materials that can better withstand higher temperatures and drier or wetter conditions and technologies that can help us better adapt to the effects of climate change.

- **Pipelines** – A more complete examination of pipeline impacts from climate change and adaptation strategies is warranted.

- **Land use and climate change interactions** – Research is required to investigate how various land use development and environmental management strategies in vulnerable areas affect the magnitude of climate change impacts on communities and transportation infrastructure. A comparative analysis of current, international best practices in land use and building codes, particularly in coastal regions, could provide useful information to U.S. transportation and planning agencies.

- **Emergency management planning/coordination/modeling** – Additional study on successful approaches in coordinating emergency management planning among public agencies and major private sector entities in at-risk areas could identify opportunities for improved coordination, public-private partnering, and risk reduction. Development and application of simulation modeling should be considered to illustrate the increasing challenges of evacuating major urban areas and evaluate mitigation strategies. Collection and evaluation of real-time data gathered during emergencies is needed to determine its possible use to first responders, operating agencies, the media, and the general public. Changes in communication and information technology infrastructure also should be explored.

- **Secondary and national economic impacts** – More in-depth research into the secondary economic impacts to the region and Nation of freight disruption would benefit understanding of national trends and vulnerabilities and inform development of appropriate policies.

- **Site-specific impacts** – This assessment considers scenarios of change for the counties that comprise the central Gulf Coast. More detailed analysis is desirable since specific transportation facilities will ultimately be affected by climate change. This will require development of climate data and information that is specific to much smaller geographic areas, in addition to detailed analyses of specific facilities.

4.5. Sources

4.5.1. References

Administration on Aging, (2006). *Aging into the 21st Century. Department of Health and Human Services.* Available on-line at <http://www.aoa.gov/prof/Statistics/future_growth/aging2 1 /demography.asp>, viewed June 2006.

Ballard, A. (2006). *Recommended Practices for Hurricane Evacuation Traffic Operations, Product 0-4962-P2.* Available on-line at: <http://tti.tamu.edu/documents/0-4962-P2.pdf>.

Breckinridge, W. L., Najafi, F. T., Rouleau, M., Ekberg, J. & Mamaghanizadeh, E. (2006). A Summary Review on the Importance of Transportation Planning, Preparation, and Execution during a Mass Evacuation. *TRB 2007 Annual Meeting CD-ROM*, Washington, D.C., 17 pages.

Bureau of Transportation Statistics (BTS), (2004). *National Transportation Atlas Database (NTAD) 2004.* U.S. Department of Transportation, Washington, D.C., CD-ROM. [Available on-line at http://www.bts.gov/publications/national_ transportation_atlas_ database/]

Budzik, Phillip, (2003). *Energy Information Administration (EIA), U.S. Natural Gas Markets: Relationship Between Henry Hub Spot Prices and U.S. Wellhead Prices*. Available on-line at: <http://www.eia.doe.gov/oiaf/analysispaper/henryhub/index.html>.

City of Biloxi, August (2005). Storm & Flood Preparedness. <http://www.biloxi.ms.us/PDF/Storm_Flood_2005 .pdf>, viewed December 2006.

City of New Orleans, (2006). Emergency Preparedness Plan: July 2006. Available on-line at: <http://www.cityofno.com/Portals/Portal46/portal.aspx?portal=46&tabid=3 8>, viewed December 2006.

City of Houston, Office of Emergency Management, (2006). Evacuation Information. http://www.houstontx.gov/oem/hcra.html>, viewed December 2006.

City of Houston, Office of Emergency Management, (2006). The City of Houston Emergency Plan. <http://www.houstontx.gov/oem/plan.html>, viewed December 2006.

City of Houston, Office of Emergency Management, (2006). Annex E Evacuation. Available on-line at: <http://www.houstontx.gov/oem/plan.html>, viewed December 2006.

City-Parish of Lafayette, Office of Emergency Preparedness, What To Do In Case of an EVACUATION. <http://brgov.com/dept/oep/plan.asp# Search%20the%20Plan>, viewed December 2006.

City-Parish of Lafayette, Office of Emergency Preparedness, (2006). Lafayette Hurricane Evacuation Routes. Available on-line at: <http://info.louisiana.edu/mahler/oep/sug-rout.html>, viewed December.

Code of Federal Regulations, (2007). Title 49, Section 192.705, Transportation of Natural and other gas by pipeline: Minimum Federal safety standards, transmission lines: patrolling.

Code of Federal Regulations, (2007). Title 49, Section 195.412, Transportation of Hazardous Liquids by pipeline: Inspection of rights-of-way and crossings under navigable waters.

Dessens, J., (1995). Severe convective weather in the context of a nighttime global warming. *Geophysical Research Letters., 22*, 1,241–1,244.

Durham, C., (2006). Hurricane Evacuation. *TRB 2007 Annual Meeting CD-ROM*, Washington, D.C., 19 pages.

East Baton Rouge, Parish of, (2006). Emergency Operations Plan. February 2006. <http://brgov.com/dept/oep/plan.asp#Search%20the%20Plan>, viewed December 2006.

Ely, J. L. (2005). 2004 Hurricane Season – Florida's Toll Agencies' Rise to the Challenge. Presented at the International Bridge, Tunnel, and Turnpike Association 73rd Annual Meeting and Exhibition. Cleveland, Ohio. Available on-line at: < http://www.ibtta.org/files/PDFs/Ely_Jim.pdf>, viewed December 2005.

Energy Information Administration (EIA), (2005). *Special Report – Hurricane Katrina 's Impact on U.S. Energy*. Department of Energy, Washington, D.C., September 2, 2005. http://tonto.eia.doe.gov/oog/special/eia1_katrina_090205.html.

First Response Coalition, April (2006). *The Imminent Storm 2006: Vulnerable Emergency Communications in Eight Hurricane Prone States*.

Gaspard, K., Martinez, M., Zhang, Z. & Wu. Z. (2007). Impact of Hurricane Katrina on Roadways in the New Orleans Area. Technical Assistance Report No. 07-2TA, LTRC Pavement Research Group, Louisiana Department of Transportation and Development, Louisiana Transportation Research Center, March 2007, 73 pages.

Graumann, A., Houston, T., Lawrimore, J., Levinson, D., Lott, N., McCown, S., Stephens, S. & Wuertz, D. (2005). Hurricane Katrina – a climatological perspective. October 2005,

Updated August 2006. Technical Report 2005-01 page 28. NOAA National Climate Data Center, available at http://www.ncdc.noaa.gov/oa/reports/tech-report200501 z.pdf.

Groisman, P. Y., Knight, R. W., Karl, T. R., Easterling, D. R., Sun, B. & Lawrimore, J. H. (2004). Contemporary changes of the hydrological cycle over the contiguous United States: Trends derived from in situ observations. *J. Hydrometeorol., 5*, 64-85.

Guillet, Jaime, (2007). 2: LA 1: $1.4 billion. *New Orleans City Business*, February 26, 2007. Available on-line at: <http://www.neworleanscitybusiness.com/viewstory.cfm? recID=1 8279>, viewed June 2007.

Hitchcock, C., Angell, M., Givler, R. & Hooper, J. (2006). *A Pilot Study for Regionally Consistent Hazard Susceptibility Mapping of Submarine Mudslides, Offshore Gulf of Mexico*. Minerals Management Service, Engineering & Research Branch, Herndon, Virginia, April 5, 2006.

Hooper, James, R. (2005). Hurricane-Induced Seafloor Failures in the Mississippi Delta. Presented to 2005 Offshore Hurricane Readiness and Recovery Conference, Houston, Texas, July 26-27, 2005.

Karl, T. R. & Trenberth, K. E. (2003). Modern global climate change. *Science, 302* (5651), 1,719-1,723.

Louis Berger Group, Inc., (2004). OCS-Related Infrastructure in the Gulf of Mexico Fact Book, OCS Study MMS 2004-027, June 2004. The reported value is believed by the author to represent only in-service oil pipelines in the MMS jurisdiction of the GOM.

Louisiana Recovery Authority, (2006). Press Release: LRA Releases Estimates of Hurricane Impact: Magnitude of Disaster to Factor into Block Grant Discussion Friday, Jan 12, 2006. Available on-line at: <www.lra.louisiana.gov/pr_1_12_impact.html>, viewed 10 July 2007.

Louisiana State Police, (2006). Louisiana Citizen Awareness and Disaster Evacuation Guide, Emergency Evacuation Guide. Available on-line at: <http://www.lsp.org/evacguide. html>, viewed December 2006.

Mississippi Gulf Coast Regional Planning Commission, (2006). Transportation Improvement Program FY 2007-2012, amended 9-28-06. Gulfport, Mississippi.

Mississippi DOT, (2007). Gulf Coast Transportation Recovery Update. 22 June 2007. Mississippi Department of Transportation.

Mississippi DOT, (2005).Hurricane Evacuation Guide. Available on-line at:
<http://www.mdot.state.ms.us/cetrp/ms _coastal _hurricane.pdf>, viewed December 2006.

Mobile County Alabama, (2006). Evacuation Procedures. Available on-line at: < http://www. mcema.net/sa_procs.html>, viewed December 2006.

National Assessment Synthesis Team (U.S.). Climate Change Impacts on the United States: The Potential Consequences of Climate Variability and Change. Cambridge University Press, Cambridge, 2001. Available on-line at: <http://www.usgcrp.gov/ usgcrp/Library/ nationalassessment/foundation.htm>, viewed December 2005.>

Paul, H. (2007). *Impacts of Climate Change and Variability on Transportation Systems and Infrastructure: Gulf Coast Study*. Presented at Impacts of Climate Change and Variability on Transportation Systems and Infrastructure: Gulf Coast Study, Phase I Federal Advisory Committee Meeting, May 16-17, 2007.

Pipeline and Hazardous Materials Safety Administration (PHMSA), (2007). *Annual Reports: National Pipeline Mapping System*. PHMSA, U.S. Department of Transportation, Washington, D.C.

Plans Branch, Louisiana Office of Homeland Security and Emergency Preparedness, (2006). *Pelican Parish Planning Guidance and Crosswalk for Parish Multi-Hazard Emergency Operations Plans*. March 2006. Available on-line at: <http://www.ohsep. louisiana.gov/ plans/PELICAN%20CROSSWALK06F.doc>, viewed December 2006.

Skinner, John, (2006). *Pipeline Damage Assessment from Hurricane Ivan*. Det Norske Veritas Technology Services, Department of Interior, Minerals Management Services report. Available on-line at: <http://www.mms.gov/tarprojects/553 .htm>.

Sloan, Kim, (2007). Bay St. Louis Bridge Open. Mississippi DOT Public Affairs. 17 May 2007. <http://www.gomdot.com/newsapp/newsDetail.aspx? referrer=list&id=5292007 10349>, viewed 10 July 2007.

State of Alabama, Emergency Operations Plan, April 20, (2006). <http://ema.alabama. gov/Misc/Alabama%20EOP%202006.pdf>, December 2006.

State of Louisiana, Office of Homeland Security and Emergency Preparedness, Emergency Operations Plan. April 2005. Available on-line at: < http://www.lsuhsc. edu/hcsd/loep. pdf>, viewed December 2006.

Transportation Research Board, Special Report 281, *Transportation Pipelines and Land Use*, Washington, D.C., 2004, page 47.

U.S. Census Bureau: Areas Affected by Hurricane Katrina (Revised). Available on-line at: <http://ftp2.census.gov/geo/maps/special/HurKat/Katrina_Pop_65+_v2.pdf>, viewed May 2006.

U.S. Department of Transportation (DOT) in Cooperation with U.S. Department of Homeland Security, 2006: Catastrophic Hurricane Evacuation Plan Evaluation: A Report to Congress. Available on-line at: <http://www.fhwa.dot.gov/ reports/ hurricane vacuation/ index.htm>.

U.S. Geological Survey (USGS), 2004: *National Elevation Dataset (NED)*. U.S. Geological Survey, EROS Data Center, Sioux Falls, South Dakota. http://ned.usgs.gov/]

Warren, C. L., et al., (2005). *Toll Suspension Plan during Hurricanes*. Transportation and Expressway Authority Membership of Florida (TEAMFL).

Wilbur Smith Associates, (2007). LA 1 Improvements.<http://www.la1project.com/ index. cfm>, viewed 3 July 2007.

End Notes

[1] Aside from introductory and summary sections, the climate drivers are not addressed in order of relative importance but rather are discussed in the same order throughout this chapter for purposes of consistency: temperature, precipitation, sea level rise, and storm activity.

[2] Bridges may be of particular interest in this regard. Phase II of this study, which will include an in-depth analysis of a single location within the study region, is expected to include a systematic analysis of the potential impacts of climate change on bridges, because bridges play a key role across multiple modes, and their failures can produce bottlenecks.

[3] As noted in chapter 2.0, this report focuses on interstates, arterials, and collectors, but not local roads.

[4] Much of the material in this section was developed through personal communication with David Read, Principal Investigator, Transportation Technology Center, Inc., an Association of American Railroads subsidiary located in Pueblo, CO.

[5] It should be noted that many existing facilities at low elevations are protected by levees and other physical structures, which provide some resistance to gradual changes in sea level and the impacts of storm surge. The effects of existing or planned protections were not addressed by this study. Even with this protection, the infrastructure described in this study is potentially still at risk.

[6] These variables affect the performance of departing aircraft in particular; landing aircraft use less runway because of decreased landing weight (from fuel usage), as compared to take-off weight, and the use of flap settings.

[7] The approach is not completely different. The FAA AC provides design guidance for both small aircraft and large aircraft by using the charts within the AC or directing the reader to obtain manufacturer performance charts for small or large aircraft. The FAA AC also stipulates what design procedure to apply, based on whether or not Federal dollars are involved, e.g., funding through the Airport Improvement Program (AIP).

[8] Runways at military airports are designed to military aircraft specifications.

[9] It should be noted that the FAA Airport Design for Microcomputer software is solely for **planning purposes** and not for design, since the software generates roughly estimated lengths.

[10] One hundred percent of all large aircraft category is seldom used in runway design since very few airports experience the entire spectrum of large general aviation aircraft operations.

[11] Visual flight rules (VFR) are a set of aviation regulations under which a pilot may operate an aircraft, if weather conditions are sufficient to allow the pilot to visually control the aircraft's attitude, navigate, and maintain separation with obstacles such as terrain and other aircraft.

[12] This includes some extended pipeline sections beyond the boundaries of the study, because GIS coding of links included segments that spanned both inside and outside the study area.

[13] The State/Federal boundary is 3 mi offshore in the study area except in Texas, where it is 10 mi offshore.

In: Impacts of Climate Change on Transportation and… ISBN: 978-1-60741-424-7
Editor: Iason Pavlopoulos

Chapter 5

HOW CAN TRANSPORTATION PROFESSIONALS INCORPORATE CLIMATE CHANGE IN TRANSPORTATION DECISIONS?

Lead Authors: Kenneth J. Leonard, John H. Suhrbier and Eric Lindquist

Contributing Authors: Michael J. Savonis, Joanne R. Potter and Wesley R. Dean

As the previous chapters have demonstrated, there is benefit to including long-term climate considerations in the development of transportation systems. In fact, climate factors are likely to affect decisions in every phase of the transportation management process: from long-range planning and investment; through project design and construction; to management and operations of the infrastructure; and system evaluation (figure 5.1). This chapter will explore how such concerns might be addressed in the continuing process of development and renewal of transportation infrastructure. To better understand this, an overview of the planning process as generally implemented today is provided, as well as specific consideration of transportation planning within the Gulf States.

To rigorously address climate concerns, new approaches may be necessary. Since climate impacts occur into the future, and there is uncertainty as to the full magnitude and the timing of the impacts, deterministic methods as currently employed are ill suited to provide the type of information that current decision makers need. Instead it may be more fruitful to consider these impacts through a risk management approach to more effectively give transportation executives, elected officials, and the general public a more complete picture of the risks and potential solutions to climate impacts. The last section of this chapter begins the process of developing an alternate approach to planning with a conceptual framework for introducing more probabilistic approaches. Once fully operational, this type of methodology could lead to better information to address the changing climate.

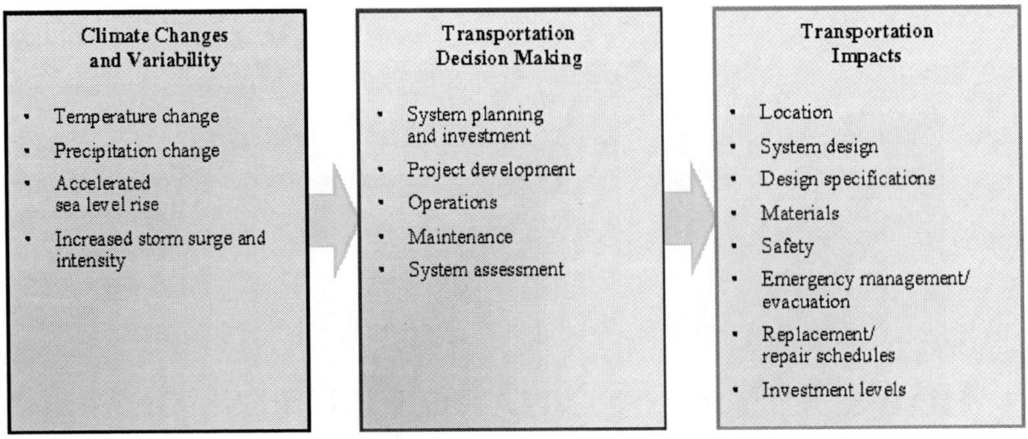

Figure 5.1. How will climate change affect transportation decisions?

5.1. Considering Climate Change in Long-Range Planning and Investment

5.1.1. Overview of the Surface Transportation Planning and Investment Processes

This section discusses how transportation planning and investment decisions are made in State and local governments and to some extent in private agencies. It reviews in particular the planning and decision making processes used by State departments of transportation (DOT) and metropolitan planning organizations (MPO). Specifically, it discusses the long-range planning taking place in the Gulf Coast study region and provides the results of a number of State DOT and MPO interviews. Finally, it discusses the challenge of how the planning process might be adapted to consider the potential impacts of climate change.

The Federal Surface Transportation Planning Process

Transportation planning processes vary with the type of agency (public or private), level of government (Federal, State, or local), mode of transportation, and other factors. This chapter will not attempt to provide an overview of all of them, but since the Federal government has specific requirements codified in law to cover the surface transportation planning process (for highways and transit investments), this chapter provides an illustrative example by using the Federal process.

Surface transportation planning and investment decision making, employed to make use of Federal transportation funding, is conducted within the framework and requirements defined by the planning provisions contained in Titles 23 and 49 of the United States Code (USC), most recently amended in August 2005 by the *Safe, Accountable, Flexible, Efficient Transportation Equity Act: A Legacy for Users* (SAFETEA-LU).

State DOTs and MPOs have lead transportation planning responsibilities, working in coordination with local governments. States and local governments may implement transportation infrastructure without Federal funding. These projects may be included within the framework of the Federal transportation process but could be implemented outside that framework.

Within the Federal process for highways and transit, State DOTs and MPOs must comply with the planning requirements to be eligible and to receive Federal transportation funds. The state DOTs within the study area are the Alabama Department of Transportation, Louisiana Department of Transportation and Development (DOTD), Mississippi Department of Transportation, and Texas Department of Transportation. Ten MPOs exist within the study area, as identified in table 5.1. Each MPO consists of one or more urbanized areas exceeding 50,000 in population, with an urban area exceeding 200,000 in population also defined as a Transportation Management Area (TMA).

The MPO's planning activities are identified in the Unified Plan Work Program which covers a 2-year period for the purpose of maintaining short- and long-term transportation plans. It is within this program that MPO staff collect data on traffic and pedestrian counts, building permits, planned developments, and accident rates, etc.; analyze trends; and evaluate potential projects. Two principal products result from the transportation planning process: a long-range transportation plan and a transportation improvement program. These two products, then, provide the basis for more detailed project development – engineering, design, and construction.

Table 5.1. Urbanized Area Metropolitan Planning Organizations (MPO) in the Gulf Coast Study Area

Urbanized Areas	2000 Population	Metropolitan Planning Organizations
Mobile, AL	354,943	Mobile Area Transportation Study
Baton Rouge, LA	516,614	Capital Regional Planning Commission
Houma, LA	108,474	Houma-Thibodaux MPO
Lake Charles, LA	183,577	Imperial Calcasieu Regional Planning and Development Commission
Lafayette, LA	215,061	Lafayette MPO
New Orleans, LA; Slidell, LA; Mandeville-Covington, LA	1,193,847	Regional Planning Commission of New Orleans
Gulfport-Biloxi, MS; Pascagoula, MS	363,987	Gulf Regional Planning Commission
Hattiesburg, MS	80,798	Hattiesburg-Petal-Forest-Lamar MPO
Houston, TX; Galveston, TX; Lake Jackson-Angleton, TX; Texas City, TX; The Woodlands, TX	4,669,571	Houston-Galveston Area Council
Beaumont, TX; Port Arthur, TX	385,090	South East Texas Regional Planning Commission MPO

Separate but coordinated long-range transportation plans are cooperatively developed on a statewide basis by a State DOT and for each urbanized area by an MPO. The long-range transportation plan is developed with a minimum of a 20-year forecast period, with many areas using a 30-year time horizon. The intent of a plan is to provide a long-range vision of the future of the surface transportation system, considering all passenger and freight modes and their interrelationships. As defined by SAFETEA-LU (23 USC 134 and 135), long- range plans, "shall provide for the development and integrated management and operation of transportation systems and facilities that will function as an intermodal transportation system." The transportation planning process for TMAs is essentially identical to that in urbanized areas having a population between 50,000 and 200,000 except that a congestion management process (CMP) also is required.

The transportation improvement program (TIP) is a separate document for the immediate future. It must be consistent with the long-range plan and provides the list of short-term (3 years) priorities for construction. A TIP must be developed for each metropolitan area, and a Statewide Transportation Improvement Program (STIP) must be developed for the State that is consistent with the TIPs. The STIP must be approved by U.S. DOT.

Environmental considerations have long played a role in the planning and development of transportation projects. Changes over time, though, have occurred in the manner in which environmental analyses have been conducted and the underlying legal framework in which these analyses are conducted. SAFETEA-LU, in section 6001, defines the following eight planning factors that should guide a transportation planning process and the development of projects, strategies, and services (figure 5.2):[1]

1. "Support the economic vitality of the United States, the States, nonmetropolitan areas, and metropolitan areas, especially by enabling global competitiveness, productivity, and efficiency;
2. Increase the safety of the transportation system for motorized and nonmotorized users;
3. Increase the security of the transportation system for motorized and nonmotorized users;
4. Increase the accessibility and mobility of people and freight;
5. Protect and enhance the environment, promote energy conservation, improve the quality of life, and promote consistency between transportation improvements and plans for state and local planned growth and economic development patterns;
6. Enhance the integration and connectivity of the transportation system across and between modes throughout the State, for people and freight;
7. Promote efficient system management and operation; and
8. Emphasize the preservation of the existing transportation system." [INSERT Figure

The SAFETEA-LU legislation requires that long-range transportation plans be developed in consultation with agencies responsible for land use management, natural resources, environmental protection, conservation, and historic preservation. Further, this consultation is to consider, where available, conservation plans or maps and inventories of natural or historic resources. This is typically a time- and labor-intensive effort requiring years to complete, with extensive public involvement that was made far more difficult by the 2005 hurricanes. The Gulfport MPO reports that in addition to the several years the overall effort took prior to

2005, the agency needed another year to reconsider the land use and demographic changes taking place as well as the Plan's regional goals to make them consistent with the Governor's Recovery Plan.

An interesting question is the manner in which the impacts of climate change can be addressed in the list of eight planning factors and the associated consultative process. As will be discussed later in this section, while climate change is not now named as part of any of the eight factors, a number of them reflect considerations that are directly related to climate change. In addition to protecting, enhancing, and mitigating impacts on the environment, these include system preservation, system management and operation, safety, and economic vitality (see especially factors 1, 2, 6, and 8).

Transportation plans, programs, and projects historically have been developed to meet the needs of future, projected, or planned land use, including population and employment patterns. In recent years, though, transportation and land use are being addressed in a much more interactive or coordinated manner. Rather than land use being viewed as driving transportation decisions, transportation investment and management decisions are increasingly being made collaboratively and in concert with growth management and economic development decisions. In this view, the manner in which transportation infrastructure is developed and managed is seen as one tool for helping to achieve desirable growth objectives.

Source: U.S. Department of Transportation.

Figure 5.2. SAFETEA-LU planning factors. Eight planning factors that should guide the development of plans, programs, and projects are identified in SAFETEA-LU

The overall transportation planning and investment process is illustrated in figure 5.3 with an emphasis that is helpful in identifying where in this transportation planning process considerations related to climate change impacts potentially could be introduced. Using terminology that is consistent with current planning and strategic management approaches, separate steps are identified for establishing a long-range vision and for establishing goals, objectives, and performance measures. Developing an *understanding of the problem* is seen as occurring on a continuing and iterative basis throughout the planning process, including the analysis of data and evaluating tradeoffs and establishing priorities among candidate policies and projects. The process culminates with development of a long-range transportation plan, a short-range transportation improvement program, and project development and implementation.

In terms of introducing climate-related changes into the long-range transportation planning and investment process, the potential exists at each step illustrated in figure 5.3. As shown, long-range environmental quality, economic development, mobility, and other desired conditions such as safety commonly are defined as part of a vision and accompanying mission statement and then translated into goals, objectives, and performance indicators. Thus, protection from climate change impacts could be introduced at these stages as well. Given these defined goals and objectives, strategies then are developed that are specifically designed to meet the agreed upon goals and objectives and are evaluated using the appropriate performance measures. Again, strategies could be developed that address climate change and variability. Similarly, climate change protection and mitigation strategies could be evaluated with respect to their potential impact on the transportation system.

Coordination in Transportation Planning

The Federal transportation planning and investment process is highly collaborative in which transportation agencies work in partnership with natural resource agencies, communities, businesses, and others throughout the period of planning, programming, developing, implementing, and operating transportation projects. Transportation agencies are charged with helping to accomplish multiple transportation; economic development; environmental, community, safety, and security objectives. Going beyond the Federally mandated process, the continued development and operation of the multimodal network requires extensive coordination.

Although planning and programming of the highway system, and its coordination with other modes of travel, are major responsibilities of the State DOT and the MPO, the actual development and operation of the transportation system is the responsibility of various levels of government and private agencies. States typically own and operate a relatively small portion of the road network but that portion (the Interstate System and arterial highways) usually accommodates the majority of the road travel. In some cases, States also own and operate local and state transit systems and freight rail lines. However, the majority of highway miles and transit systems are local responsibilities, and most of the Nation's freight system and air passenger system is owned by the private sector.

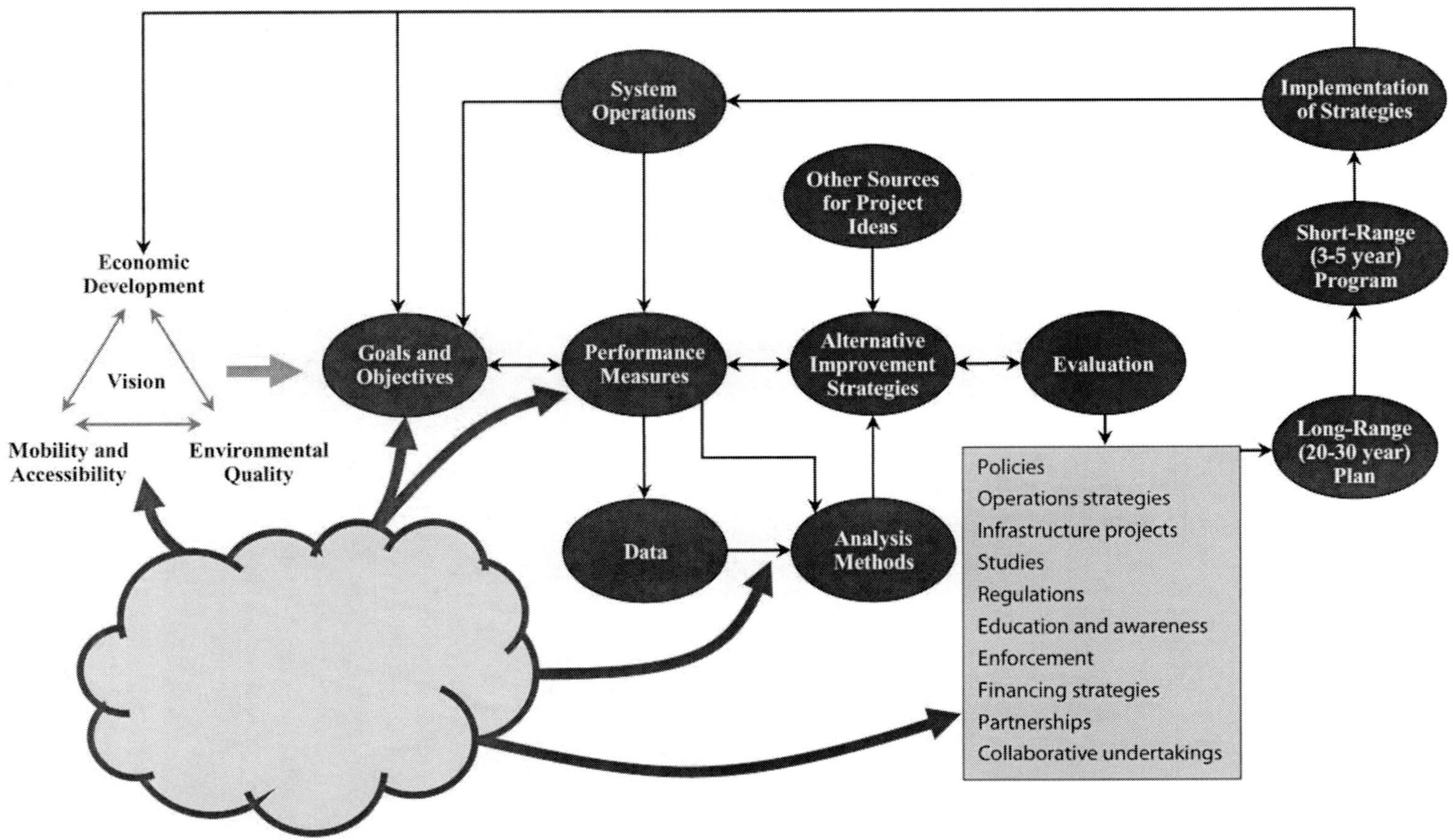

Source: Adapted from Michael Meyer, Georgia Institute of Technology.

Figure 5.3. Steps in the transportation planning process

Meeting the requirements of the Federal planning process is necessary as a condition of receiving Federal financial assistance. However, for States and MPOs the number of different organizations who have independent roles makes it important to have a collaborative decision making process, one that is based on valid and convincing information. At the MPO level, decisions are a collaboration of the individual local governments that comprise the MPO and serve on the policy board that is usually supported by the advice and analysis of a technical coordinating committee.

At the State level, the ultimate decisions are typically made by the Governor and the State legislature[2], with recommendations and advice coming from the State DOT. Decisions within the State DOT also occur at many levels and units within the organization. State DOT decisions encompass all aspects of the roadways under State jurisdiction: planning, engineering, operations, design, and construction.

Most of the freight and part of the aviation and passenger systems are owned by the private sector. State DOT and MPO plans that make recommendations for these systems must get the concurrence from the private sector for implementation. In the vast majority of the cases, the private sector invests in their current system or a new system if they feel it is cost effective to do so. The State and MPO may have some influence through the planning process or through the provision of financial assistance. For instance, a railroad will not likely move a rail line unless it improves their return on investment or because the government helps finance it.

Since the freight network is largely owned by the private sector, the long-range transportation planning process for both States and metropolitan areas ensures that the private users and providers of transportation are represented and their comments considered. In fact, the Federal planning regulations discussed above require that in developing or updating long-range transportation plans, States and MPOs shall have a process to allow freight shippers and providers of freight transportation services a reasonable opportunity to review and comment on key decision points and the proposed transportation plan. Planning agencies normally include private shippers and transportation providers on their plan advisory committees to guarantee representation early and throughout the planning process.

For these systems to be effective at efficiently moving people and goods – as well as meeting the higher needs of society in terms of economic development and environmental enhancement – a high degree of coordination is crucial. In terms of meeting the particular challenges that climate change poses, each entity, whether public agency or private firm, needs to consider how climate stressors might affect their businesses. Further, these agencies need to work together to consider how climate changes affect the efficient movement of people, goods and services as a whole to take full advantage of system redundancy and resilience, explained later in this chapter.

5.1.2. Current State of Practice in Incorporating Climate Change Considerations

In this Gulf Coast Study, representation of the private freight industry was sought during the development of the modal technical papers. For example, railroads were involved in the review of the rail technical paper, and discussions were held with the Association of

American Railroads about possible impacts to rail lines from climate change such as "sun kinks" and the importance of prestressed rail track. The CSX Railroad provided significant information on hurricane Katrina impacts and adaptation strategies through public comments and the sharing of information. The CSX Railroad reported that it cost about $250 million to repair damage from Katrina, and the damage caused them to further consider relocating the rail line. The CSX Railroad is exploring the feasibility of new construction within the existing corridor but further inland. Also, increased use of alternative Mississippi River crossings (Baton Rouge/Vicksburg) is under study. Interviews included a private toll road authority and port employees for two separate ports (Galveston and Houston) that were publicly owned but privately operated facilities. The toll road representative expressed concern about potential impacts of sea level rise since the toll facilities do approach the coastline, particularly in the Houston metropolitan area. The port representatives also were concerned about the impacts of possible sea level rise and the impacts of increased precipitation on sedimentation of port channels and port runoff that could cause local flooding. In the next phase of the Gulf Coast Study, the private sector involvement will be intensified to determine what specific climate change impacts are possible and in detailing likely adaptation strategies and costs.

Two approaches were utilized to determine how state DOTs and MPOs currently are addressing issues of climate change and also how climate change might be addressed in the future. The approaches involved:

1. Obtaining and reviewing current long-range transportation plans, transportation improvement programs, and other recent documents for the States and selected MPOs within the study area, addressing infrastructure development, operation, and management; and
2. Interviewing State DOT and representative MPO officials responsible for transportation planning within the study area.

Some MPOs within the study region currently are in the process of updating their vision statements and long-range transportation plans. In some of these cases, MPOs are actively considering issues related to the potential effects of climate change and variability, including the impacts of hurricanes such as Katrina and Rita. The two aspects of climate that are receiving the most attention in these more recent planning activities are: (1) evacuation planning and management and (2) preventing infrastructure damage resulting from storm surge-related flooding.

Long-range transportation plans, statewide transportation improvement programs, and annual reports were obtained, where available, from the Internet for the States of Alabama, Mississippi, Louisiana, and Texas. In addition, the corresponding documents were similarly obtained for the following urban areas:

- Mobile, AL (South Alabama Regional Planning Commission);
- Hattiesburg, MS (Hattiesburg-Petal-Forrest-Lamar Metropolitan Planning Organization);
- Gulfport, MS (Gulf Regional Planning Commission);
- Lake Charles, LA (Imperial Calcasieu Regional Planning and Development Commission);

- Lafayette, LA (Lafayette City-Parish Consolidated Government Metropolitan Planning Organization);
- New Orleans, LA (Regional Planning Commission for Jefferson, Orleans, Plaquemine, St. Bernard and St. Tammany parishes); and
- Houston and Galveston, TX (Houston-Galveston Area Council).

None of the existing State and MPO documents examined here, all of which date from 2000 to 2006, directly addresses or acknowledges issues of climate change and variability. This is, in part, due to their age; most were developed 2 to 4 years ago, prior to the recent increase of interest in climate change and the associated increase in the availability of climate change-related information. Also, most of these documents were prepared prior to Hurricanes Katrina and Rita, so the many actions being taken by State DOTs and MPOs in response to these two storms have only recently been included in updated and published documents.

The following observations result from a review of these planning documents, organized into the following three categories: plans including missions and goals, planning activities, and prioritization criteria.

State and Metropolitan Planning Organization Plans and Planning Activities

Most of the state and MPO plans in the region include a mission or goals that include statements about providing environmentally sound transportation systems or preserving the quality of the environment and enhancing the quality of life. There also are goals that include strategies to encourage land use planning and to incorporate public transportation, walking, and bicycles. Essentially, all of the plans recognize the environmental impacts (excluding climate change) and issues related to transportation growth and expansion. The Louisiana long-range transportation plan defines 57 "mega projects," whose evaluation criteria for development and implementation include environment, demonstrating context- sensitive design and/or sound growth management principles, and emergency evacuation capabilities. Nine of the 22 Priority "A" mega projects involve I-10, including construction of a six-lane I-10 Twin Span across Lake Pontchartrain. Other Priority "A" mega projects located in evacuation areas include upgrading I-49 south of Lafayette and construction of a new two-lane road between U.S. Highway 90 and Louisiana Highway 3127. The Houston- Galveston long-range transportation plan identifies eight distinct ecological zones within the region and pays particular attention to the wetlands, which protect shoreline areas from erosion and serve as buffers from flooding.

As mentioned above, however, some of the planning activities and infrastructure reconstruction since Hurricanes Rita and Katrina are being done to address the impacts of climate change. In Mississippi, the flooding that resulted from Hurricane Katrina has resulted in new design standards for the bridges that are being rebuilt and is serving as a catalyst for considerable debate on the interrelationships between land use and transportation investment within the coastal areas of the State.

The Regional Planning Commission for the New Orleans urbanized area and the Mandeville/Covington and Slidell urbanized areas is refining its metropolitan transportation plan (MTP) for the New Orleans region so that it can provide a framework within which the projected climate change effects can be assessed and addressed. The Houston-Galveston Area Council (H-GAC) is in the process of conducting a visioning exercise, the results of which will then guide the development of an updated regional transportation plan. Since this is occurring post-Hurricane Rita, climate change and the means of reducing the risk of flooding have been raised in the outreach sessions and working meetings.

In addition to including policies to provide maintain and improve the area's intermodal systems, the States and MPOs in the study area also are including consideration of future uncertainties and evacuation management. The Mississippi transportation plan and associated STIP both acknowledge uncertainty in future year conditions in areas such as growth, air quality, road maintenance, and congestion. The STIP contains a section on planning and research that describes planning as looking at what has to be done today to be ready for an uncertain tomorrow. While climate change and variability are not explicitly mentioned in either the current plan or the STIP, and the major effects of climate change may not occur within the plan's current 30-year timeframe, the stage certainly is set to both recognize and respond to potential issues of climate change in future planning activities.

Following Hurricane Rita, the Governor of Texas established a task force on evacuation, transportation, and logistics. The report of this task force was completed and submitted on February 14, 2006. Twenty recommendations are made, including the development of contraflow plans for major hurricane evacuation routes, including some in the study area, such as north out of the Houston-Galveston metropolitan area on I-45, U.S Highway 290, U.S. Highway 59; west out of Houston on I-10; and north out of Beaumont on U.S. Highway 69. Evacuation routes represent one element of the operations and system management portion of the long-range transportation plan for the Houston-Galveston metropolitan area, with extra points given to evacuation routes in the prioritization of projects. Short-term recommendations to improve evacuation capabilities were developed in 2006. Longer-term evacuation priorities also are being assessed, some of which may require significant public investment over a period of many years, according to the task force report. These may include new evacuation routes, reconstruction of existing evacuation routes, and reduction in the number and severity of traffic bottlenecks. The location of new development in flood- and storm-prone areas also is arising as an issue.

Site Interviews with Transportation Representatives in the Gulf Coast Region

In addition to reviewing planning documents and interviewing DOT and MPO officials as described above, another set of interviews was conducted between December 15, 2006, and January 10, 2007, to understand in more specific terms the issues facing the area selected for more intense study in Phase II of this effort. These interviews included a representative of each of the transportation modes represented in the site study area. The objective of the study site interviews was to consider the potential climate impacts at the level of the individual decision maker/planner. This information was used to develop and refine the conceptual framework for assessing potential impacts on transportation presented below. There were three general lines of inquiry used to generate a localized picture of climate change impacts and transportation decision making:

1. Interviewees' Perspectives on Climate Change – Respondents were asked about their perception of climate change, its potential impact on the respondent's specific facility or system, and whether or not the respondent currently was incorporating climate change and variability science or indicators in their decision making and planning.

2. Decision and Planning Processes in which Respondents are Involved – Interviewees were asked to describe the types of decisions they are engaged in at the facility and/or system level in their area of responsibility. The interview guide solicited responses in regard to the factors that were the most relevant to making facility or system decisions; the role of the respondent in the local decision and planning process and interactions with the State and Federal processes; what information was used for informing these decisions; and what threshold or tipping point factors would facilitate changes in policy or planning, both from the climate perspective and in general.
3. Utility of the General Project Report Findings – Respondents were asked their opinions regarding the applicability of the climate scenarios and various report concepts that might be used in their analysis. The respondents were presented with a two-page summary of study findings – including climate scenarios for the study area, and the assessment of exposure, vulnerability, and resilience – for their review and input.

The interviews were designed and conducted according to standard social science research methodologies and practices. The questions were open ended in order to solicit a range of responses as broad as possible.

The interview subjects were contacted and interviewed by using a questionnaire approved by the Texas A&M University Institutional Review Board. As such, they were informed that their expressed opinions and any information they provide would be kept confidential and that they were free to refuse to answer any questions that made them uncomfortable. Because of the size and public nature of the research area, only limited references are made to the positions of these individuals within the hierarchy of their system or institution.

Fourteen individuals were interviewed, four of whom provided general context information on climate change and variability and the Galveston County area, and 10 of whom were formal interview subjects. These included:

- An employee of Transtar, the Houston Traffic Management Center;
- An individual responsible for evacuation in the Galveston County area,;
- A representative of a toll road authority;
- Employees of the City of Houston Aviation Department;
- A county engineer;
- Employees of the Texas DOT; and
- Employees of the Ports of Galveston and Houston.

Significance of climate considerations – Although the respondents were comfortable with the idea that climate conditions would be changing in the Gulf Coast area, most respondents reported that climate was not an issue that they considered in development of the plans and TIPS. The perceptions of the respondents were that climate change is an issue that has been of limited concern to the State and Federal agencies that affect their decision making, yet responses varied. Representatives of at least one agency indicated a strong belief that climate change should be treated as an issue of importance in the transportation planning of the region. In contrast, others indicated that climate change is not an issue that has received any official treatment. Several interviewees felt that future consideration of climate change would be directed by guidelines established by the Federal government.

None of the interview subjects indicated they were using climate change data in their transportation decision making. However, the entire sample of interview subjects was convinced that climate change is a matter of some concern.

Value of climate information – Findings from the general project synthesis report were of some use to the interview subjects. At least one interview subject indicated s(he) had not been concerned with climate change until s(he) saw the predictions for sea level and storm surge in the Galveston County area. The value of the specific predictions varied from one respondent to the next. Many respondents found sea level rise and storm surge information to be useful; however, they would like the projections to be for time periods more applicable to their own decision making timeframes. At least one respondent suggested that the elevations for storm surge and sea level should be selected from a range more relevant to the Galveston County area. Much of Galveston County is at an elevation of 4.6 m (15 ft); the 5.5-m (18-ft) threshold used in the storm surge map was not as relevant as this decision maker would like.

Perceived importance of individual climate factors – The degree to which respondents considered various climate stressors to affect the transportation infrastructure modes for which they were responsible is characterized in table 5.2 with a scale of low, limited, moderate, high, and highest perceived concern.

The high degree of concern exhibited by all respondents about *storm frequency and magnitude* as a stressor betrays the strong affective power of recent hurricanes on the hazard perceptions of respondents in the Galveston and Harris County area. The majority of subjects expressed their concern for storm frequency and magnitude in regards to the capacity of their infrastructure mode of responsibility to fully function during a hurricane evacuation, or in the case of the port, to be evacuated. An exception was the flood control subject who shared this fear but was primarily concerned about the ability of the drainage system to cope with severe storms.

Temperature was of limited importance to the respondents with the exception of the Transtar subject who described his equipment as tested and hardened against temperature extremes and the airport representative who described temperature as a key variable in airport performance measures. The other airport representative was not as concerned about temperature. We account for this variation as a function of their respective roles. The second representative is involved in construction and does not directly grapple with operations logistics. Operations logistics are heavily determined by temperature because increased temperature reduces lift and results in an increase of the airport facility's average annual delay of departures.

Average precipitation was of limited importance to many of the respondents in comparison to *extreme precipitation* events. Of special interest was the flood control engineer who indicated that increases or decreases in average precipitation have limited effect on flooding. His concern was principally with precipitation events that could be categorized as high in intensity, frequency, and duration. The one interview subject who was directly and seriously concerned with overall precipitation levels was the port engineer, who linked average levels of precipitation to the sedimentation of port channels. The second port engineer and manager were concerned with precipitation as well, especially with the consequences of port runoff for local flooding.

Sea level was of high importance to many of the interview subjects. The factor that governed the strength of this concern was proximity to the coast, moderated by the relative imperviousness of the infrastructure in question. For example, the toll road authority representative expressed a potential concern about sea level because the toll facility does approach the coast; however, this facility was designed to be elevated well above the surge levels predicted in the climate and vulnerability summaries, as well as the levels to which this respondent was previously familiar. Other respondents had broader purviews of responsibility such as multiple highways, the evacuation of residents, and facilities near sea level. These respondents expressed high concern about sea level rise. The port representatives characterized their concerns about sea level rise differently. One port engineer was highly concerned about sea level rise, but this respondent noted that his concern was coupled with his concern about local subsidence. The second port interview subject could imagine sea level rise having an impact on the region; however, the infrastructure elements of concern – piers – were rebuilt often enough that only a catastrophic degree of sea level rise would have any impact. This respondent explicitly stated that such an event was highly unlikely.

The responses in regards to questions about decision making thresholds were fairly uniform. Interview subjects suggested that the impetus to make fairly radical policy shifts could only come from higher levels of government and usually in response to a disaster. Otherwise, they simply did not have the autonomy or the access to funding to adopt new policies or planning approaches.

Table 5.2. Level of Decision Maker Concern about Climate Stressors

Area of Interest of Interviewees	Sea Level	Precipitation	Temperature	Storm Frequency and Magnitude
Traffic Management – Transtar	Moderate	Moderate	High	High
Emergency Management	High	Limited	Limited	High
Toll Authority	Low	Limited	Limited	High
Aviation	Limited	Moderate	Highest	High
Aviation	Limited	Moderate	Moderate	High
County Engineer	High	Limited	Limited	Highest
Port Engineer	High	High	Limited	Highest
Port Engineer	Low	High	Limited	Highest
Flood Control – Houston	Limited	Limited	Low	Highest
State Transportation Engineer	Highest	Limited	Low	High

Since these interviews were conducted, however, there appears to have been a shift in some of the expressed opinions due to the impacts of Hurricanes Katrina and Rita, as evidenced by adaptation measures being undertaken. For example, as detailed in chapter 4.0, the rebuilding of certain facilities, like U.S. Highway 90 in Mississippi, have taken into account the likely impacts of future storms. Further, the activities and opinions expressed to the study authors by State and local authorities indicate a much greater appreciation for the potential impacts of climate change than those of the interviewees.

The involvement of private users and providers of freight transportation in these interviews was limited. Employees at two public ports that use private facilities and a private toll road authority representative were interviewed; however, the private sector's involvement in the next phase of the study will be substantially expanded to capture specific impacts and adaptation activities. Also, additional insight to private sector impacts and adaptation considerations were learned from other regions of the study area in the aftermath of Hurricane Katrina. As an example, the CSX Railroad received extensive damage on the Gulf Coast, particularly in Mississippi and Louisiana, and had to consider alternative adaptation strategies such as rerouting, rehabilitation with strengthening, or relocation further inland.

5.1.3. Challenges and Opportunities to Integrating Climate Information

Transportation agencies consider a broad range of future conditions, including demographic, environmental, economic, and other factors. It is within this broader context that it is reasonable for some agencies to address the additional consideration of climate change over the lifetimes of their transportation facilities, to the extent possible.

Over time, fundamental and significant changes may be desirable in the manner in which long-range transportation plans are developed and investment decisions are made. Similar to what transportation agencies are now doing to address freight, safety, economic development, environmental mitigation, and other emerging issues, considerations of climate change can be incorporated in each step of the transportation planning process, particularly during the earliest parts of the planning process – the formulation of a vision and the development of goals and objectives.

Timeframes

Long-range transportation plans are developed with a time horizon that typically extends 20 to 30 years into the future. Most long-range transportation plans being developed today have time horizons of 2030 or 2035. However, as illustrated in figure 5.4, individual facilities being recommended in those plans will be designed with a considerably longer service life. For instance, bridges being built today should last 60 to 80 years or more. Furthermore, bridges being proposed in the long-range plans will be designed to last beyond 2100. Although the timeframe for significant climate change might appear to be longer than most plan horizons, studies have found that the effects of climate change are being experienced today. And while climate change is typically thought of as a gradual, incremental process over many years, scientists expect that climate changes are likely to include abrupt and discontinuous change as well. To begin to adequately consider the implications of climate change, transportation planners would benefit from consideration of longer time horizons.

Climate changes over longer time periods could be addressed as part of a long-term visioning that helps determine where transportation investments are needed and should be located. This process would inform the transportation planning process with supplementary information. For example, in the planning process depicted in figure 5.3, climate change could be added to the initial visioning step, along with other factors such as economic and environmental considerations.

While it is difficult to know the planning horizons of private companies, given their proprietary nature, it is likely that their focus would benefit from an expanded time horizon as well. Since the infrastructure likely to be affected by future climate impacts is currently under development, planners and decision makers need to start now in considering how climate changes may affect them.

Land Use

Responding to the potential effects of climate change, as demonstrated by the ongoing discussions in Texas, Louisiana, and Mississippi, may involve changes in the location of transportation facilities, housing, and business. Transportation planning already attempts to forecast these types of demographic and economic shifts. Potential changes in the future climate and its resulting impacts on the existing ecology may make such forecasting far more difficult.

A further challenge for transportation planners and climate scientists is to better understand the interplay of the built environment with the local ecology toward the betterment of both. For example, barrier islands serve to protect existing infrastructure by reducing the impacts of major storms. Preservation of these ecologically sensitive coastal wetlands is one way of minimizing damage from hurricanes, by restoring critical buffer areas that absorb storm energy. Similarly, a variety of human activities are contributing to the current and projected rate of land subsidence, including but not limited to, the location and management of navigation channels. The impacts of climate change will likely make understanding and protecting these natural systems even more important, not only for their own sake but to prolong the viability of transportation infrastructure. The development of the full range of port, pipeline, and shipping facilities, and their supporting land transportation infrastructure, can be examined for their potential to either directly or indirectly affect coastal areas. In essence, this is extending the concept of "secondary and cumulative effects" (as required under the National Environmental Policy Act [NEPA]), to include coastal ecology and storm protection. Similarly, strategies proposed to protect coastal areas should be screened for potential implications on the transportation system.

Institutional Arrangements

Existing institutional arrangements may not be sufficient for transportation agencies to fully address and respond to issues of climate change. Increased collaboration may be necessary for transportation planning and investment decision makers to effectively respond to climate change issues, including their partnering with climate change specialists. State DOTs and MPOs already are consulting with resource agencies such as natural resources, conservation, and historical preservation in the planning process. Collaborating on climate change might be a natural extension of that consultation process.

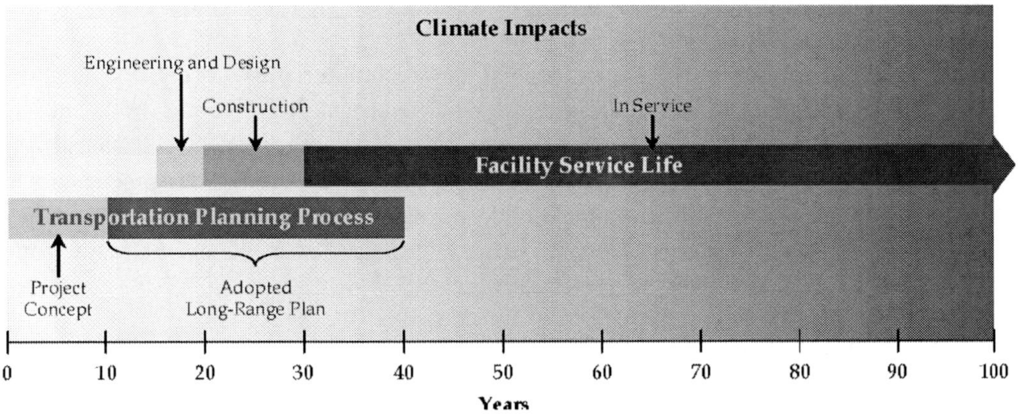

Figure 5.4. Relationship of transportation planning timeframe and infrastructure service life to increasing climate change impacts

It also will be necessary for State DOTs to collaborate within their agencies so that planning, engineering, and programming have a common understanding of the potential for climate change and the alternative responses possible. Likewise, the MPOs need to accomplish a similar effort with their local governments. Finally, for the vast amount of the transportation system owned by private agencies, climate change information must be made available to them so that their decisions can be coordinated with and compliment those of the public sector. In some cases, this may lead to public/private investment options.

A New Approach

Based on currently available climate change information, there appear to be important implications of climate change for the manner in which transportation investments are planned, developed, implemented, managed, and operated. This report shows that these implications are sufficiently significant that transportation planners should develop an improved understanding of climate change issues and reflect them in their decision making today.

The long timeframe for climate change, as compared to the existing 20-year view of most transportation plans, makes the specification of its impacts considerably more difficult. Instead of relatively precise estimates of potential impacts that are needed for many aspects of transportation planning, broad ranges are more typically what climatologists currently can provide. Given this lack of certainty, climatologists are moving toward the determination of probabilities of potential impacts.

Currently, the transportation planning process does not consider probabilities in determining future travel demand and ways to meet it.[3] Instead, transportation professionals generally rely on more deterministic methods that yield a single answer based on the inputs - such as well accepted engineering, construction and other standards - along with professional judgment.

Such methods are ill-equipped to addressing the uncertainties associated with the timing and magnitude of many climate change impacts. What is needed are new tools that can address the uncertainties associated with climate change and yet provide more useful information to the transportation community that would be used to create a more robust and resilient system.

The following section provides a conceptual approach that represents the first step toward development of such a tool. It suggests a new approach to viewing both individual transportation facilities and the system as a whole, borrowing concepts and relationships from ecology, risk management, decision theory, and transportation practice. It proposes a way to help planners, designers, and engineers think through the potential harm that changing conditions in the natural environment might cause and the ability of the existing and proposed facilities to withstand such harm.

5.2. Conceptual Framework for Assessing Potential Impacts on Transportation

While climate factors are not usually considered for transportation planning purposes, as shown in the previous section, some agencies are beginning to explore how they might be incorporated. This section attempts to provide a conceptual approach to how climate concerns – with their inherent uncertainties – might be addressed in a transportation context. This is a first step toward creation of a way to consider risk and uncertainty in transportation planning as an alternative to the largely deterministic approaches currently employed. Further refinement will be necessary in Phase II of this study to make this approach operational in a pilot test area.

While the focus of this project is on a portion of the U.S. Gulf Coast, the intent is to develop a conceptual framework that lays the groundwork for an assessment that links climate change and transportation and to focus on this nexus by using a specific case as an illustration. Climate change impacts vary by region, with some areas being more vulnerable to some aspects of exposure than others. Regardless of the specific site characteristics related to this chapter, the general framework and relationships between information, decision maker, and process will be transferable to other situations. Developing a conceptual framework at this stage in the research, rather than a static tool or model, provides the transportation sector with the basic understanding of these relationships at this early stage of recognition of the potential impacts of climate change and variability on transportation infrastructure.

This section focuses on: (1) a description of the basic factors that can be useful in an assessment of the potential impacts of climate change on transportation and (2) a description of the development of a conceptual framework incorporating these basic components.

5.2.1. Factors of Concern: Exposure, Vulnerability, Resilience, and Adaptation

There are four major conceptual factors to consider climate concerns in transportation: exposure to climate stressors, vulnerability, resilience, and adaptation. These concepts and their definitions are borrowed from, and consistent with, ecological and hazard assessment practices and represent transportation infrastructure's probable levels of exposure to damage from climate change factors, its capacity to resist such damage or disruption of service, and its ability to recover if damaged. For purposes of this project, we adapted the Intergovernmental Panel on Climate Change (IPCC) definitions of these concepts, in general, with reference to applied and theoretical applications for more specific or articulated examples. It was determined by the research team to closely approximate the IPCC terminology and methodology, as this also informs many other regional and sectoral assessments conducted in the United States and elsewhere.

With specific regard to climate change, *exposure* comprises the "nature and degree to which a system is exposed to significant climatic variations" (IPCC, 2001, p. 987). Exposure also is often articulated as the probability of occurrence (the probable range of climate change stressors, such as sea level rise or increased rainfall) and the physical characterization of the local area. In this study, *exposure* is the combination of stress associated with climate-related change (sea level rise, changes in temperature, frequency of severe storms) and the probability, or *likelihood*, that this stress will affect transportation infrastructure.

While there are different kinds of exposure (see Tobin and Montz, 1997, for a discussion), two types are applicable to this approach: perceived (based on the situational perspective of the particular decision maker) and predicted (based on "objective" measures). For predicted exposures, the following environmental impacts appear to be most relevant in the central Gulf Coast region, depending on the specific infrastructure component and location:

- Sea level rise, historical trends, and predicted range (including rates of subsidence and/or erosion;
- Temperature range, scenarios, and probability distribution functions (with special consideration to changes in extreme temperatures);
- Precipitation range, scenarios, and probability distribution functions and intensity; and
- Major storm characteristics (projected magnitude of storm surge and winds, as well as frequency).

Vulnerability, in general, refers to the "potential for loss" (Tobin and Montz, 1997) due to *exposure* to a particular hazard. The IPCC defines vulnerability as: "the degree to which a system is susceptible to, or unable to cope with, adverse effects of climate change, including climate variability and extremes. Vulnerability is a function of the character, magnitude, and rate of climate variation to which a system is exposed, its sensitivity, and its adaptive capacity" (IPCC, 2001, p. 995). More specifically for this project, vulnerability considers the structural strength and integrity of key facilities or systems and is defined as the resulting potential for damage and disruption in transportation services from climate change stressors. The vulnerability of a facility or system then depends on the level of exposure to which it is subject.

The risk that a transportation facility or a system faces can be defined from these notions of exposure and vulnerability. Risk is the product of the probability that a facility will be exposed to a climate stressor of destructive (or disruptive, at the systems level) force times the damage that would be done because of this exposure.

While transportation is frequently thought of as the built infrastructure, transportation's value to society is the service or performance this system of facilities and operations provides to move goods and people. Loss of capacity is the reduction from full performance capacity for a particular transportation system or facility. For example, Berdica (2002) defines vulnerability to the road system as a problem of reduced accessibility. System vulnerabilities to specific locational risks will vary based on the performance expectations of those specific system segments. The loss in performance would be the reduction of system capacity measured according to the relevant metrics. For example, highway capacity would be measured in volume of traffic flow; a loss in performance would be gauged by the reduction of traffic flow capacity.

It is important to note that vulnerability, like exposure, may be perceived differently among stakeholders and across modes. Key factors for the determination of transportation facility or system vulnerability may include:

- Age of infrastructure element;
- Condition/integrity;
- Proximity to other infrastructure elements/concentrations; and
- Level of service.

The concept of *resilience* is used to refer to the restoration capacity of the infrastructure at the facility and system level. In general, resilience is defined as the "amount of change a system can undergo without changing state" (IPCC, 2001, p. 993). In the climate change context, resilience also refers to regenerative capacity, the speed of response and recovery of various system elements, and mitigation and adaptation efforts. It also is generally considered to be a "multidimensional concept, encompassing biogeophysical, socioeconomic and political factors" (Klein et al., 1998, p. 260). Adger et al. (2005) define resilience more specifically as the capacity of a system to absorb disturbances and retain essential processes.

We can apply these concepts to the transportation context. System-level resilience is particularly important in the transportation sector because of the inherent connectivity of transportation facilities. Resilience can be looked at as the ability of a transportation network to maintain adequate performance levels for mobility of goods and services through redundant infrastructure and services. The fact that one component is out of service may not be crucial in areas where alternative transportation facilities or services are available. For an individual facility such as a road or bridge, resilience can be thought of as how quickly full service can be restored either through repair or replacement.

Key factors influencing resilience in our conceptual framework can be categorized across three dimensions: mode or structure (highway segment or port, for example), socioeconomic (political will and resources), and system-level factors. These factors may include:

1. Mode/structure:
    - Repair/replacement cost; and
    - Replacement timeframe.
2. Socioeconomic:
    - Public support;
    - Interorganization cooperation;
    - Economic resources; and
    - Social resources.

3. System level:
    - Redundancy among components;
    - Essential service resumption;
    - System network connectivity;
    - Institutional capacity; and
    - Relevance of existing plans for response to events (e.g., floods).

Transportation planners and decision makers may consider these factors (either formally or informally) and generate a basic perception of resilience. For example, for any given facility the relevant decision maker would have a general idea as to: (1) how much replacement would cost; (2) how long it would take; (3) the economic resources available for replacement; (4) public sentiment regarding replacement (or not); (5) how essential the facility is to system performance; and (6) whether or not plans exist for dealing with disruption of facility and/or system performance over the duration of the replacement time. This understanding of the resilience of the facility or system can be based on either the general feeling and experience of the decision maker, or it can be developed systematically with quantifiable measures.

The IPCC defines *adaptation* as the: "adjustment to natural or human systems to a new or changing environment. Adaptation to *climate change* refers to adjustment in natural or human systems in response to actual or expected climatic stimuli or their effects, which moderates harm or exploits beneficial opportunities" (IPCC, 2001, p. 982). An associated concept, *adaptive capacity*, refers to "the ability of a system to adjust to climate change (including climate variability and extremes) to moderate potential damages, to take advantage of opportunities, or to cope with the consequences" (IPCC, 2001, p. 982).

In this project, we are interested in understanding adaptation as a decision that officials can make in response to perceptions or objective measurements of vulnerability or exposure. For example, given a certain climate change scenario, a decision maker may choose to advocate for certain adaptive policy responses beyond the status quo. This can be determined through interviews by asking such questions as: what is the planning horizon for this specific area; what factors (political and resource) constrain or encourage adaptive behavior in this area of concern; and what are the stakeholder perceptions of uncertainty in regard to the data and information provided and available for informed decision making (see Jones, 2001, for an example)?

Adaptive strategies can be further delineated into three possible alternatives: protect, accommodate, and retreat. These adaptive responses are derived from the IPCC framework for assessing coastal adaptation options (Bijlsma et al., 1996). Within the context of our case study in a coastal region, the *protection* strategy might aim to protect the land from the sea by constructing hard structures (e.g., seawalls) as well as by using "soft measures" (e.g., beach nourishment, wetland restoration). *Accommodation* may call for preparing for periodic flooding by having operational plans in place to redirect traffic, for example, or cleaning up roadway obstacles to return to normal service. The *retreat* option would involve no attempt to protect the facility from the climate stressor. In an extreme case along a coastal area, for example, a facility or road segment could be abandoned under certain conditions (sea level rise, persistent storm surges that reduced the feasibility of replacement). From a system perspective, it could be determined that retreat is the best decision if the road segment could be relocated without loss of system service; if performance can be maintained through other system components; or if service is no longer required due to shifts in population and commerce.

A related concept, *threshold*, also will be considered in the framework. Threshold has been defined as the point where a stimulus leads to a significant response (Jones, 2001; Parry, Carter, and Hulme, 1996). In the case of transportation decision making, we are interested in determining at what point within an assessment or decision process change is induced. A threshold can be quantified under certain circumstances (for example, the impact of temperature on pavement construction decisions), or it may be subjective, depending on the situation. Jones (2001) suggests two general thresholds for infrastructure: (1) economic write-off, when replacement costs less than repair and (2) a standard-derived threshold, when the condition of the infrastructure component falls below a certain standard. These variables can have both quantitative and qualitative characteristics. In this phase of the research, the focus is on determining qualitative characteristics and their general utility to decision makers (see Cutter et al., 2000, for a similar approach).

In summary, the following are working definitions that were applied in this section of the research. These definitions were developed in conjunction with the research team, the Federal Advisory Committee, and other experts.

Exposure – The combination of stress associated with climate-related change (sea level rise, changes in temperature, frequency of severe storms) and the probability, or *likelihood*, that this stress will affect transportation infrastructure.

Vulnerability – The structural strength and integrity of key facilities or systems and the resulting potential for damage and disruption in transportation services from climate change stressors.

Resilience – The capacity of a system to absorb disturbances and retain essential processes.

Adaptation – A decision that stakeholders can make in response to perceptions or objective measurements of vulnerability or exposure. Included in this concept is the recognition that *thresholds* exist where a stimulus leads to a significant response.

Each of these four factors is critical in our understanding of how climate change may impact transportation in the study region. As illustrated in figure 5.5, an initial risk assessment for a facility or system will include analysis of the first three factors: exposure, vulnerability, and resilience. Once a risk assessment is conducted, choices for an appropriate adaptation strategy can be considered. The implementation of a particular adaptation strategy – to protect, accommodate, or retreat – will in turn affect subsequent risk assessments by changing one or more aspects of risk. The effectiveness of the adaptation strategy can be assessed by the degree of success in maintaining system or facility performance.

5.2.2. Framework for Assessing Local Climate Change Impacts on Transportation

Having introduced the major factors for consideration in a climate change impact assessment, this section introduces the conceptual framework and outlines the input and outputs. This is followed by a description of an approach to implementing such a framework.

In general, the objective is to illustrate how climate change/variability can be integrated into existing processes for transportation policy and decision making toward the development of adaptation strategies. Even at the conceptual level, this process can assist transportation decision makers in considering the potential impacts from climate change and variability on a wide range of transportation infrastructure components of any type, including air, rail, marine, transit, or highway, as well as the overall intermodal system. It is intended to be implemented primarily at the State or local scale, since climate impacts differ by region of the country.

The framework can help direct local decision makers in raising and to some extent answering such questions as: what are the likely changes in sea level (for example) in my area; how vulnerable is the transportation infrastructure related to this probability in my area; and at what point should decision makers seek adaptive strategies to address this? The resulting information can then be utilized for making adaptation decisions.

Needed Data

Previous chapters outlined the physical, infrastructure, and socioeconomic data that was collected and aggregated specifically for the Gulf Coast study area. This section discusses how this data serves to help assess the exposure and vulnerability of any transportation network. While not all of the data collected for this project would be available to local transportation stakeholders, much of the data is available and is being updated on a regular basis.

Within this conceptual framework, the analysis begins with an assessment of what climate impacts can be determined with a relatively high degree of confidence. This is the basis for the exposure analysis, including some idea as to the probability that transportation facilities will be exposed to particular impacts. For the Gulf Coast Study, various climate scenarios were analyzed and probable impacts identified at the regional level, including sea level rise, increased storm intensity, extreme temperature increases, and potential ranges quantified. The infrastructure and services will be exposed to these impacts.

The vulnerability of specific portions of the transportation infrastructure will depend on its location relative to the location of the impacts, as well as other characteristics. Sea level rise is a good example, as coastal infrastructure will be more vulnerable than inland facilities. Based on location, the physical characteristics of the region, and socioeconomic data, the vulnerability of transportation facilities can be assessed.

From the probability of an exposure to a climate impact and the assessment of vulnerability, some idea of the risk the facility or the system faces can be determined. In order to do this, repair or replacement costs, economic losses, or other metrics of potential damage must be developed. In addition, precise estimates of risk would require quantitative estimates of exposure. Whether risk can be quantitatively determined remains to be seen.

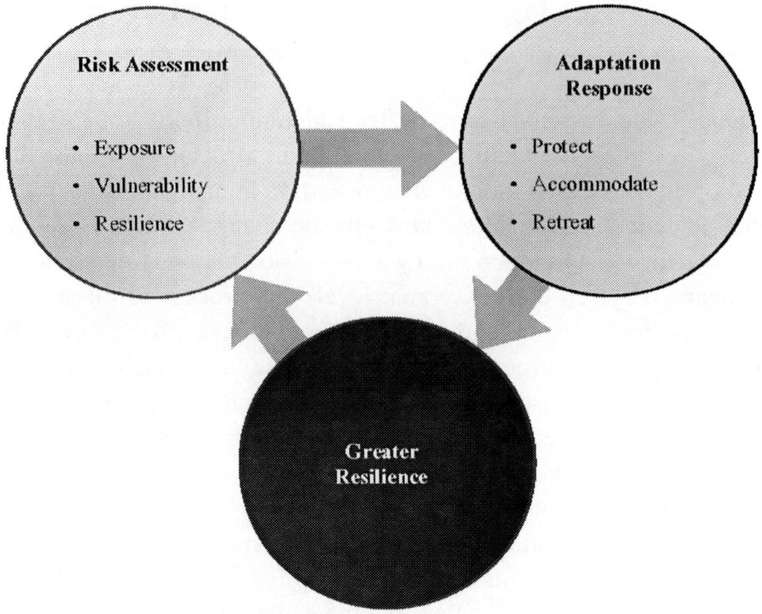

Figure 5.5. A risk-assessment approach to transportation decisions

Resilience was not addressed in the first phase of the Gulf Coast analysis, but will be in the second phase. The analysis of resilience requires different data for systems versus facility consideration. At the systems level, an in-depth knowledge of the movement of goods and people is necessary to assess the potential for redundant services that can at least minimally maintain service. For facilities, the time and cost needed to bring damaged infrastructure to full performance would be critical.

Outcomes

Having considered how transportation facilities might be exposed and having determined their vulnerability and the resilience of the network, decision makers can then consider ways to improve transportation in the region to be more robust to the climate impacts identified.

The primary outputs from the conceptual framework are policy recommendations or changes derived from the decision makers' understanding and interpretation of the major factors (exposure, vulnerability, and resilience and adaptation) associated with climate change. Where appropriate, these recommendations should lead to capital, maintenance, or operational improvements that will result in a more robust and resilient network.

The process of following the framework can be used to characterize the exposure of particular facility or system components to climate hazards; the vulnerability and resiliency of these elements; and the adaptation options available to the decision maker. Examples of potential thresholds or tipping points indicated for each of these factors targeted at each relevant transportation infrastructure element can then be used as input into the planning and decision processes available to the user. This output from the conceptual framework could be designated for the local level or State DOT level of planning. It will be up to the stakeholder or decision maker to determine how the assessment output would impact existing or proposed decision and planning processes at the relevant scale.

Figure 5.6 illustrates the relationship between risk assessment and the value of performance to the type of adaptation strategy that may be selected. As the importance of maintaining uninterrupted performance increases, the appropriate level of investment in adaptation should increase as well, taking into account the degree of risk facing the specific facility or system. For example, maintaining a specific bridge may be essential to ensure safe evacuation of a particular community, because no other feasible evacuation routes or backup strategies are available. In this instance, transportation and regional planners may recommend that more conservative (and possibly more expensive) design standards be applied to protect that bridge in the event of a low probability – but high consequence – storm event in that location. Conversely, although a road segment may be assessed to be highly at risk, it may warrant less extensive adaptation investment because alternatives to that road are available to provide access and mobility, or a moderate disruption in service performance is not considered to be critical.

Making Use of Risk Assessment in Transportation Decisions

The concepts presented in this chapter can be employed to begin the assessment of climate impacts in transportation planning and investment. Additional detail will be required for implementation, but this discussion offers an initial step toward a more complete consideration of risk and uncertainty in this type of assessment. As demonstrated, probabilities for some climate impacts are now available on a regional level, but probabilities for specific impacts on individual facilities or network components cannot yet be assigned with confidence. Furthermore, while some climate impacts can be reliably identified, data are lacking for others that may be important for transportation. Nonetheless, even at the conceptual level, this discussion may be useful for transportation planners as they begin to incorporate climate concerns in their consideration of new investments.

Consider the following example of a bridge located near the coast that is scheduled for rehabilitation in 5 years. Based on the conceptual framework, the first step is to determine its exposure to stressors that may significantly impede the service it provides.

If the bridge were located within the Gulf Coast study area, the analyses in chapters 3.0 and 4.0 indicate the four main stressors of concern: sea level rise, storm surge, temperature increases, and heavy downpours giving rise to flooding. There may be others as well, and the analyst would do well to consider other potential impacts in consultation with natural resource experts.

If the bridge falls within the area identified as likely to be flooded by a 61- to 122-cm (2- to 4-ft) rise in sea level, more specific examination of the particular terrain is warranted to assess in greater detail the likelihood of flooding. If there are no mitigating factors, there is a relatively high probability that the area will flood within a 50- to 100-year time period.

The next step is to determine the bridge's vulnerability to sea level rise. How high is the bridge? How high are the approaches? How critical is the service it provides? Based on these and other considerations, the bridge's vulnerability, in the context of its role within the larger network, can be assessed. If the bridge, or critical elements of it, are below 122 cm (4 ft), it will likely flood within its projected lifespan. While more objective measures of vulnerability to the service flowing over the bridge would be desirable, at a minimum the analyst should be able to derive a qualitative determination of the bridge's vulnerability.

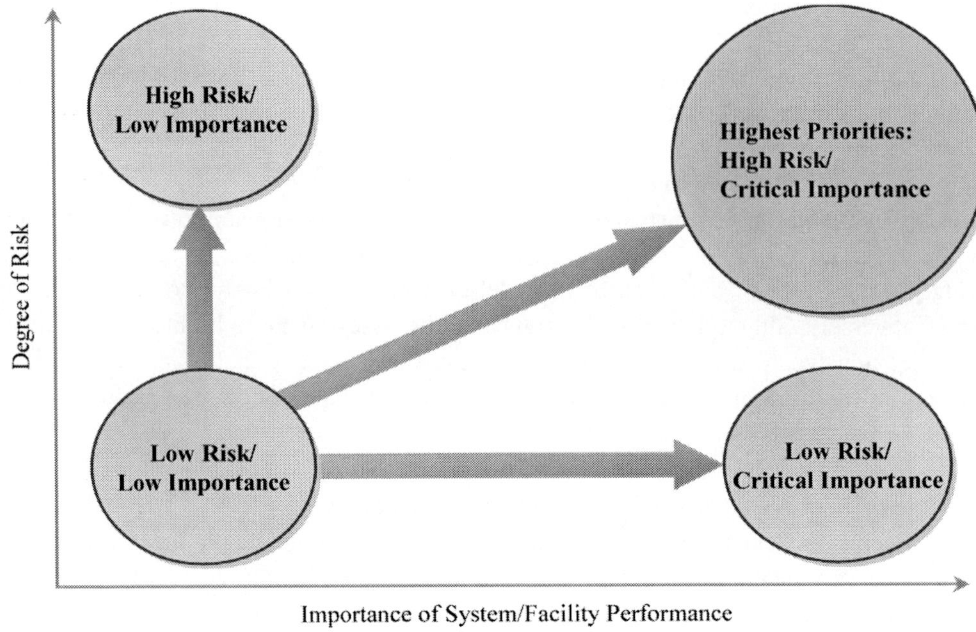

Figure 5.6. Degree of risk and importance of system or facility performance inform the level of adaptation investment

Judgment must be applied to assess the risk (probability of exposure times vulnerability) posed by flooding with current knowledge. Precise estimates of its components are not possible, but the direction and likely ranges are known, and from this a general sense of the risk can be inferred. If the bridge is heavily trafficked and vulnerable, the risk is high because the sea is rising, leading to permanent flooding, and the bridge's period of service will be cut short before it reaches the end of its useful life. Since (in the example) the bridge is scheduled for rehabilitation, now would be an appropriate time to consider options.

The adaptation options are to protect, accommodate, or retreat. Accommodation, which might include operational strategies to work around the flooding or simply live with it, does not appear to be viable since the flooding is permanent, and operational strategies like pumping the water out do not seem viable. Protection may include raising the bridge or its approaches or relocating the facility. Retreat, which in this case amounts to abandonment of the bridge, is likely the option of last choice since the bridge presumably provides a critical service. Engineering, design, landscape, and regional considerations will play crucial roles in the determination of the best option, as will the consideration of the additional resources necessary to best protect the bridge. Transportation agencies have extensive experience in exercising the judgment necessary to make these determinations.

In similar fashion, each of the stressors can be assessed for their likelihood and the bridge examined for its vulnerability. Risk can be determined and options identified to prolong the bridge's useful life and minimize disruptions to the critical service it provides. For stressors whose impacts are well understood, a higher level of analysis can and should be done to consider the potential for synergistic impacts that may be more severe than the individual effect. The end result of the analysis will be recommendations for investment whose implementation will result in a more robust and reliable transportation facility and system.

Experience indicates that the total cost to transportation agencies will probably be lower than failing to consider these impacts when the full costs – capital, operating, and economic loss due to disrupted service – are included.

5.3. CONCLUSIONS

Climate change and variability have not historically been considered in the planning and development of transportation facilities, and this was clearly expressed in the interviews conducted as part of this study. Until recently, it may not have been possible to effectively use climate data to serve as the basis of considerable capital investment due to its relative uncertainties. That appears to be changing. The destructive forces of Hurricanes Katrina and Rita have underscored the need to carefully consider the effects of the natural environment on transportation to a much higher degree. State, local, and possibly private (though less is known about their myriad approaches) transportation agencies are beginning to incorporate more information about the natural environment, including those effects wrought or exacerbated by climate change.

With the advent of increasingly greater certainty about the regional effects of climate change and better tools to assist their examination, analyzing the impacts of climate and the natural environment has become possible. Clearly there is benefit to do so. Subsidence and climate-induced sea level rise, coupled with the likely increased severity of hurricanes, threaten infrastructure, potentially causing severe disruptions to essential transportation services or cutting short the useful lives of important facilities. Transportation planners across the United States would do well to follow the lead of progressive agencies in the Gulf Coast and other places to begin immediately to consider the impacts of climate change on the natural environment and thus on transportation facilities under their purview.

This chapter introduces a taxonomy and conceptual approach toward incorporating climate change impacts in transportation planning. Standard deterministic approaches used in transportation planning will not suffice to address the timeframes and uncertainties that a changing climate poses. The approach is based on the quantitative or qualitative assessment of exposure to potentially disruptive impacts, examination of a facility's (or a network's) vulnerability, the risk of its loss, and possible adaptation strategies to mitigate these impacts and prolong service. It is premature to consider any formal changes to the established Federal transportation planning process. If for no other reason, the timeframes and other requirements such as fiscal constraint do not mesh well. Nonetheless, the consideration of climate impacts is possible and useful to transportation plans at all levels of government and the private sector. For instance, in the planning process shown in figure 5.3, climate change could be considered early on as part of a visioning process and later in the development and evaluation of alternative improvement strategies to consider future services and their location. Climate change could be considered in the project development process when design and engineering are addressed. Likewise, the concept of uncertainty and the use of risk analysis could be incorporated into the entire planning and project development process.

5.4. SOURCES

5.4.1. References

Adger, W. N., Hughes, T. P., Folke, C. S.R. Carpenter, and J. Rockström, (2005). Social-Ecological Resilience to Coastal Disasters. *Science*, *309 (5737)*, 1,036-1,039.

Berdica, K. (2002): An Introduction to Road Vulnerability: What Has Been Done, Is Done, And Should Be Done. *Transport Policy, 9(2)*, 117-127.

Bijlsma, L., Ehler, C. N., Klein, R. J. T., Khlshrestha, S. M., McLean, R. F., Mimura, N., Nicholls, R. J., Nurse, L. A., Perez Nieto, H., Stakhiv, E. Z., Turner, R. K. & Warrick, R. A. (1996). Coastal Zones and Small Islands. In: *Climate Change 1995: Impacts, Adaptations, and Mitigation of Climate Change: Scientific-Technical Analysis.* Contribution of Working Group II to the Second Assessment Report of the Intergovernmental Panel on Climate Change. [Watson, R.T., M.C. Zioyowera, and R.H. Moss, eds.] Cambridge University Press, Cambridge, United Kingdom, and New York, New York, USA, pages 289-324.

Cutter, S. L. (2003). The Vulnerability of Science and Science of Vulnerability. *Annals of the Association of American Geographers*, *93(1)*, 1-12.

Cutter, S. L., Mitchell, J. T. & Scott, M. S. (2000). Revealing the Vulnerability of People and Place: A Case Study of Georgetown County, South Carolina. *Annals of the Association of American Geographers*, *90(4)*, 7 13-737.

Intergovernmental Panel on Climate Change (IPCC), (2001). *Climate Change 2001: Impacts, Adaptation, and Vulnerability*. J.J. McCarthy, O.F. Canziani, and N.A. Leary, et al. (Eds.). Contribution of Working Group II to the Third Assessment Report of the Intergovernmental Panel on Climate Change. New York, New York: Cambridge University Press. [Available on-line at: http://www.ipcc.ch/]

Jones, R. N. (2001). An Environmental Risk Assessment/Management Framework for Climate Change Impact Assessment. *Natural Hazards*, *23,* 197-230.

Kelly, P. M. & Adger, W. N. (2000). Theory and Practice in Assessing Vulnerability to Climate Change and Facilitating Adaptation. *Climatic Change*, *47*, 325-352.

Klein, R. J. T., Smit, M. J., Goosen, H. & Hulsbergen, C. H. (1998). Resilience and Vulnerability: Coastal Dynamics or Dutch Dikes? *The Geographical Journal*, *164(3),* 259-268.

Levina, E. & Tirpak, D. (2006). *Adaptation to Climate Change: Key Terms*. Paris: Organisation for Economic Cooperation and Development/International Energy Agency.

Llort, Y. (2006). Transportation Planning in Florida. In: *Statewide Transportation Planning: Making Connections.* Transportation Research Circular E-C099, Transportation Research Board, Washington, D.C.

Omenn, G. S. (2006). Grand Challenges and Great Opportunities in Science, Technology, and Public Policy. *Science*, *314*, 1,696-1,704.

Parry, M. & Carter, T. (1998). *Climate Impact and Adaptation Assessment: Guide to the IPCC Approach*. Earthscan Publications, Ltd., London, United Kingdom.

Parry, M. L., Carter, T. R. & Hulme, M. (1996). What is Dangerous Climate Change? *Global Environmental Change.*, *6(1)*, 1-6.

Power, M. & McCarty, L. S. (1998). A comparative Analysis of Environmental Risk Assessment/Risk Management Frameworks. *Policy Analysis*, *32(9)*, 224 A-231 A.

Presidential/Congressional Commission on Risk Assessment and Risk Management, (1997). *Framework for Environmental Health Risk Management: Final Report.* The Presidential/Congressional Commission on Risk Assessment and Risk Management, Washington, D.C.

Scheraga, J. D. & Furlow, J. (2001). From Assessment to Policy: Lessons Learned from the U.S. National Assessment. *Human and Ecological Risk Assessment*, *7(5)*, 1,227-1,246.

Schneider, S., Sarukhan, J., Adejuwon, J., Azar, C., Baethgen, W., Hope, C., Moss, R., Leary, N., Richels, R., van Ypersele, J. P., Kuntz-Duriseti, K. & Jones, R. N. (2001). Overview of Impacts, Adaptation, and Vulnerability to Climate Change. In: *Climate Change: Impacts, Adaptation, and Vulnerability, the Contribution of Working Group II to the IPCC Third Assessment Report (TAR) of the Intergovernmental Panel on Climatic Change* [J.J. McCarthy, O.F. Canziani, N.A. Leary, D.J. Dokken, and K.S. White, eds.] Cambridge University Press, Cambridge, United Kingdom, 75-103.

Smit, B., Burton, I. & K Klein, J. T. (1999). The Science of Adaptation: A Framework for Assessment. *Mitigation and Adaptation Strategies for Global Change*, *4(3-4)*, 199-2 13.

Stern, P. C. & Fineberg, H. V. Eds., (1996). *Understanding Risk: Informing Decisions in a Democratic Society.* National Academies Press, Washington, D.C.

Tobin, G. A. & Montz, B. E. (1997). *Natural Hazards: Explanation and Integration*. Guilford Press, New York City.

U.S. Department of Transportation, (2006). SAP-4.7, *Prospectus for Impacts of Climate Variability and Change on Transportation Systems and Infrastructure – Gulf Cast Study.* Department of Transportation, Washington, D.C.

5.4.2. Background Sources

Baumert, K. & Pershing, J. (2004). *Climate Data: Insights and Observations.* Pew Center on Global Climate Change, Washington, D.C.

VTPI, (2006). *Online TDM Encyclopedia.* Victoria Transport Policy Institute, Victoria, B. C., Canada. [On-line at: http://www.vtpi.org/tdm/index.php]

End Notes

[1] This list represents a refinement of a similar list contained in previous intermodal surface transportation legislation.

[2] Some DOTs, such as Mississippi's, do not report to the Governor.

[3] Steps have been made in this direction with the development of TRANSIMS, a new generation transportation simulation model, which employs sampling and statistical methods to generate future travel demand; however TRANSIMS is not yet in general use.

In: Impacts of Climate Change on Transportation and... ISBN: 978-1-60741-424-7
Editor: Iason Pavlopoulos

Chapter 6

WHAT ARE THE KEY CONCLUSIONS OF THIS STUDY?

Lead Authors: Michael J. Savonis, Virginia R. Burkett and Joanne R. Potter

Contributing Authors: Thomas W. Doyle, Ron Hagelman, Stephen B. Hartley, Robert C. Hyman, Robert S. Kafalenos, Barry D. Keim, Kenneth J. Leonard, Matthew Sheppard, Claudia Tebaldi and Jessica E. Tump

The primary objectives of this phase of the Gulf Coast Study were to assemble the data needed for an analysis of the potential impacts on transportation; determine whether climate and ecological data could be usefully employed in such an assessment; identify and implement an assessment approach; and provide an overview of the potential impacts. The results are striking. They show that the data can provide useful information to transportation decision makers about the natural environment as it exists today, as well as the likely changes stemming from climate shifts. By using the historical data on the natural environment, an ensemble of climate models, a range of emissions scenarios, well- established literature on climate impacts, and a conservative approach toward interpretation, this study indicates that the potential impacts on transportation in the Gulf Coast are highly significant, as summarized below.

While further study is needed to examine in more detail the impacts on specific transportation facilities, such as individual airports or rail terminals, this preliminary assessment finds that the potential impacts on infrastructure are so important that transportation decision makers should begin immediately to assess them in the development of transportation investment strategies. Phase II of this effort will examine one small part of the Gulf Coast study region in much more detail. While the significance of climate factors will vary across regions of the United States, responsible transportation agencies in other areas would do well to consider these types of impacts as well, since the decisions they make today may result in infrastructure that will last 50 to 100 years. While the timing and pace of these impacts cannot be specified with precision, the central Gulf Coast already is vulnerable to certain impacts, as demonstrated by the 2005 hurricane season.

Given the characteristics of the climate system – especially the long periods of time greenhouse gases remain in the atmosphere and the virtually certain increases in carbon dioxide concentrations in the coming decades – some degree of impacts cannot be avoided. Based on analysis of different emissions scenarios, the magnitude of future impacts will depend on the amount of greenhouse gases emitted. While the modeled scenarios demonstrate very similar levels of climate impacts over the next 40 years, lower emission scenarios show lesser impacts in the longer term (60 to 100 years). If aggressive measures result in reduced emission levels globally, the climate impacts identified here may be on the lower end of the anticipated ranges.

The study authors believe that prudent steps can be taken to fortify the existing transportation system, as warranted, after an evaluation of impacts on critical transportation facilities and systems. Structures can be hardened, raised, or even relocated as need be, and where critical to safety and mobility, expanded redundant systems may be considered as well. What adaptive strategies may be employed, the associated costs, and the relative effectiveness of those strategies will have to be determined on a case-by-case basis, based on studies of individual facilities and systemwide considerations. As transportation agencies struggle to meet the challenges of congestion, safety, and environmental mitigation – as well as maintaining transportation infrastructure in good repair – meeting the challenges posed by a changing climate poses a new and major hurdle toward creation of a more resilient transportation network in a time of increasingly scarce resources. Phase III of this effort will examine potential response strategies and develop methods to assist local decision makers in assessing the relative merits of various adaptation options.

6.1. Trends in Climate and Coastal Change

The central Gulf Coast is particularly vulnerable to climate variability and change because of the frequency with which hurricanes strike, because much of its land is sinking relative to mean sea level and because much of its natural protection – in the form of barrier islands and wetlands – has been lost. While difficult to quantify, the loss of natural storm buffers will likely intensify many of the climate impacts identified in this report, particularly in relation to storm damage.

- **Relative Sea Level Rise –** Since much of the land in the Gulf Coast is sinking, this area is facing much higher increases in relative sea level rise (the combination of local land surface movement and change in mean sea level) than most other parts of the U.S. coast. Based on the output of an ensemble of general circulation models (GCM) run with a range of Intergovernmental Panel on Climate Change (IPCC) emissions scenarios, relative sea level in the study area is very likely to increase by at least 0.3 m (1 ft) across the region and possibly as much as 2 m (6 to 7 ft) in some parts of the study area over the next 50 to 100 years. The analysis of even a middle range of potential sea level rise of 0.3 to 0.9 m (2 to 4 ft) indicates that a vast portion of the Gulf Coast from Houston to Mobile may be inundated in the future. The projected rate of relative sea level rise for the region during the next 50 to 100 years is consistent with historical trends, region-specific analyses, and the IPCC 4th Assessment Report (2007) findings, which assume no major changes in ice-sheet dynamics.

Protective structures, such as levees and sea walls, could mitigate some of these impacts, but considerable land area is still at risk to permanent flooding from rising tides, sinking land, and erosion during storms. Subsidence alone could account for a large part of the change in land area through the middle of this century, depending on the portion of the coast that is considered. Sea level rise induced by the changing climate will substantially worsen the impacts of subsidence on the region.

- **Storm Activity** – The region is vulnerable today to transportation infrastructure damage during hurricanes and, given the potential for increases in the number of hurricanes designated as Category 3 and above, this vulnerability will likely increase. This preliminary analysis did not quantitatively assess the impact of the loss of protective barrier islands and wetlands, which will only serve to make storm effects worse. It also did not consider the possible synergistic impacts of storm activity over a sea that has risen by 0.6 to 1.2 m (2 to 4 ft). This potential would likely make a bad situation even worse, as well.

- **Average Temperature Increase** – All GCMs used by the IPCC in its Fourth Assessment Report (2007) indicate an increase in average annual Gulf Coast temperature through the end of this century. Based on GCM runs under three different IPCC emission scenarios (A1B, A2, and B1), the average temperature in the Gulf Coast region appears likely to increase by at least 1.5 °C ± 1 °C (2.7 °F ± 1.8 °F) during the next 50 years, with the greatest increase in temperature occurring in the summer.

- **Temperature Extremes** – With increases in average temperature also will come increases in extreme high temperature. Based on historical trends and model projections, it is very likely that the number of days above 32.2 °C (90 °F) will increase significantly across the study area; this has implications for transportation operations and maintenance. The number of days above 32.2 °C (90 °F) could increase by as much as 50 percent during the next 50 years.

- **Precipitation Change** – Future changes in precipitation are much more difficult to model than temperature. Precipitation trends in the study area suggest increasing values, with some climate divisions, especially those in Mississippi and Alabama, having significant long-term trends. Yet while some GCM results indicate that average precipitation will increase in this region, others indicate a decline in average precipitation during the next 50 to 100 years. Because of this ambiguity, it is difficult to reach conclusions about what the future holds regarding change in mean precipitation. Even if average precipitation increases slightly, average annual runoff in the region could decline as temperature and evapotranspiration rates increase.

- **Extreme Rainfall Events** – Average annual precipitation increased at most recording stations within the study area since 1919, and the literature indicates that a trend towards more rainfall and more frequent heavy downpours is likely. At this stage, climate modeling capacity is insufficient to quantify effects on individual precipitation events, but the potential for temporary flooding in this region is clear. In an area where flooding already is a concern, this tendency could be exacerbated by extreme rainfall events. This impact will become increasingly important as relative sea level rises, putting more and more of the study area at risk.

6.2. Transportation Impacts

Based on the trends in climate and coastal change, transportation infrastructure and the services that require them are vulnerable to future climate changes as well as other natural phenomena. While more study is needed to specify how vulnerable they are and what steps could be taken to reduce that vulnerability, it is clear that transportation planners in this region should not ignore these impacts.

- **Inundation from Relative Sea Level Rise** – While greater or lesser rises in relative sea level are possible, this study analyzed the effects of relative sea level rise of 0.6 and 1.2 m (2 and 4 ft) as realistic scenarios. Based on these levels, an untenable portion of the region's road, rail, and port network is at risk of permanent flooding.
  Twenty-seven percent of the major roads, 9 percent of the rail lines, and 72 percent of the ports are at or below 122 cm (4 ft) in elevation, although portions of the infrastructure are guarded by protective structures such as levees and dikes. This amounts to more than 3,900 km (2,400 mi) of major roadways that are at risk of total inundation for the highway system alone. While flood protection measures will continue to be an important strategy, rising sea levels in areas with insufficient protection may be a major concern for transportation planners. Furthermore, the crucial connectivity of the intermodal system in the area means that the services of the network can be threatened even if small segments are inundated.
  While these impacts are very significant, they can be addressed and adaptive strategies developed if transportation agencies carefully consider them in their decisions. The effectiveness of such strategies will depend on the strategies selected and the magnitude of the problem because scenarios of lower emissions demonstrate lesser impacts. It may be that in some cases, the adaptive strategy may be wholly successful, while in others further steps may need to be taken. Adaptive strategies that can be undertaken to minimize adverse impacts will be assessed in phase III of this study.

- **Flooding and Damage from Storm Activity** – As the central Gulf Coast is already is vulnerable to hurricanes, so is its transportation infrastructure. This study examined the potential for short-term flooding associated with a 5.5- and a 7.0-m (18- and 23-ft) storm surge. Based on these relatively common levels, a great deal of the study area's infrastructure is subject to temporary flooding. More than half (64 percent of interstates; 57 percent of arterials) of the area's major highways, almost half of the rail miles, 29 airports, and virtually all of the ports are subject to flooding.

The nature and extent of the flooding depends on where a hurricane makes landfall and its specific characteristics. Hurricanes Katrina and Rita demonstrated that that this temporary flooding can extend for miles inland.

This study did not examine in detail the potential for damage due to storm surge, wind speeds, debris, or other characteristics of hurricanes since this, too, greatly depends on where the hurricane strikes. Given the energy associated with hurricane storm surge, concern must be raised for any infrastructure in its direct path that is not designed to withstand the impact of a Category 3 hurricane or greater.

Climate change appears to worsen the region's vulnerability to hurricanes, as warming seas give rise to more energetic storms. The literature indicates that the intensity of major storms may increase 5 to 20 percent. This indicates that Category 3 storms and higher may return more frequently to the central Gulf Coast and thus cause more disruptions of transportation services.

The impacts of such storms need to be examined in greater detail; storms may cause even greater damage under future conditions not considered here. If the barrier islands and shorelines continue to be lost at historical rates and as relative sea level rises, the destructive potential of tropical storms is likely to increase.

- **Effects of Temperature Increase –** As the average temperature in the central Gulf Coast is expected to rise by 0.5 °C to 2.5 °C (0.9 °F to 4.5 °F), the daily high temperatures, particularly in summer, and the number of days above 32.2 °C (90 °F) also will likely increase. These combined effects will raise costs related to the construction, maintenance, and operations of transportation infrastructure and vehicles. Maintenance costs will increase for some types of infrastructure because they deteriorate more quickly at temperatures above 32 °C (90 °F). Increase in daily high temperatures could increase the potential for rail buckling in certain types of track. Construction costs could increase because of restrictions on days above 32 °C (90 °F), since work crews may be unable to be deployed during extreme heat events and concrete strength is affected by the temperature at which it sets. Increases in daily high temperatures would affect aircraft performance and runway length because runways need to be longer when daily temperatures are higher (all other things being equal). While potentially costly and burdensome, these impacts may be addressed by transportation agencies by absorbing the increased costs and increasing the level of maintenance for affected facilities.

- **Effects of Change in Average Precipitation –** It is difficult to determine how transportation infrastructure and services might be impacted by changes in average precipitation since models project either a wetter or a drier climate in the southeastern United States. In either case, the changes in average rainfall are relatively slight, and the existing transportation network may be equipped to manage this.

- **Effects of Increased Extreme Precipitation Events –** Of more concern is the potential for short-term flooding due to heavier downpours. Even if average precipitation declines, the intensity of those storms can lead to temporary flooding as culverts and other drainage systems are overloaded. Further, Louisiana Department of Transportation and Development reports that prolonged flooding of 1 to 5 weeks can damage the pavement substructure and necessitate rehabilitation (Gaspard et al., 2007). The central Gulf Coast already is prone to temporary flooding, and transportation representatives struggle with the disruptions these events cause. As the climate changes, flooding will probably become more frequent and more disruptive as the intensity of these downpours will likely increase. As relative sea level rises, it appears likely that even more infrastructure will be at risk because overall water levels already will be so much higher. While these impacts cannot be quantified at present, transportation representatives can monitor where flooding occurs and how the sea is rising as an early warning system about what facilities are at immediate risk and warrant high-priority attention. In a transportation system that already is under stress due to congestion, and with people and freight haulers increasingly dependent on justin-time delivery, the economic, safety, and social ramifications of even temporary flooding may be significant.

6.3. Implications for Planning

The network in the study area provides crucial service to millions of people and transports enormous quantities of oil, grain, and other freight. It is a network under increasing strain to meet transportation demand as the American public's desire for travel and low-cost goods and services continues to grow. Even minor disruption to this system causes ripple effects that erode the resources of transportation agencies as well as the good will and trust of the public. Good stewardship requires that the transportation network be as robust and resilient as possible within available resources.

This preliminary assessment raises clear cause for concern regarding the vulnerability of transportation infrastructure and services in the central Gulf Coast due to climate and coastal changes. These changes threaten to cause both major and minor disruptions to the smooth provision of transport service through the study area. Transportation agencies – bearing the responsibility to be effective stewards of the network and future investments in it – need to consider these impacts carefully.

Steps can be taken to address the potential impacts to varying degrees. This study demonstrates that there is benefit to examining the long-term impacts of climate change on transportation. Climate data and model scenarios may be productively employed to better plan for transportation infrastructure and services, even if there is not as much information or specificity as transportation planners might prefer. State and local planners need to examine these potentialities in greater detail within the context of smaller study areas and specific facilities. But to effectively consider them, changes are likely necessary in the timeframes and approaches taken.

- **Planning Timeframes –** Current practice limits the ability of transportation planners to examine potential conditions far enough into the future to adequately plan for impacts on transportation systems resulting from the natural environment and climate change. As such, insufficient attention is paid to longer-term impacts in some cases. The longevity of transportation infrastructure argues for a long timeframe to examine potential impacts from climate change and other elements of the natural environment. The current practice for public agencies of examining 20 to 30 years in the future to plan for transportation infrastructure may represent the limits of our sight for social, economic, and demographic assessments, as well as for consideration of fiscal constraint and other Federal planning requirements. However, the natural environment, including the climate, changes over longer time periods and warrants attention – perhaps as part of a long-term visioning process that helps to determine where transportation investments are needed and should be located. Such an approach would inform the long-range planning process with valuable supplementary information.
  This study could not examine transportation decision making in the private sector in detail due to proprietary concerns and the numerous companies involved. Clearly, some companies, such as CSX Railroad, have responded to issues posed by the 2005 hurricane season and made contingency plans to reroute service. Since the concerns are every bit as real for the private sector, these companies also would do well to plan for and implement adaptive strategies related to climate and other natural environment impacts.

- **Connectivity –** In addition to analysis at the level of particular facilities – such as an airport, bridge, or a portion of rail line – it would be useful for planners to examine the connectivity of the intermodal system for vulnerability assessed at the local, regional, national, and international levels to long-term changes in the natural environment, including changes induced by climate. This helps to identify critical links in the system and ways to buttress them against exposures to climate factors or other variables, or to create redundancies to maintain critical mobility for directly and indirectly affected populations alike.

- **Integrated Analysis –** From a transportation planning perspective, it is unnecessary and irrelevant to separate impacts due to climate change from impacts occurring from other naturally occurring phenomena like subsidence or storm surge due to hurricanes. In fact, such impacts are integrally related. Climate change is likely to increase the severity or frequency of impacts that already are occurring. Any impact that affects the structural integrity, design, operations, or maintenance that can be reasonably planned for should be considered in transportation planning. Efforts to restore ecological systems to redevelop protective buffers and reverse land loss may likewise help to protect transportation infrastructure from future climate impacts.

6.4. FUTURE NEEDS

The analysis of how a changing climate might affect transportation is in its infancy. While there is useful information that can be developed, the continued evolution of this type of study will serve to enhance the type of information that planners, engineers, operators, and maintenance personnel need to create an even more robust and resilient transportation system, ultimately at lower cost. This study begins to address the research needs identified in chapter 1.0 based on the current literature, but much more investigation is required. Based on the experience gained in conducting this study, research gaps are indicated in several chapters and specifically identified in chapter 4.0. Taken together, they indicate the following areas where more information is critical to the further estimation of the impacts of a changing climate on transportation infrastructure and services.

- **Climate Data and Projections** – The transportation community would benefit from the continued development by climatologists of more specific data on projected future impacts. Higher resolution of climate models for regional and subregional studies would be useful. More information about the likelihood and extent of extreme events, including temperature extremes, storms with associated surges and winds, and precipitation events could be utilized by transportation planners.

- **Risk Analysis Tools** – In addition to more specific climate data, transportation planners also need new methodological tools to address the uncertainties that are inherent in projections of climate phenomena. Such methods are likely to be based on probability and statistics as much as on engineering and material science. The approaches taken to address risk in earthquake-prone areas may provide a model for developing such tools.
  This study proposes a conceptual framework that may provide one way of approaching the development of new tools. More effort is needed to make the concepts presented here operational and thus useful to planners in the region. Specifically, more effort is needed to identify thresholds at which adaptive actions are warranted and taken. Monitoring short-term flooding due to increased downpours; relative sea level rise; and operating, maintenance, and construction costs serves as a good first step toward the identification of these thresholds. Eventually, it would be most useful to have standards of transportation service based on societal needs to guide future investments, and in some instances, changes in design standards may be indicated to ensure the desired levels of service.

- **Region-Based Analysis** – Future phases of this study will examine in more detail the potential impacts specific to the Gulf Coast and determine possible adaptation strategies. In addition, information developed either in this or subsequent studies would be valuable on freight, pipelines, and emergency management in particular. Additional analysis on demographic responses to climate change, land use interactions, and secondary and national economic impacts would help elucidate what impacts climate will have on people and the Nation as a whole, should critical transportation services in the region be lost. However, the impacts that a changing climate might have depends on where a region is and the specific characteristics of its natural environment. The research conducted in this study should be replicated in other areas of the country to determine the possible impacts of climate change on transportation infrastructure and services in those locations. Transportation in northern climates will face much different challenges than those in the south. Coastal areas will similarly face different challenges than interior portions of the country.

- **Interdisciplinary Research** – This study has demonstrated the value of cross-disciplinary research that engages both the transportation and climate research communities. Continued collaboration will benefit both disciplines in building methodologies and conducting analyses to inform the Nation's efforts to address the implications of climate change.

6.5. References

Gaspard, K., Martinez, M., Zhang, Z. & Wu, Z. (2007). *Impact of Hurricane Katrina on Roadways in the New Orleans Area.* Technical Assistance Report No. 07-2TA, LTRC Pavement Research Group, Louisiana Department of Transportation and Development, Louisiana Transportation Research Center, March 2007, 73 pages.

Intergovernmental Panel on Climate Change (IPCC), (2007). *Climate Change 2007: The Physical Science Basis, Summary for Policy-Makers.* Contribution of Working Group I to the Fourth Assessment Report of the Intergovernmental Panel on Climate Change. Geneva, Switzerland, 21 pages.

Appendix A

GULF COAST STUDY GIS DATASETS

Table A.1. Datasets in the Geographic Information System for the Gulf Coast Study Area

Topic	Dataset
Elevation/Subsidence	National Elevation Dataset (NED) for Gulf Coast study area
	Lidar for coastal Louisiana
	LSRC/NGS Subsidence Measurement Network for LMRV (NOAA Tech Report 50)
Transportation Infrastructure	Pipeline data obtained from the U.S. DOT OPS National Pipeline Mapping System
	Road networks from TIGER data, 2003
	Individual State evacuation routes
	Railway network from TIGER data
	Rail networks (Source: Federal Railroad Administration, 2005)
	Amtrak stations (Source: Federal Railroad Administration, 2005)
	Fixed guideway transit facilities (Source: Bureau of Transportation Statistics, Federal Transit Administration)
	Airports (Source: Federal Aviation Administration, 2005)
	Ports (Source: U.S. Army Corps of Engineers, 2005)
	Intermodal freight terminals (Source: Bureau of Transportation Statistics, 2004)
	Navigable waterways (Source: U.S. Army Corps of Engineers)
Imagery and Topographic Maps	Landsat 5 Thematic Mapper (TM) satellite data at 90-m resolution
	Aerial Photography at 1-m resolution from the 1998 DOQQ
	Topographic Maps (DRG) at 1:24,000, 1:100,000, and 1:250,000 scales

Table A.1. (Continued)

Topic	Dataset
Earth Sciences	Geology at 1:2,000,000 covering the study area
	1:500,000 for Louisiana
	State Soil Geographic Database (STATSGO) for Gulf Coast study area
	Soil Survey Geographic Database (SSURGO) in tabular form available for Gulf Coast study area
	National Land Cover Dataset (NLCD)
	EcoRegions
Hydrology	National Hydrographic Dataset (NHD)
	Federal Emergency Management Agency (FEMA) Q3 flood data
	Hydrologic unit watershed coverage of Gulf Coast drainage
Administrative Geography and Other Infrastructure	
Political boundaries (Source: U.S. Census)	
Demographic data (Source: U.S. Census)	
Urbanized areas (Source: U.S. Census)	
MPO planning boundaries (Source: BTS)	
Coastal and hazard planning districts (Source: FEMA)	
Petrochemical and energy resources (Source: EPA/CENSUS)	
Industrial centers (Source: EPA/CENSUS)	
Employment centers (Source: CENSUS)	
Government/Federal facilities (Source: USGS)	
Military bases (Source: MTMCTEA)	
Public health, education, service facilities (Source: CENSUS)	
Emergency response and safety facilities (Source: FEMA)	

Appendix B

ADDITIONAL DATA ON SOCIAL AND ECONOMIC SETTING

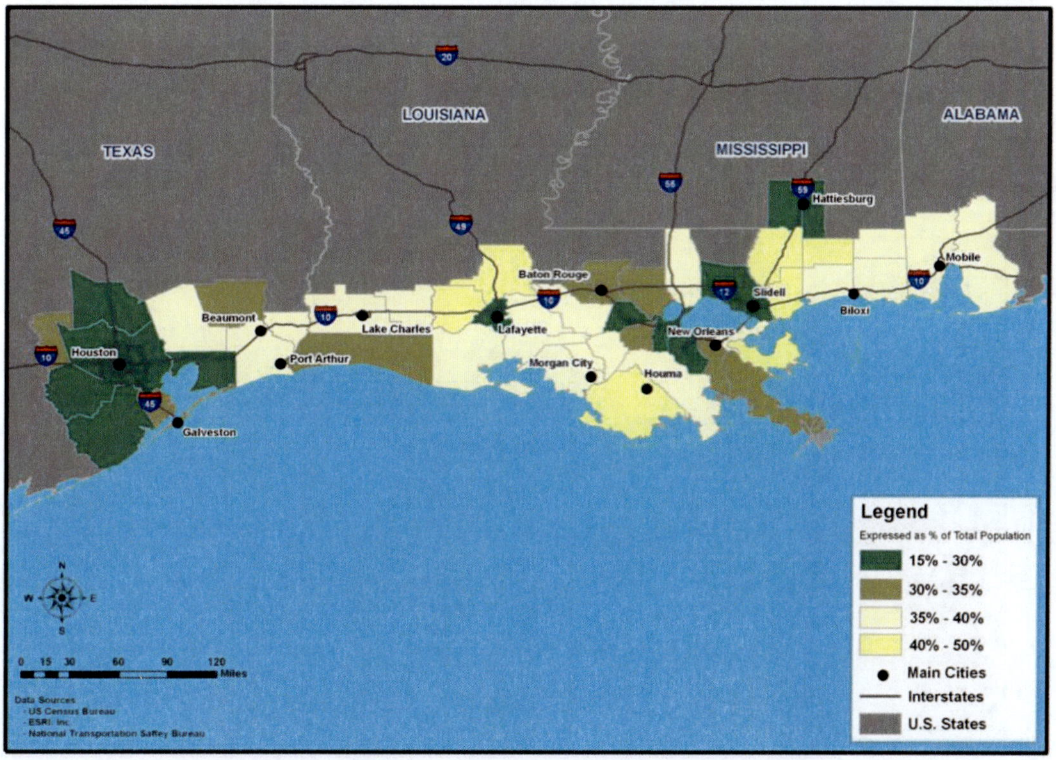

Source: U.S. Census Bureau; ESRI, Inc.; National Transportation Safety Bureau.

Figure B.1. Persons reporting disabilities

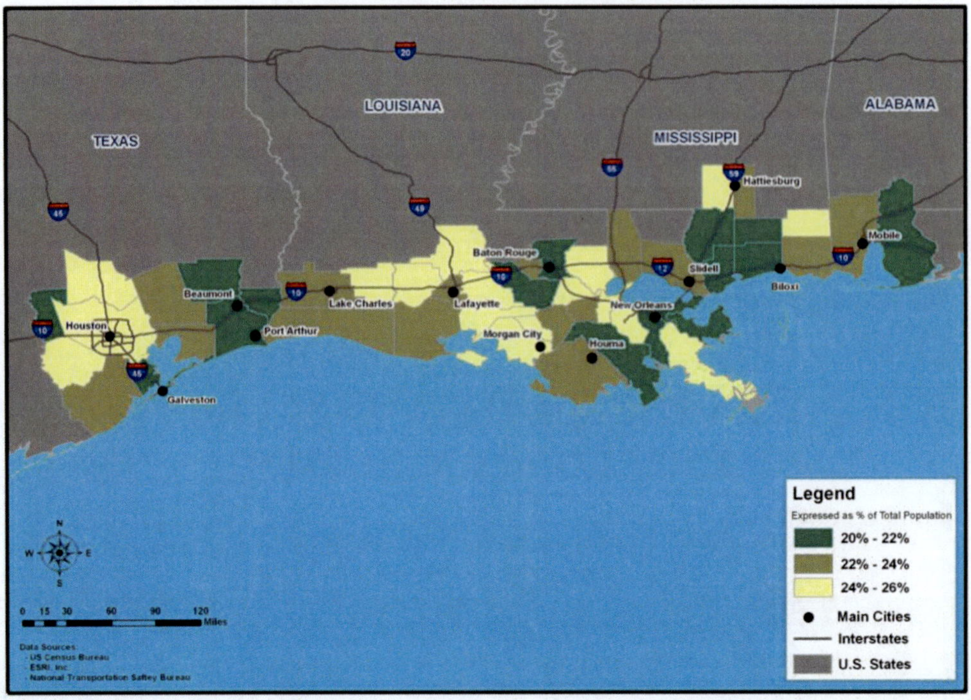

Source: U.S. Census Bureau; ESRI, Inc.; National Transportation Safety Bureau.

Figure B.2. Children age 14 and under

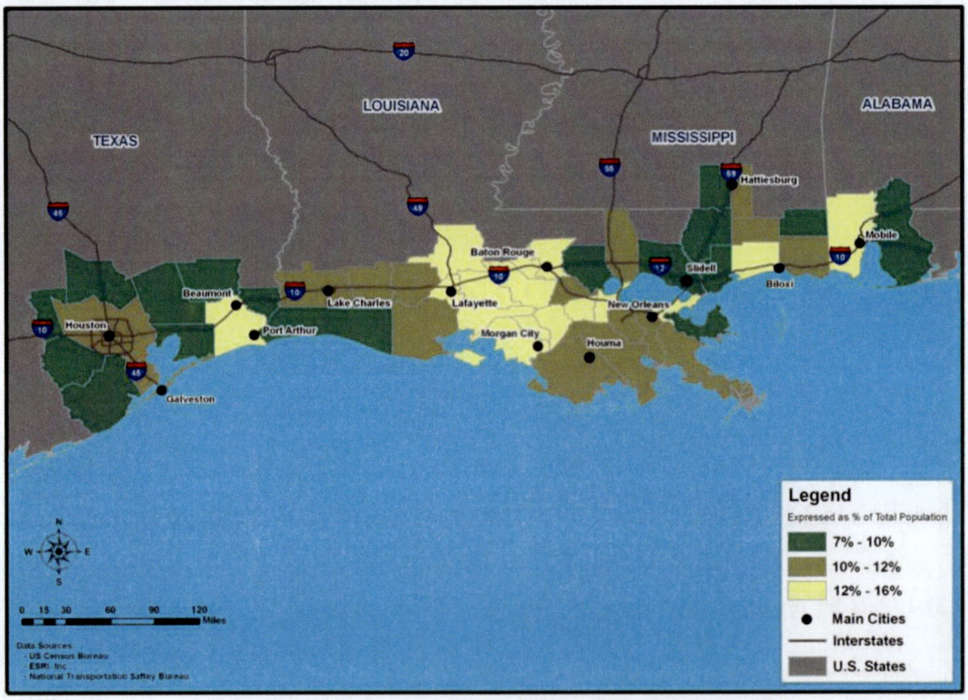

Source: U.S. Census Bureau; ESRI, Inc.; National Transportation Safety Bureau.

Figure B.3. Single adult households with children

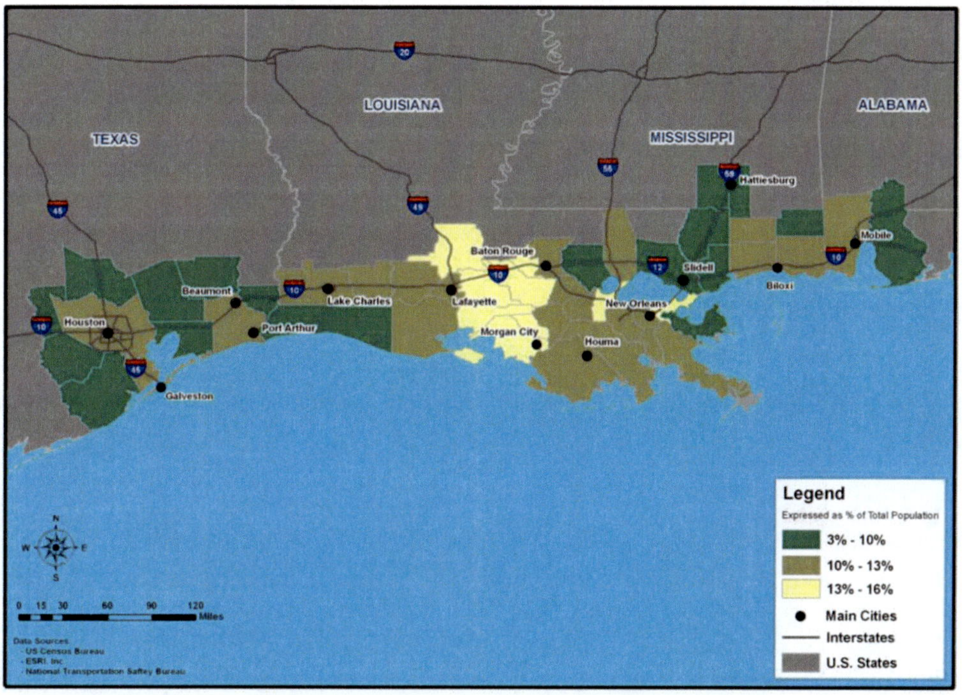

Source: U.S. Census Bureau; ESRI, Inc.; National Transportation Safety Bureau.

Figure B.4. Linguistically isolated households

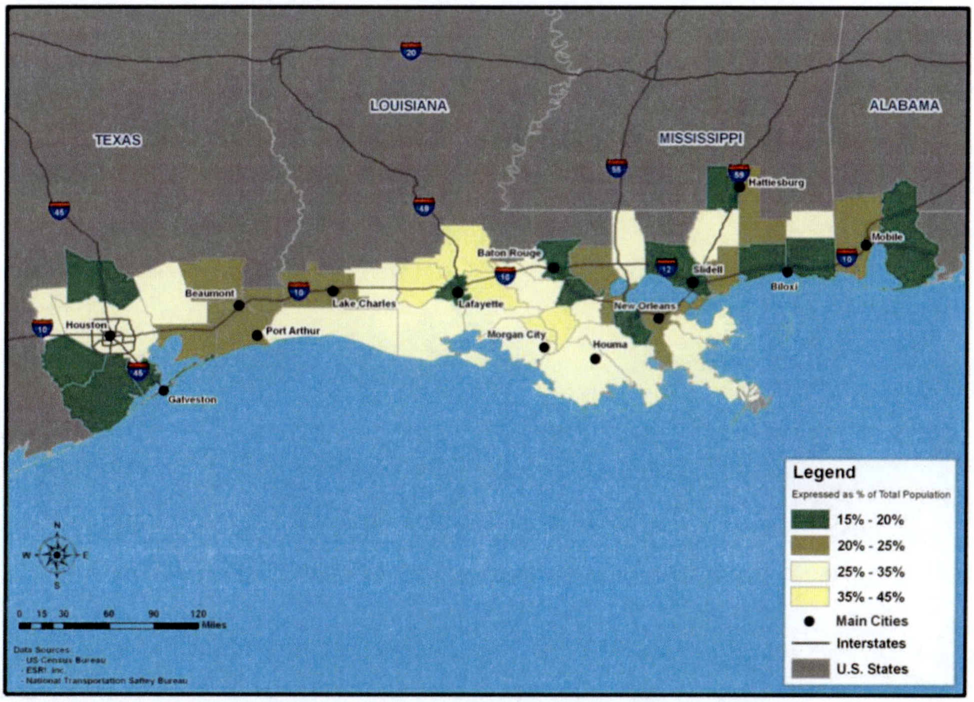

Source: U.S. Census Bureau; ESRI, Inc.; National Transportation Safety Bureau.

Figure B.5. Percent of population with no high school diploma or equivalent

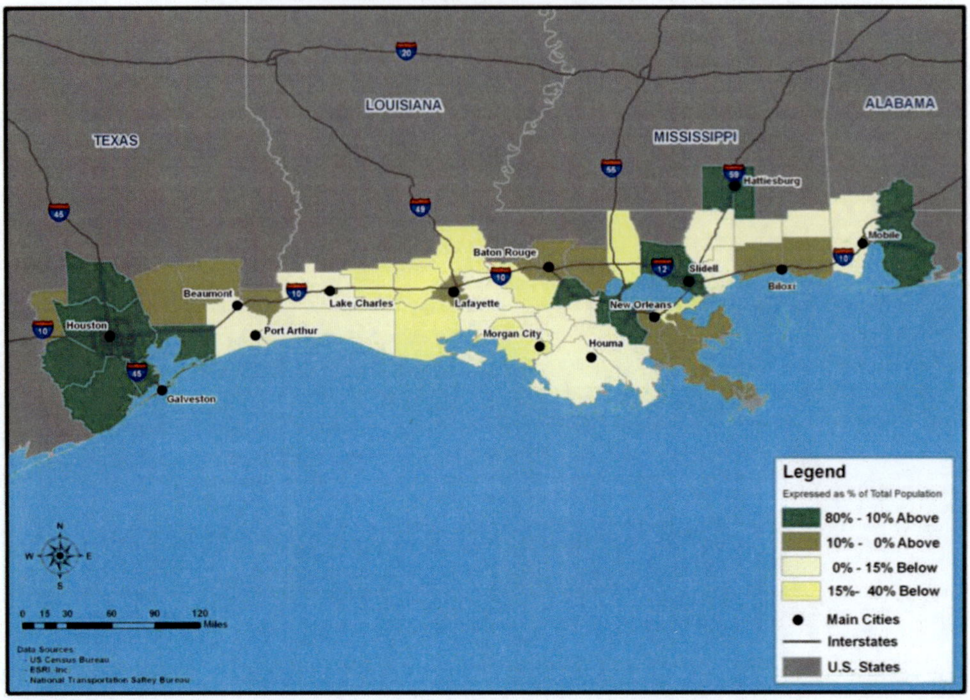

Source: U.S. Census Bureau; ESRI, Inc.; National Transportation Safety Bureau.

Figure B.6. Percent of population below/above median income of study area

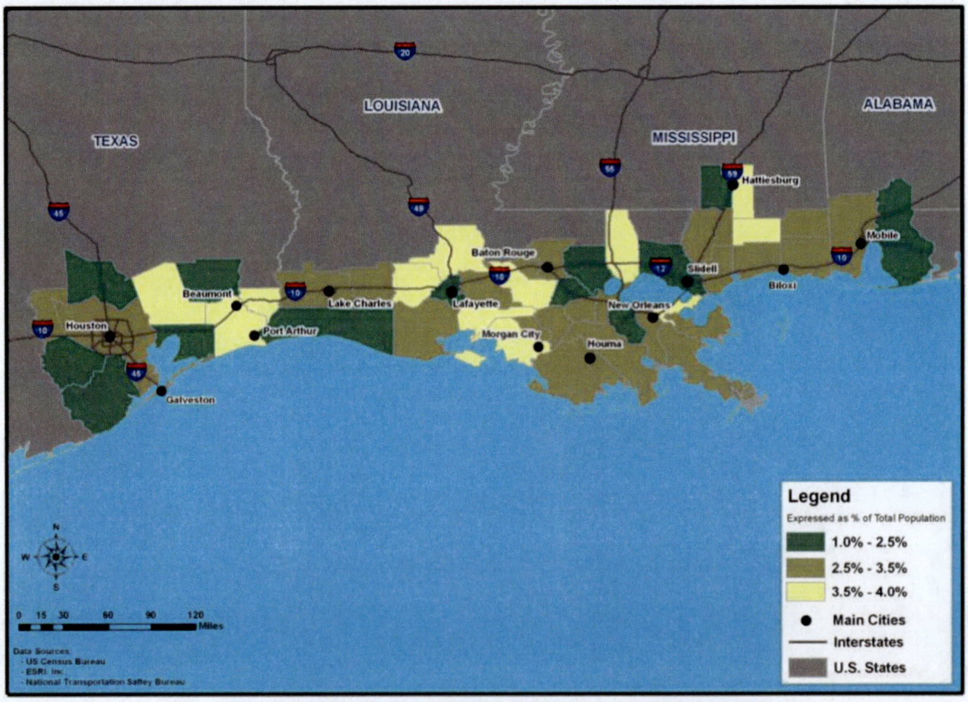

Source: U.S. Census Bureau; ESRI, Inc.; National Transportation Safety Bureau.

Figure B.7. Percentages of persons receiving public assistance income

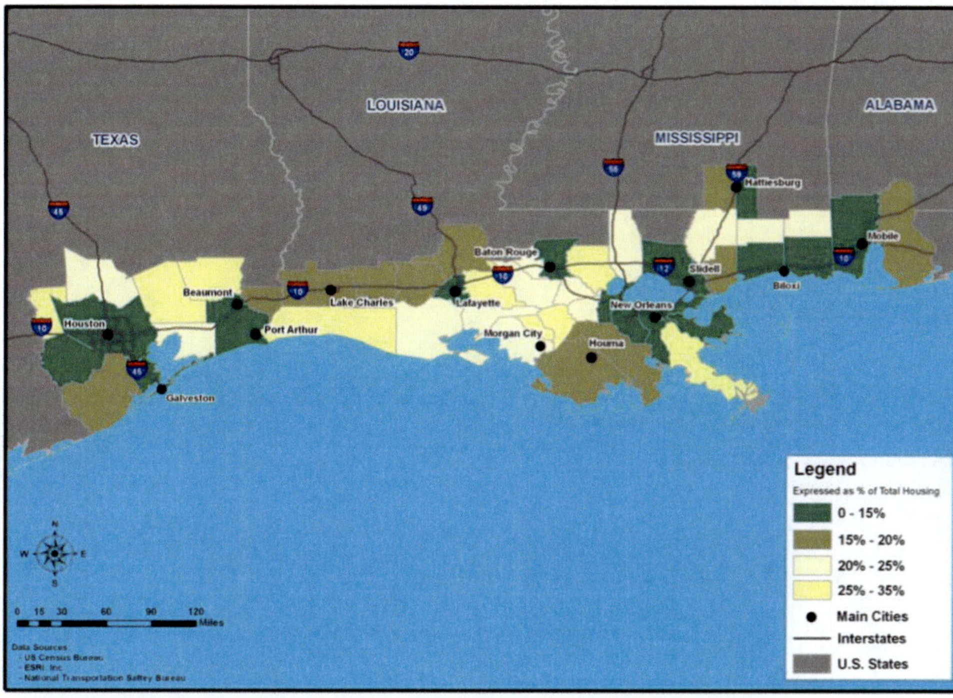

Source: U.S. Census Bureau; ESRI, Inc.; National Transportation Safety Bureau.

Figure B.8. Percent of housing units that are mobile homes

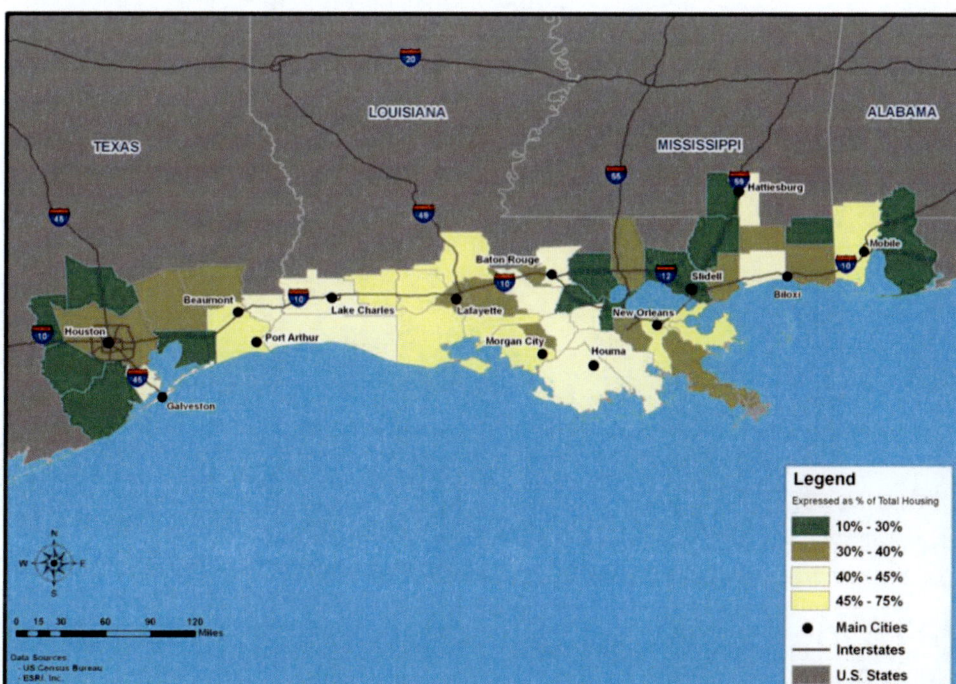

Source: U.S. Census Bureau; ESRI, Inc.; National Transportation Safety Bureau.

Figure B.9. Percent of housing built before 1970

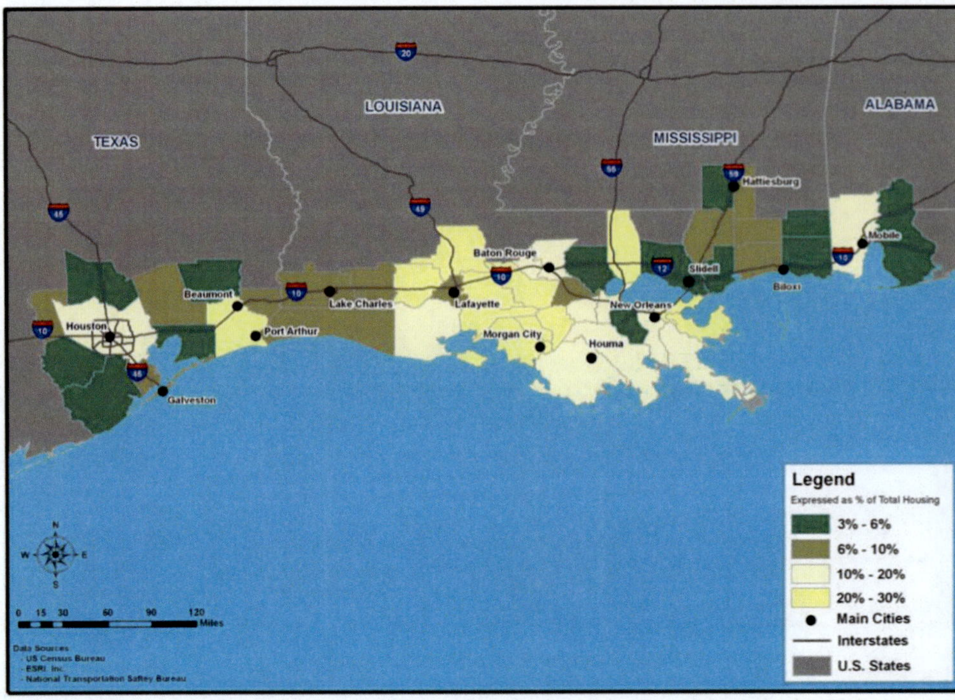

Source: U.S. Census Bureau; ESRI, Inc.; National Transportation Safety Bureau.

Figure B.10. Percent of housing units reporting no vehicle

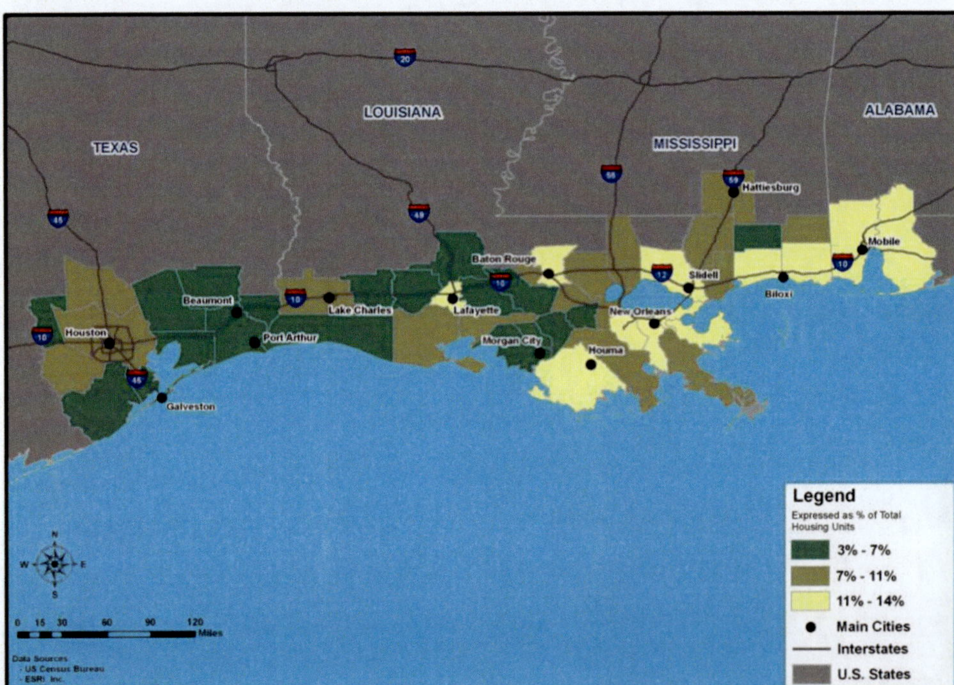

Source: U.S. Census Bureau; ESRI, Inc.; National Transportation Safety Bureau.

Figure B.11. Percent of housing units with a second mortgage or home equity loan

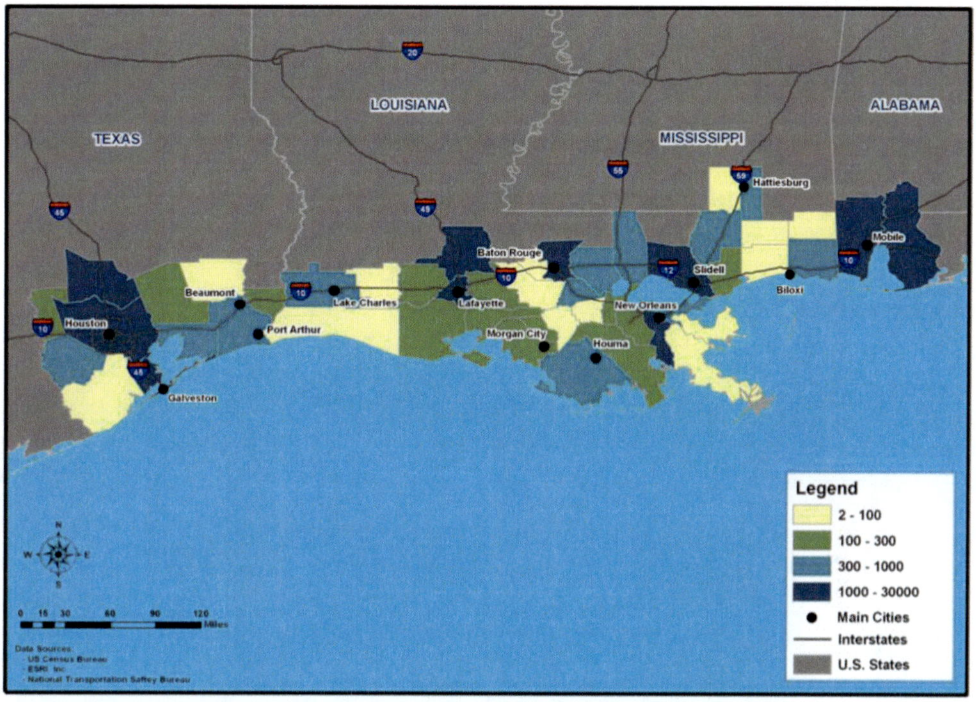

Source: U.S. Census Bureau; ESRI, Inc.; National Transportation Safety Bureau.

Figure B.12. Number of building permits issued, 2002

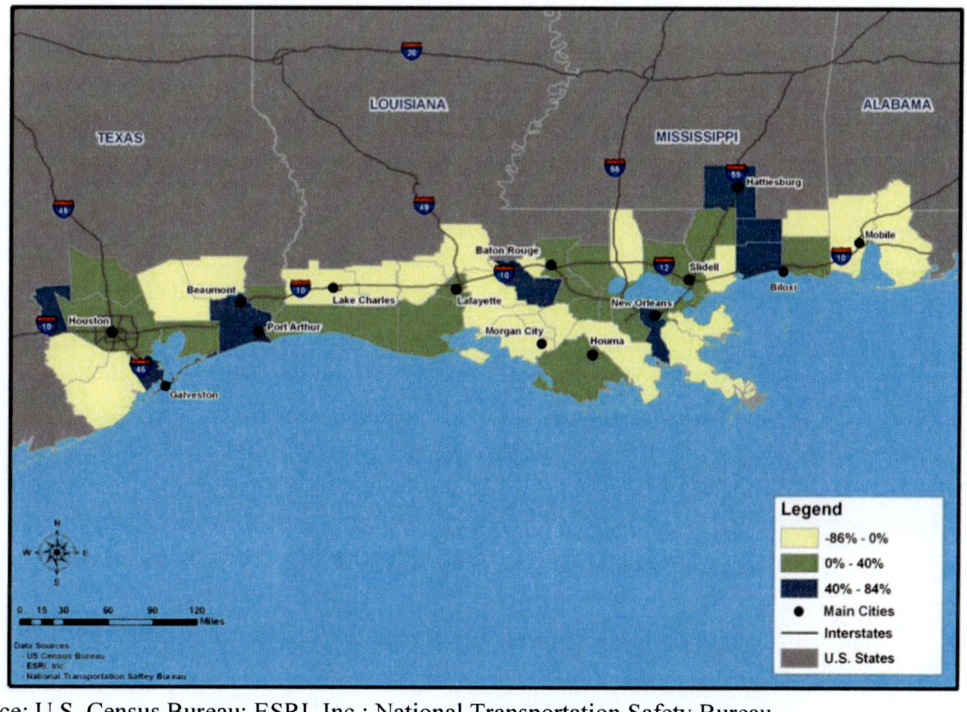

Source: U.S. Census Bureau; ESRI, Inc.; National Transportation Safety Bureau.

Figure B.13. Percent change in building permits issued, 1997 to 2002

Appendix C

ADDITIONAL RAIL DATA

Table C.1. Freight Rail Facilities in the Gulf Coast Study Area

Name	Mode Type	City	State	Gridcode
Kansas City Southern-Metairie-LA	Rail & Truck	Metairie	LA	0
Larsen Intermodal, Inc.-Metairie-LA	Rail & Truck	Metairie	LA	0
New Orleans Cold Storage & Warehouse Company, Ltd	Rail & Truck	Metairie	LA	0
Port of Gulfport	Truck - Port – Rail	Gulfport	MS	0
Port of Galveston	Truck - Port – Rail	Galveston	TX	0
NS-New Orleans-LA	Rail & Truck	New Orleans	LA	1
Up-Avondale-La-Intermodal Facility	Rail & Truck	Avondale	LA	1
Port of Freeport	Truck - Port – Rail	Freeport	TX	1
Dry Storage Corporation of Louisiana	Rail & Truck	Kenner	LA	2
DSC Logistics-Kenner-LA	Rail & Truck	Kenner	LA	2
Yellow-New Orleans-LA Terminal	Rail & Truck	New Orleans	LA	2
BNSF-New Orleans-LA	Rail & Truck	Westwego	LA	3
BNSF-New Orleans-LA-539 Bridge	Rail & Truck	Westwego	LA	3
BNSF-New Orleans-LA-Intermodal Facility	Rail & Truck	New Orleans	LA	3
Intermodal Cartage Company-New Orleans-LA	Truck - Port – Rail	New Orleans	LA	3
Transflo-New Orleans-LA	Rail & Truck	New Orleans	LA	3
BNSF-Avondale-LA-101 Avonda	Rail & Truck	Avondale	LA	4
Down South Transfer, Inc.-Avondale-LA	Rail & Truck	Avondale	LA	4
Port of New Orleans	Truck - Port – Rail	New Orleans	LA	4
CSX Intermodal-New Orleans-LA	Rail & Truck	New Orleans	LA	5
Dupuy Storage and Forwarding Corporation	Rail & Truck	New Orleans	LA	5

Table C.1. (Continued)

Name	Mode Type	City	State	Gridcode
Port of Iberia	Truck - Port – Rail	New Iberia	LA	5
BNSF-Mobile-AL	Rail & Truck	Mobile	AL	6
CN-Mobile-AL	Truck - Port – Rail	Mobile	AL	6
Miller Transporters, Inc.-Mobile-AL	Rail & Truck	Prichard	AL	6
CN-New Orleans-LA	Truck - Port – Rail	New Orleans	LA	6
Continental Grain Co.-Westwego-LA	Rail & Truck	Westwego	LA	6
Hayes Dockside, Inc.-New Orleans-LA	Rail & Truck	New Orleans	LA	6
Illinois Central Railroad-New Orleans-LA	Rail & Truck	New Orleans	LA	6
Lake Charles Harbor and Terminal District-Lake CHA	Rail & Truck	Lake Charles	LA	6
LST (Floating Elevator)-Belle Chasse-LA	Rail & Truck	Belle Chasse	LA	6
NS Independent Bulk Transfer Terminal-Arabi-LA	Rail & Truck	Arabi	LA	6
Port of Lake Charles	Rail & Port	Lake Charles	LA	6
Port of South Louisiana	Rail & Port	Laplace	LA	6
Port of Pascagoula	Truck - Port – Rail	Pascagoula	MS	6
Houston Fuel Oil Terminal Co.-Houston-TX	Rail & Truck	Houston	TX	6
Miller Transporters, Inc.-Beaumont-TX	Rail & Truck	Beaumont	TX	6
Port Arthur-TX	Rail & Truck	Port Arthur	TX	6
Port of Orange	Truck - Port – Rail	Orange	TX	6
Port of Port Arthur	Truck - Port – Rail	Port Arthur	TX	6
Mobile Moving and Storage	Rail & Truck	Mobile	AL	7
Yellow-Mobile-AL Terminal	Rail & Truck	Mobile	AL	7
Aimcor Galveston Marine Terminal-Texas City-TX	Rail & Truck	Texas City	TX	7
ASW Supply Chain Services, LLC-Beaumont-TX	Truck - Port – Rail	Beaumont	TX	7
GATX Terminals Corporation-Galena Park-TX	Rail & Truck	Galena Park	TX	7
PCI Transportation, Inc.-Houston-TX	Rail & Truck	Houston	TX	7
Port of Beaumont	Truck - Port – Rail	Beaumont	TX	7
UP-Laporte-TX	Rail & Truck	Strane	TX	7
Walton Barge Terminal-Houston-TX	Truck - Port – Rail	Houston	TX	7
Wilson Warehouse Co. of Texas, Inc.-Beaumont-TX	Rail & Truck	Beaumont	TX	7
Yellow-Beaumont-TX Terminal	Truck - Port – Rail	Beaumont	TX	7
Cargill Marketing Co. Inc.	Truck - Port – Rail	Eight Mile	AL	8
Meador Warehousing and Distribution, Inc.-Mobile-A	Rail & Truck	Mobile	AL	8
The Finch Companies-Mobile-AL	Rail & Truck	Mobile	AL	8
Acme Transfer-Baton Rouge-LA	Rail & Truck	Baton Rouge	LA	8
Agway Systems Inc.	Rail & Truck	Baton Rouge	LA	8
Branch Warehousing and Distribution Center, Inc.-l	Rail & Truck	Lafayette	LA	8

Table C.1. (Continued)

Name	Mode Type	City	State	Gridcode
Innovative Waste Systems, Inc-Baton Rouge-LA	Rail & Truck	Baton Rouge	LA	8
Miller Transporters, Inc.-Baton Rouge-LA	Rail & Truck	Baton Rouge	LA	8
Port Manchac Distribution Center	Rail & Truck	Ponchatoula	LA	8
Yellow-Layfayette-La Terminal	Rail & Truck	Broussard	LA	8
Miller Transporters, Inc.-Hattiesburg-MS	Rail & Truck	Hattiesburg	MS	8
Miller Transporters, Inc.-Lumberton-MS	Rail & Truck	Lumberton	MS	8
Yellow-Gulfport-MS Terminal	Rail & Truck	Gulfport	MS	8
Adams Distribution Center, Inc.-Houston-TX	Rail & Truck	Houston	TX	8
BNSF-Houston-TX	Rail & Truck	Houston	TX	8
BNSF-Houston-TX-10000 Wal	Rail & Truck	Houston	TX	8
BNSF-Houston-TX-5500 Walli	Rail & Truck	Houston	TX	8
Care Terminal	Truck - Port – Rail	Houston	TX	8
Charles Emmons Pulpwood Co.-Cleveland-TX	Rail & Truck	Cleveland	TX	8
CSX Intermodal-Houston-TX	Rail & Truck	Houston	TX	8
General Stevedores, Inc.-Houston-TX	Rail & Truck	Houston	TX	8
Gulf Winds International, Inc.-Mykawa-TX	Rail & Truck	Mykawa	TX	8
Guthrie Lumber Sales Inc.-Houston-TX	Rail & Truck	Houston	TX	8
Intercontinental Terminals Co.-Houston-TX	Rail & Truck	Houston	TX	8
Intermodal Cartage Company-Houston-TX	Truck - Port – Rail	Houston	TX	8
International Distribution Corp.-Houston-TX	Rail & Truck	Houston	TX	8
MCC Transport, Inc.-Houston-TX	Rail & Truck	Houston	TX	8
Miller Transporters, Inc.-Houston-TX	Rail & Truck	Houston	TX	8
Oil Tanking Houston, Inc.-Houston-TX	Rail & Truck	Houston	TX	8
Palmer Logistics-Houston-TX	Rail & Truck	Houston	TX	8
Port of Houston Authority	Truck - Port – Rail	Houston	TX	8
Quality Carriers-Houston-TX	Rail & Truck	Houston	TX	8
South Coast Terminals, l.P.-Houston-TX-3730fm 196	Rail & Truck	Houston	TX	8
Southern Warehouse Corporation-Houston-TX	Rail & Truck	Houston	TX	8
Tencon Industries, Inc.-Houston-TX	Rail & Truck	Houston	TX	8
Texas Rice Inc.	Rail & Truck	Houston	TX	8
Thompson Cargo Specialists, Inc.-Houston-TX	Rail & Truck	Houston	TX	8
Union Pacific Bulk Tainer Service-Spring-TX	Rail & Truck	Spring	TX	8
United DC, Inc Corporate Headquaters-Houston-TX894	Rail & Truck	Houston	TX	8

Table C.1. (Continued)

Name	Mode Type	City	State	Gridcode
United DC, Inc.-Houston-TX-1200 Lathr	Rail & Truck	Houston	TX	8
UP-Houston-TX-5500	Rail & Truck	Houston	TX	8
Western Intermodal Services, Ltd.-Houston-TX	Rail & Truck	Houston	TX	8
Yellow-Houston-TX Terminal	Truck - Port – Rail	Houston	TX	8

Source: National Transportation Atlas Database (BTS, 2004).

Table C.2. Amtrak Facilities in the Gulf Coast Study Area

Station/City	State	Gridcode
Mobile	AL	6
Lake Charles	LA	6
New Orleans	LA	6
Schriever	LA	6
Slidell	LA	6
Pascagoula	MS	6
Beaumont	TX	6
Galveston	TX	6
Lamarque	TX	6
New Iberia	LA	7
Bay St. Louis	MS	7
Biloxi	MS	7
Bay Minette	AL	8
Baton Rouge	LA	8
Hammond	LA	8
Lafayette	LA	8
Gulfport	MS	8
Hattiesburg	MS	8
Picayune	MS	8
Houston	TX	8
South Houston	TX	8

Source: National Transportation Atlas Database (BTS, 2004).

Appendix D

WATER BALANCE MODEL PROCEDURES

Temperature and precipitation data were either the calculated historical values for the study area or the forecast-modified values under Special Report on Emissions Scenarios (SRES) scenario A1B. Reference evapotranspiration (ET_o), was calculated by using the Turc (1961) model (Jensen et al., 1997; Fontenot, 2004). Turc was selected for use over the original Thornthwaite model (as described in Dingman 2002) because of its ability to more closely simulate FAO-56 Penman-Monteith ET_o with a limited set of meteorological data (Fontenot, 2004). Allen (2003) defined the Turc equation for operational use:

$$ET_O = a_T 0.013 \frac{T_{mean}}{T_{mean} + 15} \frac{23.8856 R_S + 50}{\lambda} \tag{1}$$

where ET_o is evapotranspiration (mm day^{-1}), T_{mean} is the mean daily air temperature ($^{\circ}C$), R_s is solar radiation (MJ m^{-2} day^{-1}), and λ is the latent heat of vaporization (MJ kg^{-1}). The coefficient a_T is a humidity-based value. If the mean daily relative humidity (RH_{mean}) is greater than or equal to 50 percent, then $a_T = 1.0$. If the mean daily relative humidity is less than 50 percent, then a_T has the value of:

$$a_T = 1 + \frac{50 - RH_{mean}}{70} \tag{2}$$

Humidity data (historical or forecasted) were not available for the study area, so the assumption was made that the dew point temperature was equal to the mean monthly minimum temperature. This procedure is recommended by Allen et al. (1998) for approximating daily humidity values when measured values are not available. Solar radiation (R_s) was estimated by using the Hargreaves model as described by Allen et al. (1998):

$$R_s = k_{RS} \sqrt{(T_{MAX} - T_{MIN})} R_a \tag{3}$$

where R_s is the solar radiation as stated above, k_{RS} is an adjustment coefficient, T_{MAX} and T_{MIN} are the mean daily maximum and minimum air temperatures ($^{\circ}C$), and R_a is extraterrestrial radiation (MJ m^{-2} day^{-1}). A value of 0.19 was used for k_{RS} as suggested by Allen et al. (1998)

for use in coastal locations. The Turc model was run by using the monthly temperature data and radiation data for the 15th – the midpoint – of each month. The values were then multiplied by the appropriate number of days in each month to create a monthly value for ET_o. For simplicity, leap days were not included.

After the basic input variables were prepared, the data were entered into the water balance model. First, by using the temperature data, the monthly precipitation was partitioned into rain and snow components, where:

$$RAIN_M = F_M \bullet P_M \tag{4}$$

$$SNOW_M = (1 - F_M) \bullet P_M \tag{5}$$

Where P_M is the monthly precipitation and F_M is a melt factor that is computed by using the following method:

$$\begin{aligned} &\text{If } T_M \leq 0° \text{ C:} \quad F_M = 0 \\ &\text{If } 0° \text{ C} < T_M < 6° \text{ C:} \quad F_M = 0.167 \cdot T_M \\ &\text{If } T_M \geq 6° \text{ C:} \quad F_M = 1 \end{aligned} \tag{6}$$

where T_M is the mean monthly temperature (Dingman, 2002). F_M also is used to determine the monthly snowmelt amount:

$$MELT_M = F_M \bullet (PACK_{m-1} + SNOW_m) \tag{7}$$

with $PACK_{m-1}$ being the water equivalent of the snow pack at the end of the previous month and $SNOW_m$ being the snow fall total of the current month. The previous month's pack amount is calculated as:

$$PACK_m = (1 - F_M)^2 \bullet P_M + (1 - F_M) \bullet PACK_{m-1} \tag{8}$$

The overall hydrological input into the model is defined by W_M as:

$$W_M = RAIN_m + MELT_m \tag{9}$$

In this study, the probability of the study region having any significant snow amounts is low, but the variable was included to provide for the possibility in the forecasted model runs.

Changes in soil moisture are calculated by using the following logic. If WM $\geq ET_o$, monthly evapotranspiration (ET_M) occurs at the ET_o rate. If ET_M equals ET_o, then soil moisture would increase or remain steady if the soil moisture already is at field capacity (Dingman, 2002). For the purposes of this study, field capacity ($SOIL_{MAX}$) has been set to 150 mm (5.9 in). The monthly value for soil moisture is therefore:

$$SOIL_M = \min\{[(W_M - ET_O) + SOIL_{m-1}], SOIL_{MAX}\} \tag{10}$$

where the soil moisture value is the lesser of the two values in the equation (Dingman, 2002). If WM is less than ET_o, then ET_M is equal to the hydrological input (W_M) and a drying factor:

$$ET_M = W_M + \left\{ SOIL_{m-1} \bullet \left[1 - \exp\left(-\frac{ET_{OM} - W_M}{SOIL_{MAX}} \right) \right] \right\} \tag{11}$$

where ET_{OM} is the monthly Turc ET_o value (Dingman, 2002).

After computing soil moisture change, any excess water in the budget was declared as surplus. The monthly surplus parameter is synonymous with runoff in these wetland environments, as long lags are not common between the generation of surplus water and the resultant streamflow. If W_M does not meet the environmental demand, then a deficit is created until W_M meets the environmental demand. In this study, we retained surplus as an index for runoff and dismissed the modeled runoff term as invalid.

REFERENCES

Allen, R. G. (2003). *REF-ET User's Guide*. University of Idaho Kimberly Research Stations: Kimberly, Idaho.

Allen, R. G., Pereira, L.S., Raes, D. & Smith, M. (1998). *Crop Evapotranspiration – Guidelines for Computing Crop Water Requirements.* FAO Irrigation and Drainage Paper 56. Food and Agriculture Organization. Rome.

Dingman, S. L. (2002). *Physical Hydrology, 2nd Ed.*: Upper Saddle River, New Jersey. Prentice Hall.

Fontenot, R. (2004). *An Evaluation of Reference Evapotranspiration Models in Louisiana.* M.N.S. Thesis, Department of Geography and Anthropology, Louisiana State University.

Jensen, D. T., Hargreaves, G. H., Temesgen, B. & Allen. R. G. (1997). Computation of ETo under Nonideal Conditions. *Journal of Irrigation and Drainage Engineering, 123(5)*,394-400

Turc, L. (1961). Evaluation des besoins en eau d'irrigation, evapotranspiration potentielle, formule climatique simplifee et mise a jour. (In French). *Annales Agronomiques, 12(1)*,13-49.

Appendix E

HURASIM MODEL DESCRIPTION

HURASIM is a spatial simulation model of hurricane structure and circulation for reconstructing estimated windforce and vectors of past hurricanes. The HURASIM model generates a matrix of storm characteristics (i.e., quadrant, windspeed, and direction) within discrete spatial units and time intervals specified by the user for any specific storm or set of storms. HURASIM recreates the spatial structure of past hurricanes based on a tangential wind function, inflow angle offset, forward speed, and radius of maximum winds. Figure E.1 (below) shows the graphic user interface of the HURASIM model for a windfield reconstruction of Hurricane Katrina (2005) making landfall southeast of New Orleans, LA. Data input for the model includes tracking information of storm position, latitude and longitude, and maximum sustained wind speed every six hours or less. The model offers a suite of mathematical functions and parameter sets for the tangential wind profile taken from other hurricane studies (Harris, 1963; Bretschneiger and Tamaye, 1976; Neumann, 1987; Kjerfve et al., 1986; Boose et al., 1994). The user can specify the set of functions that provide more or less robust constructions of the range and extent of estimated winds.

Model output is user-specified for given geographic locations assigned by a given point or boundary area. Latitude and longitude for each study site location was supplied to the model to create a log of hurricane activity at 15-minute intervals for predicted winds above 30 mi/h for the period of record (1851-2003). The model estimates a suite of storm characteristics (i.e., quadrant, wind speed, and direction) within discrete spatial units and time intervals specified by the user for designated storms, years, and study site locations. Profiles of estimated wind conditions for a given site application are stored by year and storm. Time intervals of storm reposition and speed for this study were generated every 15 minutes. Minimum conditions of windspeed or distance can be set to parse the data output if warranted. In this study, windspeed estimates for any point or grid location were retained for further analysis if greater than 30 mi/h or tropical depression status.

HURASIM has been used extensively for field and modeling studies to relate biological response to hurricane forcing. HURASIM model output from Hurricane Andrew was correlated with field data to construct data tables of damage probabilities by site and species and to determine critical windspeeds and vectors of tree mortality and injury (Doyle et al., 1995a, 1995b). HURASIM also has been applied to reconstruct probable windfields of past hurricanes for remote field locations and correlated with tree-ring growth patterns and direction of leaning trees and downed logs (Doyle and Gorham, 1996). HURASIM also has been used to construct landscape templates of past hurricane activity that are linked with landscape simulation models of coastal habitat (Doyle and Girod, 1997).

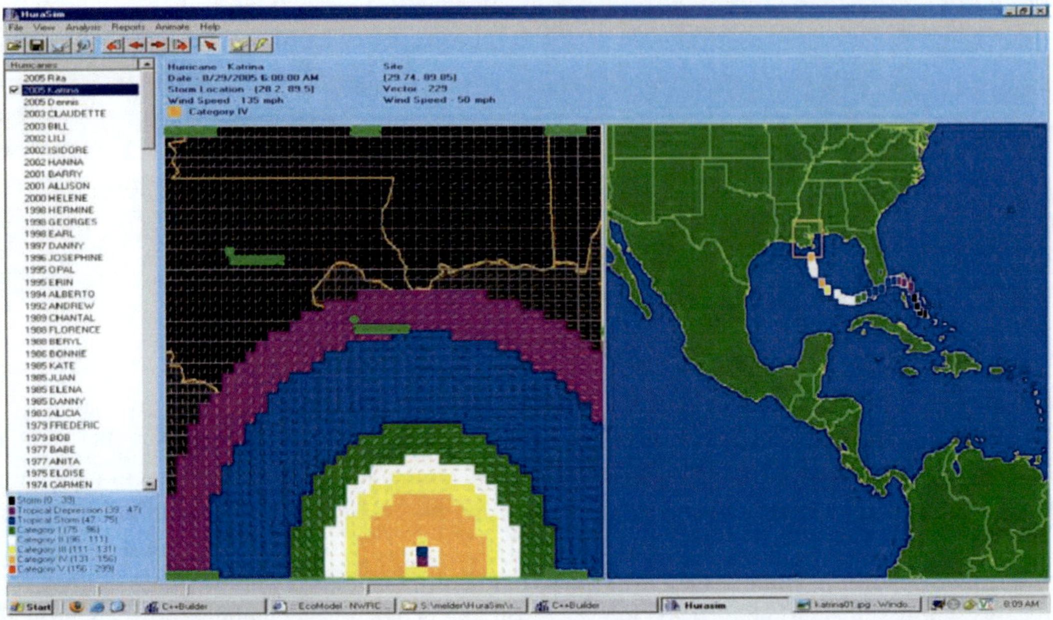

Figure E.1. Graphic user interface of the HURASIM model displaying storm track and windfield reconstruction of Hurricane Katrina (2005) at landfall south of New Orleans, LA. Grid cell color schemes represent different categories of storm strength and a directional line of wind direction

References

Boose, E. R. (1994). Hurricane impacts to tropical and temperate forest landscapes. *Ecological Monographs, 64*, 369-400.

Bretschneigder, C. L. & Tamaye, E. E. (1976). Hurricane wind and wave forecasting techniques. In: *Proceedings of the 15th Coastal Engineering Conference*, held Honolulu, Hawaii, July 11-17, (ASCE), Chapter *13*, pages 202-237.

Doyle, T. W., Keeland, B. D., Gorham, L. E. & Johnson, D. J. (1995a). Structural impact of Hurricane Andrew on forested wetlands of the Atchafalaya Basin in coastal Louisiana. *Journal of Coastal Research, 18*, 354-364.

Doyle, T. W., Smith, III, T. J. & Robblee, M. B. (1995b). Wind damage effects of Hurricane Andrew on mangrove communities along the southwest coast of Florida, USA. *Journal of Coastal Research, 18*, 144-159.

Doyle, T. W. & Gorham, L. E. (1996). Detecting hurricane impact and recovery from tree rings. In: J.S. Dean, Meko, D.M., and Swetnam, T.W., *Tree-Rings, Environment and Humanity,* Radiocarbon Press, Tucson, Arizona. Radiocarbon 1996, pages 405-412.

Doyle, T. W. & Girod, G. (1997). The frequency and intensity of Atlantic hurricanes and their influence on the structure of south Florida mangrove communities. Chapter 7. pages 111-128. In: H. Diaz and R. Pulwarty (eds.), *Hurricanes, Climatic Change and Socio economic Impacts: A Current Perspective*, Westview Press, New York, New York, pages 325.

Harris, D. L. (1963). *Characteristics of the hurricane storm surge*. U. S. Department of Commerce, Technical Paper No. 48, Washington, D.C., page 138.

Neumann, C. J. (1987). *The National Hurricane Center risk analysis program (HURISK)*. NOAA Technical Memorandum NWS NHC 38, Washington D.C., page 56.

Appendix F

PROJECTING FUTURE SEA LEVEL RISE WITH THE SLRRP MODEL

The Sea Level Rise Rectification Program (SLRRP) is a software package designed with a user-friendly interface to generate a suite of future sea level projections from various GCM models and scenario outputs obtained from the Intergovernmental Panel on Climate Change (IPCC) (2001). The SLRRP model allows the user to select a region-based tide station, GCM model, and SRES emissions scenario to generate a graph and output file of future sea level change. SLRRP rectifies the historical tide record and future eustatic sea level rise into a common datum (default = North American Vertical Datum of 1988 [NAVD88]) to facilitate comparison with landbased features and elevations. The SLRRP model generates a sea level prediction by wrapping the historical mean monthly records for the period of record for all future years up to year 2100. Because the historical record retains the long-term trend of local subsidence and historical eustatic change, an adjustment of removing the historical eustatic rate is accomplished before adding the predicted eustatic sea level rise based on a selected IPCC model and scenario. SLRRP uses a historical eustatic sea level rate of 1.8 mm/year (0.071 in) conferred by several sources as the best estimate for the global-mean since 1963 (IPCC, 2001; Douglas, 1997).

The SLRRP model uses a series of sequential pop-up windows to facilitate user selection of GCM models, scenarios, and manual entries for projecting future sea levels (figures F.1-F.3). The SLRRP and CoastCLIM models generate similar eustatic projections, but SLRRP retains the local tidal fluctuations that will contribute to short-term flooding above mean tides. The advantage of using the historical record includes the retention of the local variability and seasonality of sea level heights and the interannual variability and long-term climatic autocorrelation.

The program gives the user options for saving graphical and digital formats of SLRRP predictions and generating a supplemental graph to visualize the timing and extent of yearly flooding potential for a given elevation (NAVD88) for a transportation feature. After generating a future sea level projection, the user can execute a seawater inundation option that builds another graph that plots the timing and rate of flooding for a selected land elevation (figure F.4). In effect, the model shows the prospective data and time period for which sea level will overtop a given landscape feature under a future changing climate. Flooding potential is the percentage of months within a year when there is inundation by seawater at a select land elevation determined by the user.

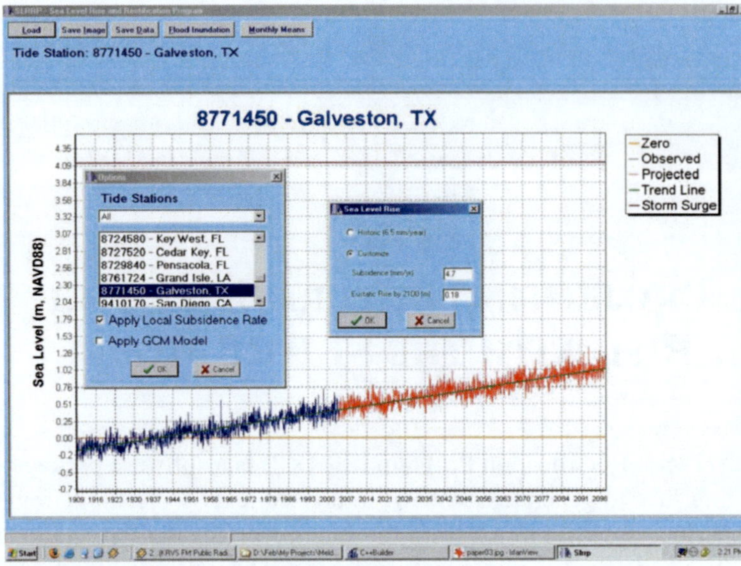

Figure F.1. User interface and simulated graph of historical sea level rise from a sample SLRRP model application displaying the pop-up windows for selecting tide gauge stations and constructing a sea level function based on local subsidence

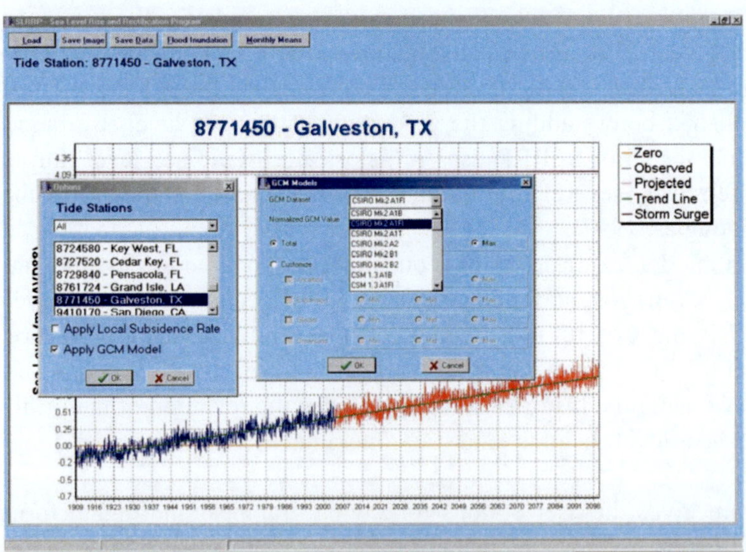

Figure F.2. User interface and simulated graph of historical sea level rise from a sample SLRRP model application displaying the pop-up window for selecting a GCM model and SRES emissions scenario

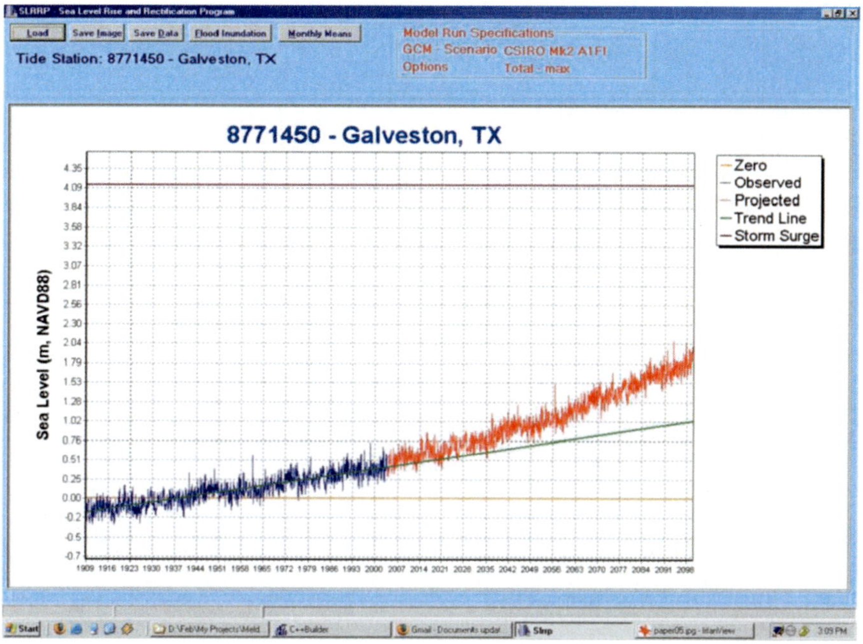

Figure F.3. User interface and simulated graph of future sea level rise from a sample SLRRP model application displaying the historical trend line, datum relationship, and maximum historical storm surge stage for the selected tide gauge location

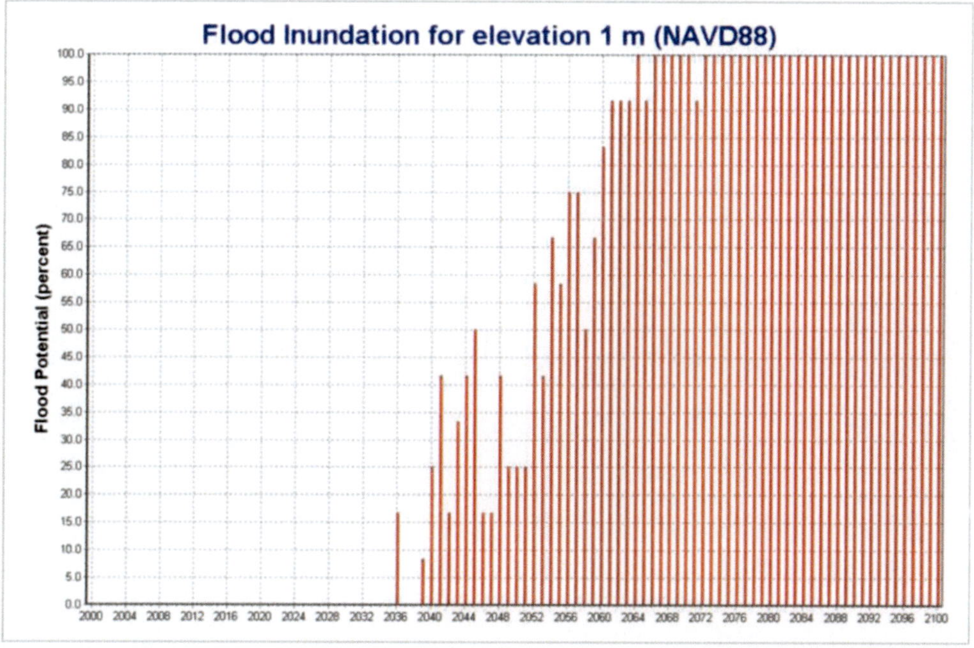

Figure F.4. Sample flood graph displaying flood timing and extent based on the hydroperiod or percent of days within a calendar year that flooding is likely to occur for a given land elevation and sea level rise projection

REFERENCES

Douglas, B. C. (1997). Global sea rise: A redetermination. *Surveys in Geophysics*, Volume *18*, pages 279-292.

IPCC, (2001): *Climate Change 2001: Impacts, Adaptation, and Vulnerability.* Contribution of Working Group II to the Third Assessment Report of the Intergovernmental Panel on Climate Change [J.J. McCarthy, O.F. Canziani, N.A. Leary, et al. (Eds.)], New York, New York: Cambridge University Press. Page 944 (Available on-line at http://www.ipcc.ch/).

List of Selected Acronyms

AASHTO	American Association of State and Highway Transportation Officials
APTA	American Public Transit Association
BRT	Bus Rapid Transit
CCSP	Climate Change Science Program
CDIAC	Carbon Dioxide Information Analysis Center
CVI	Coastal Vulnerability Index
DEM	Digital Elevation Model
DOT	Department of Transportation
ENSO	El Nino-Southern Oscillation
FAA	Federal Aviation Administration
FHWA	Federal Highway Administration
FRA	Federal Railroad Administration
FTA	Federal Transit Administration
GCM	General Circulation Model
HURDAT	HURricane DATabase
IPCC	Intergovernmental Panel on Climate Change
MARAD	Maritime Administration
MPO	Metropolitan Planning Organization
NAVD88	North American Vertical Datum 88
NBI	National Bridge Inventory
NED	National Elevation Dataset
NGVD	National Geodetic Vertical Datum
NHC	National Hurricane Center
NHS	National Highway System
NLCD	National Land Cover Dataset
NTD	National Transit Database
PDF	Probability Density Function
PHMSA	Pipeline and Hazardous Materials Safety Administration
SLOSH	Sea, Lake, and Overland Surges from Hurricanes
SLRRP	Sea Level Rise Rectification Program
SRES	Special Report on Emissions Scenarios
SST	Sea Surface Temperature

USCDD	U.S. Climate Division Dataset
USGS	United States Geological Survey
USHCN	U.S. Historic Climate Network
VMT	Vehicle Miles of Travel
WGNE-CMIP3	Working Group on Numerical Experimentation Coupled Model Intercomparison Project Phase 3

Glossary of Terms

Accretion

The process of soil buildup, generally through deposition.

Adaptation

Actions taken to reduce the vulnerability of natural and human systems against actual or expected climate change effects. Various types of adaptation can be distinguished, including anticipatory, autonomous, and planned adaptation.

- **Anticipatory Adaptation** – Adaptation that takes place before impacts of climate change are observed. Also referred to as proactive adaptation.

- **Autonomous Adaptation** – Adaptation that does not constitute a conscious response to climatic stimuli but is triggered by ecological changes in natural systems and by market or welfare changes in human systems. Also referred to as spontaneous adaptation.

- **Planned Adaptation** – Adaptation that is the result of a deliberate policy decision, based on an awareness that conditions have changed or are about to change and that action is required to return to, maintain, or achieve a desired state.

Adaptation Assessment

The practice of identifying options to adapt to climate change and evaluating them in terms of criteria such as availability, benefits, costs, effectiveness, efficiency, and feasibility.

Adaptation Benefits

The avoided damage costs or the accrued benefits following the adoption and implementation of *adaptation* measures.

Adaptation Costs

Costs of planning, preparing for, facilitating, and implementing *adaptation* measures, including transition costs.

Adaptive Capacity

The ability of a system to adjust to climate change (including climate variability and extremes) to moderate potential damages, to take advantage of opportunities, or to cope with the consequences.

Alluvium

Sand, gravel, and silt deposited by rivers and streams in a valley bottom.

Anthropogenic

Resulting from or produced by human beings.

Arterials

Major streets or highways, many with multilane or freeway design, serving high-volume traffic corridor movements that connect major generators of travel. While they may provide access to abutting land, their primary function is to serve traffic moving through the area.

Atmosphere

The gaseous envelope surrounding the earth. The dry atmosphere consists almost entirely of nitrogen and oxygen, together with trace gases, including carbon dioxide and ozone.

Baseline/Reference

The baseline (or reference) is the state against which change is measured. It might be a "current baseline," in which case it represents observable, present-day conditions. It might also be a "future baseline," which is a projected set of conditions, excluding the driving factor of interest. Alternative interpretations of the reference conditions can give rise to multiple baselines.

Basin

The drainage area of a stream, river, or lake.

Bus Rapid Transit (BRT)

A rapid mode of bus transportation that can combine the quality of rail transit and the flexibility of buses. There are a broad range of features that can be considered elements of a BRT system, including a dedicated bus-only right-of-way, bus lane reserved for buses on a major arterial road or freeway, on-line stops or stations (like light rail stations), other forms of giving buses priority in traffic, faster passenger boarding, faster fare collection, and a system image that is uniquely identifiable.

Carbon Cycle

The term used to describe the flow of carbon (in various forms, e.g., carbon dioxide) through the atmosphere, ocean, terrestrial biosphere, and lithosphere.

Carbon Dioxide (CO_2)

A naturally occurring gas fixed by photosynthesis into organic matter. A by-product of fossil fuel combustion and biomass burning, it is also emitted from land use changes and other industrial processes. It is the principal *anthropogenic greenhouse gas* that affects the earth's radiative balance. It is the reference gas against which other greenhouse gases are measured, thus having a global warming potential of one.

Climate

Climate in a narrow sense is usually defined as the "average weather," or more rigorously, as the statistical description in terms of the mean and variability of relevant quantities over a period of time ranging from months to thousands or millions of years. These quantities are most often surface variables such as temperature, precipitation, and wind.

Climate in a wider sense is the state, including a statistical description, of the *climate system*. The classical period of time is 30 years, as defined by the World Meteorological Organization (WMO).

Climate Change

A change in the mean state or variability of the climate, whether due to natural variability or as a result of human activity, that persists for an extended period, typically decades or longer. This usage differs from that in the *United Nations Framework Convention on Climate Change (UNFCCC)*, which defines "climate change" as: "a change of climate which is attributed directly or indirectly to human activity that alters the composition of the global atmosphere and which is in addition to natural climate variability observed over comparable time periods." Also see *climate variability*.

(Climate Change) Impact Assessment

The practice of identifying and evaluating, in monetary and/or nonmonetary terms, the effects of climate change on natural and human systems.

(Climate Change) Impacts

The effects of climate change on natural and human systems. Depending on the consideration of adaptation, one can distinguish between potential impacts and residual impacts.

- **Potential Impacts –** All impacts that may occur given a projected change in climate, without considering adaptation.

- **Residual Impacts –** The impacts of climate change that would occur after adaptation.

Climate Model

A numerical representation of the climate system based on the physical, chemical, and biological properties of its components, their interactions and feedback processes, and accounting for all or some of its known properties. The climate system can be represented by models of varying complexity (i.e. for any one component or combination of components a hierarchy of models can be identified, differing in such aspects as the number of spatial dimensions, the extent to which physical, chemical, or biological processes are explicitly represented, or the level at which empirical parameterisations are involved). Coupled atmosphere/ocean/sea-ice general circulation models (*AOGCM or GCM*) provide a comprehensive representation of the climate system. More complex models include active chemistry and biology. Climate models are applied, as a research tool, to study and simulate the climate, but also for operational purposes, including monthly, seasonal, and interannual climate predictions.

Climate Prediction

A climate prediction or climate forecast is the result of an attempt to produce an estimate of the actual evolution of the climate in the future; e.g., at seasonal, interannual, or long-term time scales. Also see *climate projection* and *climate scenario*.

Climate Projection

The calculated response of the climate system to emission or concentration scenarios of *greenhouse gases* and aerosols, or *radiative forcing* scenarios, often based on simulations by climate models. Climate projections are distinguished from *climate predictions* in that the former critically depend on the emission/concentration/radiative forcing scenario used, and therefore, on highly uncertain assumptions of future socioeconomic and technological development.

Climate Scenario

A plausible and often simplified representation of the future *climate*, based on an internally consistent set of climatological relationships and assumptions of *radiative forcing*, typically constructed for explicit use as input to climate change impact models. A "climate change scenario" is the difference between a climate scenario and the current climate.

Climate System

The climate system is defined by the dynamics and interactions of five major components: atmosphere, hydrosphere, cryosphere, land surface, and biosphere. Climate system dynamics are driven by both internal and external forcing, such as volcanic eruptions, solar variations, or human-induced modifications to the planetary radiative balance, for instance via anthropogenic emissions of greenhouse gases and/or land use changes.

Climate Variability

Climate variability refers to variations in the mean state and other statistics (such as standard deviations, statistics of extremes, etc.) of the climate on all temporal and spatial scales beyond that of individual weather events. Variability may be due to natural internal processes within the climate system (internal variability*)* or to variations in natural or anthropogenic external forcing (external variability). Also see *climate change*.

Collectors

In urban areas, streets providing direct access to neighborhoods as well as direct access to arterials. In rural areas, routes serving intracounty, rather than statewide travel.

Commercial Service Airport

Airport that primarily accommodates scheduled passenger airline service.

Convection

Generally, transport of heat and moisture by the movement of a fluid. In meteorology, the term is used specifically to describe vertical transport of heat and moisture in the atmosphere, especially by updrafts and downdrafts in an unstable atmosphere. The terms "convection" and "thunderstorms" often are used interchangeably, although thunderstorms are only one form of convection.

Datum

A reference point or surface against which position measurements are made. A vertical datum is used for measuring the elevations of points on the Earth's surface, while a horizontal datum is used to measure positions on the Earth.

Downscaling

A method that derives local- to regional-scale (10 to 100 km) information from larger-scale models or data analyses.

Drought

The phenomenon that exists when precipitation is significantly below normal recorded levels, causing serious hydrological imbalances that often adversely affect land resources and production systems.

El Niño-Southern Oscillation (ENSO)

El Niño, in its original sense, is a warmwater current that periodically flows along the coast of Ecuador and Peru, disrupting the local fishery. This oceanic event is associated with a fluctuation of the intertropical surface pressure pattern and circulation in the Indian and Pacific Oceans, called the Southern Oscillation. This coupled atmosphere-ocean phenomenon is collectively known as El Niño-Southern Oscillation. During an El Niño event, the prevailing trade winds weaken and the equatorial countercurrent strengthens, causing warm surface waters in the Indonesian area to flow eastward to overlie the cold waters of the Peruvian current. This event has great impact on the wind, sea surface temperature, and precipitation patterns in the tropical Pacific. It has climatic effects throughout the Pacific region and in many other parts of the world. The opposite of an El Niño event is called La Niña.

Emissions Scenario

A plausible representation of the future development of emissions of substances that are potentially radiatively active (e.g., *greenhouse gases*, aerosols), based on a coherent and internally consistent set of assumptions about driving forces (such as demographic and socioeconomic development, technological change) and their key relationships. In 1992, the IPCC presented a set of emissions scenarios that were used as a basis for the climate projections in the Second Assessment Report (IPCC, 1996). These emissions scenarios are referred to as the IS92 scenarios. In the IPCC *Special Report on Emissions Scenarios (SRES)* (Nakićenović et al., 2000), new emissions scenarios – the so-called *SRES* scenarios – were published.

Enplanements

The total number of passengers boarding an aircraft, including both originating and connecting passengers.

Ensemble

A group of parallel model simulations used for *climate projections*. Variation of the results across the ensemble members gives an estimate of uncertainty. Ensembles made with the same model but different initial conditions only characterize the uncertainty associated with internal climate variability, whereas multimodel ensembles, including simulations by several models also include the impact of model differences.

Erosion

The process of removal and transport of soil and rock by weathering, mass wasting, and the action of streams, glaciers, waves, winds, and underground water.

Evaporation

The transition process from liquid to gaseous state.

Evapotranspiration

The combined process of water *evaporation* from the Earth's surface and *transpiration* from vegetation.

Exposure

The combination of stress associated with climate-related change (sea level rise, changes in temperature, frequency of severe storms) and the probability, or likelihood, that this stress will affect transportation infrastructure.

Extreme Weather Event

An event that is rare within its statistical reference distribution at a particular place. Definitions of "rare" vary, but an extreme weather event would normally be as rare as or rarer than the 10th or 90th percentile. By definition, the characteristics of what is called "extreme weather" may vary from place to place. Extreme weather events may typically include floods and droughts.

Fixed-Route Bus Service

Service provided on a repetitive, fixed-schedule basis along a specific route with vehicles stopping to pick up and deliver passengers to specific locations; each fixed-route trip serves the same origins and destinations, unlike demand response and taxicabs.

Fixed Transit Guideway

A system of vehicles that can operate only on its own guideway constructed for that purpose (e.g., rapid rail, light rail). Federal usage in funding legislation also includes exclusive right-of-way bus operations, trolley coaches, and ferryboats as "fixed guideway" transit.

Freight Handling Facility

Marine facilities or terminals that handle freight. A given port or port area may contain multiple freight-handling facilities.

General Aviation Airport

Airport that primarily accommodates aircraft owned by private individuals and businesses.

General Circulation Model (GCM)

See *climate model.*

Greenhouse Effect

The process in which the absorption of infrared radiation by the atmosphere warms the Earth. In common parlance, the term "greenhouse effect" may be used to refer either to the natural greenhouse effect, due to naturally occurring greenhouse gases, or to the enhanced (anthropogenic) greenhouse effect, which results from gases emitted as a result of human activities.

Greenhouse Gas

Greenhouse gases are those gaseous constituents of the atmosphere, both natural and anthropogenic, that absorb and emit radiation at specific wavelengths within the spectrum of infrared radiation emitted by the Earth's surface, the atmosphere, and clouds. This property causes the *greenhouse effect.* Water vapor (H_2O), carbon dioxide (CO_2), nitrous oxide (N_2O), methane (CH_4), and ozone (O_3) are the primary greenhouse gases in the Earth's atmosphere. Besides CO_2, N_2O, and CH_4, the Kyoto Protocol deals with the greenhouse gases sulfur hexaflouride (SF_6), hydrofluorocarbons (HFC), and perfluorocarbons (PFC).

Gross-Ton Mile

One ton of equipment or freight moved one mile.

Hazardous Liquid

Petroleum, petroleum products, liquefied natural gas (LNG), anhydrous ammonia, or a liquid that is flammable or toxic.

Humidity

Generally, a measure of the water vapor content of the air. Popularly, it is used synonymously with relative humidity.

Hurricane

A tropical cyclone in the Atlantic, Caribbean Sea, Gulf of Mexico, or eastern Pacific, in which the maximum one-minute sustained surface wind is 64 knots (74 mph) or greater.

Industrial Airport

Airports which can accommodate both commercial and privately owned aircraft and are typically used by aircraft service centers, manufacturers, and cargo companies, as well as general aviation aircraft.

Infrastructure

The basic equipment, utilities, productive enterprises, installations, and services essential for the development, operation, and growth of an organization, city, or nation.

Integrated Assessment

An interdisciplinary process of combining, interpreting, and communicating knowledge from diverse scientific disciplines so that all relevant aspects of a complex societal issue can be evaluated and considered for the benefit of decision making.

Intermodal Connector

Highway providing access to intermodal facilities and designated as a National Highway System (NHS) Intermodal Connector.

Intermodal Passenger Terminal

A passenger terminal that accommodates several modes of transportation, such as intercity rail service, intercity bus, commuter rail, intracity rail transit and bus transportation, airport limousine service and airline ticket offices, rent-a-car facilities, taxis, private parking, and other transportation services.

Intermodal Transportation

Use of more than one type of transportation; e.g., transporting a commodity by barge to an intermediate point and by truck to destination. Often specifically refers to the use of cargo containers that can be interchanged between transport modes (i.e., motor, water, and air carriers) and where the equipment is compatible within the multiple systems.

Interstate Highways

Limited access, divided facility of at least four lanes designated by the Federal Highway Administration as part of the Interstate System, a system of freeways connecting and serving the principal cities of the continental United States.

Invasive Species

An introduced species that invades natural habitats.

Land Use

The total of human activities implemented in a certain land-cover type (a set of human actions). The social and economic purposes for which land is managed (e.g., grazing, timber extraction, conservation).

Lidar (Light Detection and Ranging)

A remote sensing technology that determines the distance to an object or surface by using laser pulses.

Linguistically Isolated Household

A household in which no person aged 14 and over speaks English at least "very well."

Local Road

Roads that provide access to private property or low-volume public facilities.

Long-Range Transportation Plan (LRTP)

A 20- to 30-year plan that provides a long-range vision of the future of the surface transportation system, considering all passenger and freight modes and their interrelationships. LRTPs are developed by *MPOs* as part of the Federally mandated planning process.

Metropolitan Planning Organization (MPO)

The forum for cooperative transportation decision making for a metropolitan planning area. Formed in cooperation with the state, it develops transportation plans and programs for the metropolitan area. For each urbanized area, an MPO must be designated by agreement between the Governor and local units of government representing 75 percent of the affected population (in the metropolitan area), including the central cities or cities as defined by the Bureau of the Census or in accordance with procedures established by applicable state or local law (23 U.S.C. 134(b)(1)/Federal Transit Act of 1991 Section 8(b)(1)).

Mitigation

An anthropogenic intervention to reduce the anthropogenic forcing of the climate system; it includes strategies to reduce *greenhouse gas* sources and emissions and to enhance *greenhouse gas* sinks.

Morphology

The form and structure of an organism or land form, or any of its parts.

Nonfreight Marine Facility

Marine facilities not used for transporting or handling freight. Includes unused berths; commercial fishing facilities; vessel construction, repair, and servicing facilities; marine construction services; etc.

Nonlinearity

A process is called "nonlinear" when there is no simple proportional relation between cause and effect.

Paratransit

Comparable transportation service required by the American Disabilities Act (ADA) for individuals with disabilities who are unable to use fixed-route transportation systems. Usually involves the use of demand-response systems, in which passengers or their agents contact a transit operator, who then dispatches a car, van, or bus to pick up the passengers and transport them to their destinations (also called "Dial-a-Ride").

Partial Duration Series (PDS)

A series composed of all events during the period of record that exceed some set criterion; for example, all floods above a selected base, or all daily rainfalls greater than a specified amount.

Probability Density Function

A statistical function that shows how the density of possible observations in a population is distributed.

Projection

The potential evolution of a quality or set of quantities, often computed with the aid of a model. Projections are distinguished from predictions in order to emphasize that projections involve assumptions – concerning, for example, future socioeconomic and technological developments that may or may not be realized – and are, therefore, subject to substantial uncertainty. Also see *climate projection* and *climate prediction*.

Radiative Forcing

Radiative forcing is the change in the net vertical irradiance (expressed in Watts per square meter (Wm^{-2})) at the tropopause due to an internal or external change in the forcing of the climate system, such as a change in the concentration of CO_2 or the output of the sun.

Relative Humidity

A dimensionless ratio, expressed in percent, of the amount of atmospheric moisture present relative to the amount that would be present if the air were saturated. Since the latter amount is dependent on temperature, relative humidity is a function of both moisture content and temperature. As such, relative humidity by itself does not directly indicate the actual amount of atmospheric moisture present.

Resilience

The capacity of a system to absorb disturbances and retain essential processes.

Runoff

That part of precipitation that does not *evaporate* and is not *transpired.*

Saffir-Simpson Scale

A scale from 1 to 5 that describes a hurricane's strength, where Category 1 is the weakest and Category 5 is the strongest hurricane. The categories are defined by wind speed. The scale of numbers is based on actual conditions at some time during the life of the storm; as the hurricane intensifies or weakens, the scale number is reassessed accordingly.

Scenario

A plausible and often simplified description of how the future may develop based on a coherent and internally consistent set of assumptions about driving forces and key relationships. Scenarios may be derived from projections but are often based on additional information from other sources, sometimes combined with a "narrative storyline." Also see *climate scenario* and *emissions scenario* and *Special Report on Emissions Scenarios (SRES).*

Sea Level Rise

An increase in the mean level of the ocean. Eustatic sea level rise is a change in global average sea level brought about by an increase in the volume of the world ocean. Relative sea level rise occurs where there is a local increase in the level of the ocean relative to the land, which might be due to ocean rise and/or land-level subsidence. In areas subject to rapid land level uplift, relative sea level can fall.

Sea Surface Temperature

The mean temperature of the ocean in the upper few meters.

Socioeconomic Scenarios

Scenarios concerning future conditions in terms of population, Gross Domestic Product, and other socioeconomic factors relevant to understanding the implications of climate change. Also see *Special Report on Emissions Scenarios (SRES).*

Specific Humidity

In a system of moist air, the ratio of the mass of water vapor to the total mass of the system.

Special Report on Emissions Scenarios (SRES)

The storylines and associated population, Gross Domestic Product, and emissions scenarios associated with the Special Report on Emissions Scenarios (SRES) (Nakićenović, 2000), and the resulting climate change and sea level rise scenarios. Four families of socioeconomic scenario (A1, A2, B1, and B2) represent different world futures in two distinct dimensions: a focus on economic versus environmental concerns and global versus regional development patterns.

Storm Surge

An abnormal rise in sea level accompanying a hurricane or other intense storm, whose height is the difference between the observed level of the sea surface and the level that would have occurred in the absence of the cyclone. Storm surge is usually estimated by subtracting the normal or astronomic tide from the observed storm tide.

Subsidence

A sinking down of part of the Earth's crust, generally due to natural compaction of sediments or from underground excavation (such as the removal of groundwater).

Surface Runoff

The water that travels over the soil surface to the nearest surface stream; *runoff* of a drainage basin that has not passed beneath the surface since precipitation.

Thermal Expansion

In connection with *sea level rise*, this refers to the increase in volume (and decrease in density) that results from warming water. A warming of the ocean leads to an expansion of the ocean volume and hence an increase in sea level.

Threshold

The level of magnitude of a system process at which sudden or rapid change occurs. A point or level at which new properties emerge in an ecological, economic, or other system, invalidating predictions based on mathematical relationships that apply at lower levels.

Transpiration

The evaporation of water vapor from the surfaces of leaves through stomates.

Transportation Improvement Program (TIP)

A prioritized program of transportation projects to be implemented in appropriate stages over several years (i.e., 3 to 5 years). The projects are recommended from those in the transportation systems management element and the long-range element of the planning process. This program is required as a condition for a locality to receive Federal transit and highway grants.

Tropical Storm

A tropical cyclone in which the maximum 1-minute sustained surface wind ranges from 34 to 63 knots (39 to 73 mph) inclusive.

Uncertainty

An expression of the degree to which a value (e.g., the future state of the climate system) is unknown. Uncertainty can result from lack of information or from disagreement about what is known or even knowable. It may have many types of sources, from quantifiable errors in the data to ambiguously defined concepts or terminology, or uncertain projections of human behavior. Uncertainty can therefore be represented by quantitative measures (e.g., a range of values calculated by various models) or by qualitative statements (e.g., reflecting the judgment of a team of experts).

United Nations Framework Convention on Climate Change (UNFCCC)

The UNFCCC was adopted on May 9, 1992, in New York and signed at the 1992 Earth Summit in Rio de Janeiro by more than 150 countries and the European Community. Its ultimate objective is the "stabilization of greenhouse gas concentrations in the atmosphere at a level that would prevent dangerous anthropogenic interference with the climate system." It contains commitments for all Parties. Under the Convention, Parties included in Annex I aim to return greenhouse gas emissions not controlled by the Montreal Protocol to 1990 levels by the year 2000. The Convention entered in force in March 1994.

Urbanization

The conversion of land from a natural state or managed natural state (such as agriculture) to cities; a process driven by net rural-to-urban migration through which an increasing percentage of the population in any nation or region come to live in settlements that are defined as "urban centers."

Vehicle Miles of Travel (VMT)

A unit to measure vehicle travel made by a private vehicle, such as an automobile, van, pickup truck, or motorcycle. Each mile traveled is counted as one vehicle mile, regardless of the number of persons in the vehicle. Generally, vehicle miles of travel are reported on an annual basis for a large area.

Vulnerability

The structural strength and integrity of key facilities or systems and the resulting potential for damage and disruption in transportation services from climate change stressors.

References

American Meteorological Society, accessed (2007). *American Meteorological Society Glossary of Meteorology*. Washington, D.C. Accessible at http:// amsglossary.allenpress. com/glossary.

Bureau of Transportation Statistics (BTS), accessed (2007). *BTS Dictionary*. U.S. Department of Transportation, Washington, D.C. Accessible at http://www.bts.gov/ dictionary/ index. xml.

IPCC, (2007). *Climate Change 2007: The Physical Science Basis, Summary for Policy-Makers*. Contribution of Working Group I to the Fourth Assessment Report of the Intergovernmental Panel on Climate Change. Geneva, Switzerland, 21 pages.

National Weather Service, accessed (2007). *National Weather Service Glossary*. National Oceanic and Atmospheric Administration, Silver Spring, Maryland. Accessible at http://www.weather.gov/glossary/.

USGS, accessed (2007). *USGS Geologic Glossary*. Reston, Virginia. Accessible at http:// wrgis.wr.usgs.gov/docs/parks/misc/glossaryAtoC.html.

INDEX

A

absorption, 309
accessibility, 53, 236, 252
accidents, 29, 34, 173, 225
accounting, 11, 70, 101, 193, 224, 304
accuracy, 125, 127, 139, 160
ACI, 71, 90
acute, 225
ADA, 313
adaptability, 39
adaptation, xiv, xv, xx, xxii, 1, 7, 9, 10, 11, 12, 15, 17, 19, 20, 21, 22, 23, 24, 74, 96, 166, 208, 212, 218, 226, 241, 247, 250, 252, 253, 254, 255, 256, 257, 259, 264, 270, 301, 302, 304
adjustment, 194, 253, 287, 295
advisory committees, 240
aerosols, 3, 305, 307
Africa, 3, 207
afternoon, 97, 142
age, 2, 84, 92, 131, 209, 215, 242, 276
agents, 313
aging, 165
agricultural, xvi, 50, 54, 64, 74, 83, 163, 220, 225
agricultural exports, 220
agriculture, 78, 79, 85, 220, 317
aid, 313
AIP, 231
air, 12, 21, 25, 26, 50, 54, 72, 97, 114, 115, 117, 142, 146, 152, 153, 172, 179, 193, 200, 211, 212, 216, 220, 238, 243, 255, 287, 310, 311, 314, 315
air carriers, 311
Air Force, 71
air quality, 21, 243
air traffic, 193, 200, 220
Aircraft, 194, 195, 197, 201, 202
Airplanes, 195, 197
airports, xi, xiii, xiv, xvi, xix, 4, 13, 14, 35, 54, 60, 70, 71, 72, 73, 118, 141, 159, 160, 161, 162, 163, 166, 193, 194, 197, 198, 199, 200, 203, 204, 218, 220, 221, 222, 225, 231, 263, 267
Alabama, xi, xiii, xx, 49, 52, 63, 64, 65, 68, 69, 70, 72, 76, 79, 99, 100, 123, 132, 138, 143, 145, 147, 150, 158, 205, 218, 229, 230, 235, 241, 265
Alaska, 8, 37, 41, 45, 46, 207
algorithm, 112
alluvial, 74, 78
alternative, 14, 15, 29, 53, 63, 68, 100, 101, 174, 187, 188, 241, 247, 249, 250, 252, 259
alternatives, 187, 253, 257
alters, 304
ambiguity, 265
American Community Survey, 84, 92
American Public Transportation Association, 61, 90
ammonia, 310
amplitude, 177
AMS, 152
Amtrak, 53, 60, 63, 65, 66, 67, 179, 182, 183, 185, 186, 214, 273, 286
analysts, 15, 18, 114
anthropogenic, 303, 305, 306, 309, 312, 317
appendix, 64, 66, 74, 87, 101, 119, 124, 134, 181, 183, 184
application, xiv, 22, 131, 135, 192, 227, 291, 296, 297
Arabia, 207
Arctic, 7, 8, 11, 12, 13, 15, 16, 17, 20, 28, 29, 32, 41, 42, 44, 45, 46, 47, 97
argument, 117
Arizona, 292
Arkansas, 74
Army, 27, 42, 66, 68, 69, 70, 91, 189, 191, 192, 273
Army Corps of Engineers, 27, 42, 66, 68, 69, 70, 91, 189, 191, 192, 273
Asia, 3
asphalt, 12, 25, 29, 60, 156, 164, 170, 219

assessment, xiii, xiv, xvi, xx, xxi, 5, 8, 9, 10, 18, 19, 20, 21, 24, 39, 98, 126, 131, 138, 146, 161, 209, 225, 227, 244, 250, 254, 255, 256, 257, 259, 263, 268
assessment tools, 24
assets, 13, 50, 140, 188, 209
Association of American Railroads, 183, 230, 241
assumptions, xviii, 3, 17, 130, 194, 196, 209, 305, 307, 313, 314
Atlantic, xvii, 45, 74, 118, 119, 121, 123, 138, 141, 146, 147, 148, 150, 152, 153, 213, 292, 310
Atlas, 53, 90, 227, 286
atmosphere, xvi, xx, 2, 75, 100, 117, 118, 122, 141, 142, 150, 200, 264, 302, 303, 304, 305, 306, 307, 309, 317
Australia, 9, 28, 44, 134
authority, 59, 211, 241, 244, 246, 247
autocorrelation, 295
autonomy, 246
availability, xiv, 150, 172, 186, 187, 215, 242, 301
averaging, 111, 196
aviation, 1, 7, 8, 33, 39, 53, 54, 71, 156, 157, 163, 193, 194, 195, 196, 199, 200, 203, 221, 231, 240, 310
awareness, xi, xxi, 7, 18, 76, 97, 301

B

ballast, 179
barges, 12, 65, 68, 186, 220
barrier, xv, 18, 29, 74, 75, 77, 78, 142, 145, 152, 161, 162, 191, 248, 264, 265, 267
barriers, 14, 34, 219
base year, 196
Bayesian, 101, 152
beaches, 217
behavior, 112, 118, 125, 253
benchmarks, 133, 152
benefits, 10, 12, 16, 21, 226, 301, 302
benign, 220
bias, 100
biological processes, 304
biological systems, 145
biomass, 303
biosphere, 303, 305
Boeing, 194, 197, 198
borrowing, 250
Boston, 10, 15, 19, 43, 45
bottlenecks, 223, 230, 243
breakdown, 12
BSR, 148
bubbles, 2
building code, 227
buildings, 60, 76, 93, 156, 191
Bureau of the Census, 312
burning, 303
buses, 62, 63, 172, 173, 174, 303
business model, 163
bypass, 187

C

cables, 14, 35
calibration, 138
Canada, 9, 10, 13, 14, 15, 16, 28, 37, 38, 40, 41, 42, 43, 45, 46, 64, 261
canals, 78
capital cost, 188
Carbon, 65, 98, 116, 149, 196, 299, 303
carbon dioxide, 2, 264, 302, 303, 309
carbon emissions, 146, 156, 211
cargo, xvi, 12, 26, 29, 54, 65, 71, 72, 188, 191, 199, 221, 310, 311
Caribbean, 310
Caribbean Sea, 310
carrier, 15, 72
case study, xiv, 1, 2, 22, 162, 223, 253
catalyst, 242
category a, 125, 203
CCC, 153
CCSP, xiv, 39, 299
cell, 128, 129, 135, 215, 292
cement, 65, 170
Census, 51, 52, 80, 81, 82, 83, 84, 86, 88, 89, 92, 93, 215, 230, 274, 275, 276, 277, 278, 279, 280, 281, 312
Census Bureau, 51, 52, 80, 81, 82, 83, 84, 86, 88, 89, 92, 93, 215, 230, 275, 276, 277, 278, 279, 280, 281
CH4, 309
changing environment, xviii, 226, 253
channels, 26, 74, 78, 169, 190, 191, 241, 246, 248
chaos, 4
chemical industry, 85
chemicals, 16, 62, 64, 65, 68
childcare, 53
children, 53, 81, 87, 88, 276
China, 28
chlorine, 64
circulation, xvi, 96, 100, 119, 122, 123, 124, 144, 156, 188, 264, 291, 304, 307
citizens, 50, 187
civilian, 71, 84, 86, 193
classical, 304
classification, 58, 78, 92
clay, 64, 65, 74

cleaning, 253
cleanup, 170, 219
Climate Change Science Program, xiii, xiv, 3, 5, 39, 43, 45, 150, 151, 299
climatology, 98, 100, 105, 151, 152
closure, 30, 209, 221
clouds, 142, 309
CO2, 2, 9, 97, 98, 106, 153, 303, 309, 314
coaches, 309
coal, 64, 65, 69, 70, 189, 193, 211
Coastal areas, xxiii, 271
coastal communities, 95, 118, 216
coastal zone, 8, 50, 75, 77, 132, 144, 145, 200
codes, 51, 181, 183, 184
coding, 231
collaboration, xxiii, 39, 240, 248, 271
collisions, 26, 161
Columbia, 44, 73, 207
combined effect, xvii, xviii, 77, 131, 134, 158, 193, 267
combustion, 2, 303
commerce, 30, 188, 254
commodity, 311
communication, 57, 215, 216, 227, 230
communication systems, 57, 216
communities, xiv, xvi, xx, xxii, xxiii, 1, 14, 76, 83, 95, 118, 146, 157, 162, 187, 216, 218, 219, 226, 238, 271, 292
compaction, 132, 133, 316
competitiveness, 163, 236
complex interactions, xxi
components, xxii, 101, 141, 250, 253, 254, 255, 256, 257, 258, 288, 304, 305
composition, 304
computing, 289
concentration, xvi, 2, 55, 96, 98, 129, 305, 314
concrete, 22, 28, 29, 60, 114, 164, 170, 178, 188, 219, 267
conditioning, 25, 157, 172, 211, 212, 216
confidence, 2, 3, 96, 131, 255, 257
Congress, 230
connectivity, xix, 50, 55, 159, 161, 163, 173, 177, 191, 212, 225, 236, 252, 253, 266, 269
consensus, xvii, 101, 146, 156, 212
conservation, 76, 236, 248, 311
construction, xi, xiii, xviii, 13, 14, 17, 25, 28, 31, 70, 84, 95, 114, 156, 158, 164, 165, 172, 188, 191, 194, 196, 209, 211, 226, 233, 235, 236, 240, 241, 242, 245, 249, 254, 267, 270, 313
consulting, 39, 248
consumer goods, 64, 187, 226
consumers, 192
consumption, 73, 131, 155, 157, 172, 188, 189, 190, 193, 207
consumption patterns, 188, 190
continental shelf, 78, 137
contingency, 29, 174, 269
continuous data, 100
control, 10, 26, 57, 145, 159, 177, 192, 220, 231, 245
convective, 141, 142, 148, 157, 162, 200, 228
convergence, 101, 125
conversion, 317
cooling, xviii, 13, 27, 28, 103, 112, 157, 188
Coping, 46
copper, 65
coral, 92
correlation, 209
corridors, 12, 55, 68, 133
corrosion, 16, 36, 168
cost-effective, xi, xiv
covering, 207, 273
cracking, 12, 165
creep, 76, 91
critical infrastructure, 54, 55, 217
crust, 74, 316
Cuba, 122
customers, 1, 93, 163, 181, 183, 186, 187
cycles, 12, 13, 31, 118, 124, 125, 129
cycling, 126
cyclones, 4, 117, 118, 123, 138, 148

D

Dallas, 112, 113
danger, 159, 177, 187
data collection, 20, 39
data distribution, 36
data set, 111
database, 58, 59, 61, 70, 90, 112, 114, 121, 135, 138, 139, 150, 189, 190, 227
debt, 87
decision makers, xiv, xxi, xxii, 2, 7, 17, 19, 20, 22, 23, 24, 166, 225, 226, 233, 248, 253, 254, 255, 256, 263, 264
decision support tool, 17, 20, 23
decisions, xi, xiv, xviii, xx, xxi, xxii, 1, 2, 5, 6, 18, 21, 57, 181, 187, 225, 226, 233, 234, 237, 240, 244, 247, 249, 254, 255, 256, 263, 266
decomposition, 132
defense, xv, 14, 35, 77
defense mechanisms, 14, 35
deficit, 105, 106, 110, 111, 289
definition, 308
degradation, 156, 180
delivery, 4, 30, 37, 71, 146, 163, 268

demographic change, 237
demographics, 21, 84, 86, 211, 212
density, 12, 64, 67, 82, 100, 106, 109, 187, 226, 313, 316
Department of Commerce, 91, 152, 293
Department of Energy, 5, 151, 228
Department of Health and Human Services, 227
Department of Homeland Security, 216, 230
Department of Interior, 230
Department of State, 7, 26, 29, 31, 32, 45
Department of the Interior, 71, 204
Department of Transportation, xiv, 2, 17, 41, 42, 44, 45, 46, 54, 55, 56, 57, 58, 59, 61, 67, 71, 73, 90, 91, 164, 227, 228, 229, 230, 235, 237, 261, 268, 271, 299, 318
dependent variable, 115
deposition, 75, 132, 301
depression, 60, 122, 291
deprivation, 78
designers, 1, 22, 158, 164, 250
destruction, xv, 77, 128
detection, 32, 123, 178
dew, 287
diesel, 27, 61, 63
diesel engines, 27
diffusion, 97
disabled, 61, 62, 87, 215, 223
disaster, 218, 246
dispersion, 35
displacement, 152, 208
distribution, xvi, 18, 36, 54, 80, 81, 85, 96, 100, 105, 108, 119, 125, 163, 204, 206, 251, 308
division, 99, 100, 103, 104, 132
doors, 191
DOT, xiv, xx, 2, 5, 8, 22, 23, 39, 57, 72, 160, 165, 167, 168, 170, 171, 173, 174, 175, 176, 180, 181, 184, 185, 186, 195, 197, 204, 205, 213, 214, 219, 229, 230, 234, 236, 238, 240, 241, 243, 244, 256, 273, 299
drainage, xviii, 29, 76, 144, 223, 245, 268, 274, 303, 316
drought, 3, 86, 96, 110, 143, 144, 207
Drought, 306
droughts, 112, 144, 206, 308
drying, 3, 289
DSC, 182, 283
durability, 87
duration, 75, 105, 123, 153, 245, 253

E

earth, 302, 303
Earth Science, 273
earthquake, xxiii, 206, 270
ecological, 6, 78, 91, 157, 161, 242, 250, 263, 269, 301, 316
Ecological Society of America, 153
ecology, 16, 248, 250
economic activity, 3, 21, 80, 186
economic development, xxi, 15, 97, 118, 236, 237, 238, 240, 247
economic losses, 218, 255
economic resources, 79, 253
economics, 39
ecosystem, 21, 165
ecosystems, 39, 95, 123, 145, 150
Ecuador, 307
elderly, 61, 62, 214
electrical power, 212, 219
electricity, 193, 211, 221
emergency calls, 209
emergency management, 7, 39, 53, 56, 86, 164, 211, 212, 214, 215, 217, 221, 224, 227, 270
emergency planning, 223
emergency response, 186, 212
emerging issues, 247
emission, xvii, xviii, xx, 20, 97, 137, 151, 157, 264, 265, 305
employees, 85, 241
employment, 84, 237
energy, xix, 3, 4, 5, 54, 71, 75, 85, 97, 119, 137, 155, 157, 166, 188, 189, 193, 204, 208, 211, 224, 236, 248, 267, 274
energy consumption, 155, 157, 188, 189
energy efficiency, 189
Energy Information Administration, 228
Energy Information Administration (EIA), 228
energy supply, 71, 204, 224
engines, 27, 172, 194
England, 46, 47
environment, xxi, xxiii, 5, 11, 16, 53, 71, 73, 146, 153, 161, 165, 209, 226, 236, 237, 242, 248, 250, 259, 263, 269, 271
environmental awareness, 97
environmental conditions, xxi, 86
environmental impact, 15, 16, 21, 242, 251
environmental influences, 78
environmental protection, 236
Environmental Protection Agency, 5, 43, 145, 153
EPA, 153, 274
equilibrium, 9
equilibrium state, 9
equity, 87, 280
erosion, xv, 73, 77, 78, 90, 118, 139, 177, 179, 200, 207, 210, 224, 242, 251, 265
estimating, 10, 148

European Community, 317
European Space Agency, 122
evacuation, 34, 50, 56, 86, 138, 164, 211, 212, 213, 214, 216, 217, 223, 224, 241, 242, 243, 244, 245, 246, 257, 273
evaporation, 12, 143, 144, 157, 165, 316
evapotranspiration, xvii, 101, 110, 146, 153, 191, 265, 287, 288, 289
evening, 164
Everglades, 220
evolution, 270, 305, 313
examinations, 8
exercise, xxi, 242
expertise, 39
exports, 193, 220
exposure, xvii, xxii, 128, 146, 161, 207, 208, 210, 244, 250, 251, 252, 253, 254, 255, 256, 257, 258, 259
Exposure, xxii, 250, 251, 254, 308
extraction, 84, 85, 311
extrapolation, xvi, 146

F

FAA, 72, 194, 195, 196, 197, 199, 205, 231, 299
fabric, 79
failure, 30, 34, 76, 163, 209
family, 1, 87
FAO, 147, 287, 289
faults, 76
Federal Aviation Administration, 194, 195, 197, 273, 299
Federal Emergency Management Agency (FEMA), 137, 274
Federal Highway Administration, 7, 26, 32, 34, 38, 45, 54, 91, 299, 311
Federal Railroad Administration, 56, 273, 299
Federal Transit Administration, 61, 91, 273, 299
feedback, 152, 304
feet, xiii, xvii, 125, 137, 195, 219, 220
FEMA, 76, 137, 205, 274
finance, 217, 240
Finland, 9
fire, 223
fires, 212
first responders, 227
fishing, 70, 84, 85, 208, 313
flexibility, 303
flight, 4, 155, 199, 200, 231
float, 37
flood, 6, 14, 29, 35, 49, 74, 138, 145, 159, 200, 224, 243, 245, 257, 266, 274, 297
flow, 4, 13, 26, 32, 131, 144, 155, 182, 208, 209, 252, 303, 307
fluctuations, 100, 147, 188, 295
fluid, 76, 132, 210, 306
focusing, 15
food, 64, 65, 68, 70
forecasting, 33, 118, 119, 248, 292
Forest Service, 90
forests, 75, 78
fossil, 2, 97, 131, 189, 211, 303
fossil fuel, 2, 97, 131, 189, 211, 303
FRA, 299
fracture, 208
freezing, 97
fresh water, 146
friction, 195
frost, 29
fruits, 220
FTA, 61, 62, 91, 299
fuel, 11, 26, 30, 50, 68, 131, 163, 172, 186, 194, 202, 211, 212, 231, 303
full capacity, 223
funding, xx, 61, 163, 216, 231, 234, 235, 246, 309
futures, 6, 18, 20, 22, 73, 97, 165, 207, 315

G

gas, xvi, 50, 54, 68, 71, 72, 73, 76, 78, 85, 91, 93, 96, 97, 133, 152, 193, 204, 207, 209, 210, 211, 224, 225, 228, 303, 312, 317
gasoline, 63, 68, 223
gauge, 131, 132, 133, 134, 145, 148, 296, 297
generation, 117, 194, 261, 289
generators, 58, 186, 209, 302
Geneva, 149, 150, 151, 271, 318
Geographic Information System, 273
geography, xv, 50, 188
geology, 73, 75, 132, 208
geophysical, 76
Georgia, 152, 239
Germany, 134
GIS, xv, 23, 74, 203, 206, 221, 226, 231, 273
glaciers, 308
global warming, 119, 124, 129, 148, 150, 228, 303
goals, xxi, 237, 238, 242, 247
goods and services, 4, 240, 252, 268
government, 9, 39, 70, 187, 234, 238, 240, 245, 246, 249, 259, 312
grain, 64, 189, 192, 268
graph, 295, 296, 297
grazing, 311
Great Lakes, 12, 15, 16, 37, 41, 44, 45, 132

greenhouse, xviii, xx, 2, 3, 6, 96, 97, 105, 123, 136, 145, 158, 213, 264, 303, 305, 307, 309, 312, 317
greenhouse gas, xviii, xx, 2, 3, 6, 96, 97, 105, 123, 136, 145, 158, 213, 264, 303, 305, 307, 309, 312, 317
Greenland, 131, 158
Gross Domestic Product, 315
groundwater, 10, 43, 76, 133, 148, 316
groups, 187, 199
growth, 2, 12, 80, 81, 83, 85, 88, 97, 118, 131, 146, 181, 187, 227, 236, 237, 242, 243, 291, 310
growth rate, 81, 83
guidance, 231
guidelines, 39, 245
Gulf of Mexico, xvi, xvii, 50, 54, 55, 68, 69, 74, 76, 77, 78, 80, 85, 92, 95, 97, 112, 119, 121, 122, 123, 124, 127, 128, 131, 138, 141, 143, 146, 151, 152, 153, 154, 162, 204, 209, 210, 211, 224, 229, 310

H

habitat, 21, 292
handling, xvi, 36, 55, 66, 68, 188, 189, 191, 192, 309, 313
harm, 16, 223, 250, 253
harvest, 220
Hawaii, 292
hazards, 14, 86, 92, 153, 204, 256
haze, 200
health, 26, 216, 274
Health and Human Services, 227
health services, 216
heat, xviii, 2, 3, 12, 26, 96, 119, 122, 143, 144, 147, 156, 164, 267, 287, 306
heat exhaustion, 26
heating, 97, 112, 178
height, 78, 122, 137, 138, 139, 141, 142, 143, 145, 169, 222, 226, 315
high risk, 19, 182
high temperature, xvii, xviii, 114, 116, 144, 146, 172, 188, 196, 265, 267
highways, xiii, xv, xix, 1, 4, 7, 20, 32, 39, 53, 58, 59, 60, 76, 92, 95, 155, 157, 159, 163, 164, 165, 166, 169, 170, 190, 191, 211, 212, 213, 218, 219, 223, 234, 235, 238, 246, 267, 302
histogram, 120, 124
Holland, 119, 129, 153
Holocene, 74, 75, 150
Homeland Security, 216, 230
horizon, 236, 247, 248, 253
hospitals, 93
House, 40
household, 80, 87, 312
household income, 80, 87
households, 87, 223, 224, 276, 277
housing, 81, 83, 85, 87, 88, 216, 248, 279, 280
human, xv, 2, 3, 39, 76, 78, 95, 97, 114, 121, 123, 224, 248, 253, 301, 302, 304, 305, 309, 311, 317
human actions, 311
human activity, 78, 97, 304
human behavior, 224, 317
human development, 76, 95
humanitarian, 221
humidity, 117, 142, 148, 178, 287, 310, 314
Hurricane Andrew, 209, 291, 292
Hurricane Camille, 125, 140
Hurricane Katrina, xv, xix, 11, 66, 77, 83, 84, 122, 125, 140, 148, 152, 161, 162, 163, 167, 168, 169, 179, 183, 185, 186, 187, 214, 215, 218, 220, 221, 222, 223, 224, 228, 230, 242, 247, 271, 291, 292
hydrocarbon, 151, 153
hydrological, 106, 149, 229, 288, 289, 306
hydrological cycle, 149, 229
hydrology, xiv, 23, 39, 106, 143, 144, 192
hydrosphere, 305
hypothesis, 122, 148

I

IAP, 107
ice, xvii, 2, 4, 7, 10, 11, 12, 13, 15, 16, 32, 36, 37, 38, 95, 122, 131, 158, 265
Idaho, 147, 289
identification, 270
idiosyncratic, 85
Illinois, 64, 284
imagery, 139
images, 223
imbalances, 306
impact assessment, 8, 17, 19, 23, 255
implementation, xxii, 138, 238, 240, 242, 254, 257, 258, 302
imports, xvi, 50, 54, 68, 85, 224
inactive, 123, 185
incentive, 187
incidence, 3, 10, 36, 121, 125, 127, 130, 157, 177
income, 53, 61, 80, 84, 87, 88, 90, 278
incomes, 84
independent variable, 115
Indian, 307
indication, 159, 161
indirect effect, 95, 160
industrial, 2, 55, 65, 71, 76, 197, 303
industrialization, 80

industry, 9, 16, 19, 64, 65, 85, 91, 93, 152, 188, 209, 210, 240
infancy, xxii, 18, 270
Information Quality Act, 39
information technology, 227
infrared, 96, 122, 309
injury, 34, 291
innovation, xviii, 194
insight, 247
inspection, 60, 179, 207, 208, 210
institutional change, 17, 21
institutions, 23
instruments, 193, 200
insulation, 28
insurance, 9, 10, 19, 118
integration, xxiii, 23, 138, 226, 236
integrity, xxii, 12, 26, 210, 251, 252, 254, 269, 317
interactions, xxi, xxiii, 212, 226, 244, 270, 304, 305
interdisciplinary, xiv, 5, 39, 310
interface, 77, 135, 211, 291, 292, 295, 296, 297
interference, 317
Intergovernmental Panel on Climate Change, xvii, 29, 30, 34, 35, 36, 42, 96, 149, 150, 151, 251, 260, 264, 271, 295, 298, 299, 318
Intergovernmental Panel on Climate Change (IPCC), xvii, 42, 96, 150, 251, 260, 264, 271, 295
International Energy Agency, 260
international trade, 62
interrelationships, 236, 242, 312
interstate, 58, 59, 60, 73, 80, 165, 166, 169, 207, 213, 219
interval, 125, 127
intervention, 312
interview, 210, 244, 245, 246
interviews, xxi, 234, 243, 244, 247, 253, 259
intrastate, 73, 207
inventories, 163, 236
Investigations, 148
investment, xiv, xx, xxi, xxii, 4, 5, 6, 17, 21, 39, 162, 166, 196, 200, 212, 214, 217, 225, 226, 233, 234, 237, 238, 240, 242, 243, 247, 248, 249, 257, 258, 259, 263
IPCC, xvii, xviii, 2, 3, 4, 7, 8, 9, 20, 40, 42, 76, 96, 97, 98, 100, 105, 106, 107, 112, 114, 117, 123, 129, 131, 132, 134, 137, 141, 142, 143, 144, 146, 149, 150, 151, 153, 154, 156, 251, 252, 253, 260, 261, 265, 295, 298, 299, 307, 318
Ireland, 45
irrigation, 153, 289
island, 74, 77, 146
isolation, 6, 50, 87, 208

J

Japan, 151
joints, 164, 173
judgment, xxi, 19, 249, 258, 317
jurisdiction, 17, 204, 229, 240

K

kinetic energy, 4, 119
kinks, 172, 177, 241
knots, 310, 316
Kyoto Protocol, 309

L

L1, 148
labor, 84, 190, 236
labor-intensive, 236
Lafayette, 52, 63, 72, 99, 179, 183, 202, 228, 235, 242, 284, 286
Lake Pontchartrain, 83, 183, 219, 221, 242
land use, xiv, xxiii, 2, 23, 34, 79, 141, 226, 236, 237, 242, 270, 303, 305
landscapes, 83, 87, 292
large-scale, 56, 78, 79, 212
law, 37, 223, 234, 312
legislation, 236, 261, 309
levees, xix, 76, 78, 145, 159, 166, 173, 189, 203, 226, 230, 265, 266
life-cycle cost, 29
light rail, 53, 61, 160, 172, 173, 177, 303, 309
likelihood, xxiii, 19, 35, 40, 172, 251, 254, 257, 258, 270, 308
limitations, 56
linear, 115, 156
linguistically, 87
links, 55, 71, 168, 224, 225, 231, 250, 269
liquefied natural gas, 193, 310
liquids, 210
lithosphere, 303
LNG, 193, 211, 310
local authorities, 247
local government, xx, 234, 240, 249
location information, 53
logistics, 55, 187, 243, 245
London, 7, 9, 17, 26, 34, 42, 45, 46, 260
long distance, 54
long period, 114, 264
long-distance, 53, 66, 170
longevity, xxi, 269
long-term impact, 80, 84, 268

Los Angeles, 66
losses, 10, 118, 145, 153
Louisiana State University, 39, 91, 148, 152, 289
low risk, 19, 76, 183
lying, xv, 14, 50, 75, 77, 90, 131, 140, 145, 161, 166, 220, 221, 223

M

machinery, 65, 224
Maintenance, 45, 164, 267
manufacturer, 85, 194, 197, 199, 231
manufacturing, 84, 85
mapping, 23
MARAD, 299
Maritime Administration, 299
marshes, 77, 78, 144, 161
Maryland, 153, 318
Massachusetts, 153
materials science, xxiii
matrix, 291
MCC, 285
measures, 11, 12, 13, 14, 17, 18, 26, 86, 87, 106, 132, 164, 238, 245, 247, 251, 253, 257, 264, 266, 302, 317
median, 84, 87, 278
melt, 11, 15, 21, 288
melting, xvii, 3, 131, 158
metals, 64, 65, 188
meteorological, 124, 145, 158, 210, 287
methane, 2, 309
metropolitan area, 56, 223, 236, 240, 241, 243, 312
Mexico, 50, 65, 68, 85, 122, 141, 209, 224
Miami, 10, 217
migration, 83, 317
military, 71, 197, 203, 221, 231
million barrels per day, 222
Minerals Management Service, 71, 78, 91, 204, 229, 230
Minerals Management Service (MMS), 91, 204
mirror, 84
missions, 98, 242, 264
Mississippi River, xvi, 10, 12, 55, 62, 65, 66, 68, 69, 70, 74, 78, 83, 85, 87, 91, 132, 133, 144, 145, 158, 165, 166, 188, 191, 192, 213, 220, 241
MMS, 78, 91, 208, 209, 210, 229
MOB, 72, 201, 202
mobility, xxi, xxii, 36, 50, 157, 212, 213, 236, 238, 252, 257, 264, 269
modeling, xxi, 9, 19, 39, 40, 123, 129, 139, 144, 145, 225, 227, 266, 291
moderates, 253
modernization, 131
moisture, xviii, 95, 97, 101, 104, 142, 144, 146, 157, 165, 177, 179, 288, 289, 306, 314
moisture content, 314
Monte Carlo, 10
mortality, 291
movement, xv, 28, 36, 49, 54, 79, 90, 124, 134, 153, 163, 181, 189, 211, 212, 219, 226, 240, 256, 264, 306
MPI, 134
MRI, 107
multidimensional, 252
multiple factors, 226

N

NAS, 203, 205
NASA, 122, 169
nation, xi, xiii, 53, 80, 90, 164, 310, 317
National Academy of Sciences, 152
National Bridge Inventory, 61, 91, 299
national economies, 54
National Hurricane Center, 122, 137, 293, 299
National Oceanic and Atmospheric Administration, 91, 111, 318
National Research Council, 41, 42, 43, 44
National Weather Service, 137, 318
natural environment, xxi, xxiii, 53, 73, 165, 226, 250, 259, 263, 269, 271
natural gas, xvi, 54, 68, 72, 73, 85, 193, 204, 207, 210, 211, 225, 310
natural habitats, 157, 311
natural hazards, 14
natural resources, 118, 145, 236, 248
Navy, 16, 37, 38, 44, 122
negative consequences, 178
Netherlands, 9, 27, 32, 43, 221
New Jersey, 46, 129, 148, 289
New York, 14, 42, 43, 44, 46, 66, 149, 260, 261, 293, 298, 317
New Zealand, 9, 10, 19, 43, 44
NGO, 9
NGOs, 9
NHC, 137, 138, 293, 299
NHS, 59, 91, 166, 168, 169, 171, 299, 311
nitrous oxide, 2, 309
NOAA, 3, 91, 111, 116, 119, 120, 125, 129, 132, 137, 138, 140, 143, 148, 151, 152, 153, 229, 273, 293
nongovernmental, 9
nongovernmental organization, 9
Norfolk, 64, 65, 180
normal, 20, 57, 125, 137, 178, 182, 187, 206, 221, 253, 306, 315

North America, 16, 90, 92, 111, 134, 151, 158, 295, 299
North Atlantic, 118, 119, 121, 123, 124, 128, 141, 144, 147, 149, 153
North Carolina, 42, 45, 74
Northern Hemisphere, 144
Northern Ireland, 18, 45
Norway, 41
NOS, 91, 132, 152, 153

O

observations, 112, 115, 122, 123, 125, 149, 150, 229, 242, 313
obstruction, 27, 35, 60
oceans, xiii, xvii, 4, 134, 142, 158
Office of Emergency Preparedness, 228
Office of Management and Budget, 39
offshore, 70, 137, 138, 204, 206, 207, 208, 209, 210, 211, 224, 231
offshore oil, 70, 209, 224
Ohio, 69, 74, 228
oil, xvi, 16, 50, 54, 68, 70, 72, 76, 78, 85, 91, 93, 101, 152, 204, 208, 209, 211, 224, 229, 268
OMB, 39
Oregon, 149
organic, 76, 78, 303
organic matter, 303
organism, 312
orientation, 128
oscillations, 123
outliers, 100, 101
overlay, 25, 138
ownership, 16, 58
oxide, 2, 309
oxygen, 302
ozone, 302, 309

P

Pacific, 64, 65, 92, 123, 138, 152, 180, 213, 285, 307, 310
Panama, 15
parameter, 153, 289, 291
parents, 87
Paris, 260
partnership, xiv, xx, 238
partnerships, 23
passenger, xx, 4, 53, 59, 60, 62, 63, 65, 70, 71, 159, 164, 172, 177, 178, 182, 183, 187, 194, 199, 212, 220, 221, 226, 236, 238, 240, 303, 306, 311, 312
patients, 54
PCM, 107, 134
pedestrian, 174, 235
per capita, 88, 97
percentile, 101, 106, 108, 110, 112, 196, 308
perception, 243, 253
perceptions, xxii, 244, 245, 253, 254
performance indicator, 238
periodic, 132, 253
periodicity, 123
permafrost, 7, 8, 9, 10, 11, 13, 15, 21, 28
permit, 14, 81
personal communication, 230
perturbations, 97
pest management, 28
Petal, 235, 241
petrochemical, 85
petroleum, xvi, 65, 68, 70, 71, 72, 85, 93, 163, 193, 204, 207, 210, 222, 224, 225, 310
Petroleum, 64, 68, 310
petroleum products, xvi, 65, 68, 93, 207, 210, 222, 310
PFC, 309
Philadelphia, 66
photosynthesis, 303
pilots, 193, 199, 200
Pipeline and Hazardous Materials Safety Administration, 71, 91, 204, 229, 299
pipelines, xi, xiii, xvi, 7, 31, 35, 39, 50, 54, 71, 72, 73, 118, 163, 193, 204, 206, 207, 208, 209, 210, 211, 218, 219, 222, 229, 270
planetary, 305
planning decisions, 21
plants, 193, 211
plastic, 93
play, 122, 144, 177, 206, 230, 258
Pleistocene, 151
plug-in, 188
polar ice caps, 4
policy makers, xiv, 5
pollution, 36, 145
polymer, 164
poor, 87
population, xiv, 11, 16, 23, 50, 53, 56, 63, 79, 80, 81, 82, 83, 84, 85, 86, 87, 90, 118, 131, 146, 187, 190, 212, 214, 216, 226, 235, 236, 237, 254, 277, 278, 312, 313, 315, 317
poverty, 50, 80, 87, 89, 90
power, 61, 177, 186, 189, 193, 208, 209, 211, 212, 216, 219, 222, 245
power plant, 189
prediction, 112, 117, 295, 305, 313
pressure, 137, 146, 193, 208, 307
preventive, 203

private, xiv, xvi, xx, 4, 5, 8, 9, 39, 55, 58, 66, 71, 85, 155, 156, 162, 163, 187, 218, 226, 227, 234, 238, 240, 247, 248, 249, 259, 269, 309, 311, 312, 317
private investment, xiv, xx, 249
private property, 58, 312
private sector, 4, 8, 9, 39, 155, 156, 162, 163, 226, 227, 238, 240, 241, 247, 259, 269
proactive, 211, 214, 301
probability, xvii, xxii, xxiii, 10, 17, 19, 105, 112, 119, 123, 127, 130, 142, 146, 153, 157, 162, 251, 254, 255, 257, 258, 270, 288, 308
probability density function, 105
probability distribution, 153, 251
producers, 4, 50, 192
production, xvi, 50, 54, 73, 84, 85, 133, 151, 153, 163, 181, 188, 189, 193, 207, 211, 224, 306
productivity, 85, 164, 236
professions, 5, 84
profitability, 163
program, 122, 235, 236, 238, 293, 295, 316
programming, 238, 249
property, 29, 80, 118, 177, 223, 224, 309
protection, 29, 33, 152, 173, 191, 230, 236, 238, 248, 253, 264, 266
protocols, 56
prototype, 22, 23
proxy, 106
public safety, 215
pumping, 34, 133, 210, 258

Q

quality assurance, 100
quality of life, 4, 216, 236, 242
Quebec, 42
questionnaire, 244

R

radiation, 96, 287, 309
radio, 57
radius, 291
rain, 13, 14, 28, 29, 101, 165, 288
rainfall, xv, xviii, 76, 95, 97, 103, 104, 105, 117, 119, 137, 139, 143, 144, 146, 157, 162, 165, 179, 182, 189, 200, 251, 266, 267
raw materials, 65, 163
recognition, 7, 22, 250, 254
reconstruction, 219, 226, 242, 243, 291, 292
recovery, 170, 178, 191, 215, 252, 292
recreational, 58, 70
recurrence, 125
redundancy, 34, 219, 240
refineries, 85, 222
refining, 85, 222, 224, 242
refrigeration, xviii, 28, 157, 188
regenerative capacity, 252
regression, 111, 115, 116
regression analysis, 115
regular, 15, 16, 210, 255
regulations, 207, 231, 240
regulatory requirements, 189, 210
rehabilitation, 13, 28, 247, 257, 258, 268
relationship, 110, 115, 116, 117, 123, 152, 154, 257, 297
relationships, 14, 115, 116, 121, 127, 250, 305, 307, 314, 316
relevance, 2
reliability, 163, 209
remote sensing, 179, 312
repair, xx, 25, 29, 70, 170, 208, 218, 219, 220, 225, 241, 252, 254, 255, 264, 313
Research and Development, 153
residential, 83, 190
resilience, xiv, xxii, 23, 145, 225, 226, 240, 244, 250, 252, 253, 254, 256
resins, 65
resistance, 159, 189, 230
resolution, xxiii, 78, 117, 122, 123, 135, 144, 160, 270, 273
resources, 50, 79, 118, 145, 193, 214, 236, 248, 252, 253, 258, 264, 268, 274, 306
response time, 209
retention, 189, 295
revenue, xxi, 186, 217
Reynolds, 119, 121, 151, 152
rice, 65
rights-of-way, 217, 228
rings, 292
risks, xiv, xv, xvi, xviii, xx, xxi, xxii, 7, 19, 20, 21, 22, 23, 24, 26, 29, 34, 37, 74, 96, 179, 183, 204, 206, 207, 225, 233, 252
Rita, xix, xxi, 4, 42, 56, 80, 122, 139, 147, 161, 170, 206, 208, 209, 210, 216, 218, 219, 221, 222, 223, 224, 241, 242, 243, 247, 259, 267
river ports, 192, 219
rivers, 68, 74, 79, 144, 145, 157, 165, 191, 302
Roads, 13, 29, 42, 53, 57, 58, 219, 312
Rouleau, 227
routing, 186
R-squared, 116, 154
RTA, 61, 62, 63, 173, 174
rubber, 65
runoff, xvii, 32, 60, 75, 95, 101, 104, 105, 106, 110, 141, 143, 144, 146, 157, 165, 179, 191, 207, 241, 246, 265, 289, 316

S

sabotage, 208
SAFETEA-LU, xx, 234, 236, 237
safety, xi, xiii, xiv, xx, xxi, xxii, 1, 16, 18, 21, 26, 29, 36, 50, 157, 172, 178, 179, 186, 207, 225, 228, 236, 237, 238, 247, 264, 268, 274
sales, 85
saline, 28
salinity, 9, 79
salt, 16, 28, 36
sample, 70, 85, 138, 245, 296, 297
sampling, 261
sand, 74, 207, 210, 219
SAP, 261
satellite, 3, 78, 122, 123, 131, 141, 273
saturation, 117, 207
Saudi Arabia, 207
savings, 14, 194
scaling, 135
scarce resources, 264
school, 1, 53, 87, 277
scientific community, 5
sea floor, 211
sea ice, 15
seabed, 207, 210, 211
seafood, 224
sea-level rise, 151
seasonality, 97, 141, 295
Seattle, 9, 17, 45, 147
seawater, 295
Secretary of Transportation, 39
security, xxi, 11, 13, 14, 16, 37, 236, 238
sediment, 74, 75, 78, 132, 133, 145, 150
seismic, xxi, 76
selecting, xiv, 22, 24, 296
sensing, 179, 312
sensitivity, 24, 39, 71, 251
sensors, 25, 30, 36, 208
separation, 200, 231
service provider, 4
settlements, 32, 317
severity, 19, 200, 210, 211, 217, 243, 259, 269
shade, 178
shape, xv, 77, 101, 108
sharing, 241
Shell, 96, 152
shelter, 216
shipping, 12, 16, 26, 29, 37, 38, 41, 50, 93, 181, 187, 189, 220, 248
short period, 118
short-term, xviii, 77, 125, 165, 186, 208, 236, 266, 268, 270, 295
signaling, 14, 36
signs, 35, 60, 62, 128, 172, 177
similarity, 106
simulation, 10, 97, 118, 124, 138, 227, 261, 291, 292
simulations, 97, 100, 132, 138, 139, 145, 146, 152, 154, 156, 193, 305, 307
sites, 125, 132, 159
SO2, 97, 98
social group, 86
socioeconomic, 19, 252, 255, 305, 307, 313, 315
sodium, 64
software, 194, 196, 199, 231, 295
soil, xviii, 9, 11, 101, 104, 132, 144, 146, 157, 165, 206, 207, 208, 210, 211, 288, 289, 301, 308, 316
solar, 97, 287, 305
South America, 3, 220
South Carolina, 260
South Dakota, 92, 230
soybean, 220
spatial, xv, 74, 78, 83, 85, 86, 100, 115, 124, 125, 139, 144, 291, 304, 305
spawning, 119
species, 123, 291, 311
specificity, 268
spectrum, 199, 231, 309
speed, 4, 14, 18, 25, 129, 137, 138, 161, 172, 173, 203, 252, 291, 314
speed of response, 252
spills, 16, 26, 30, 209
St. Louis, 55, 140, 168, 169, 170, 184, 185, 186, 187, 219, 220, 230, 286
stability, 35, 206, 210
stabilization, 317
stabilize, 28
stages, 20, 238, 316
stakeholders, xiv, 6, 8, 23, 252, 254, 255
standard deviation, 305
standards, 6, 19, 168, 177, 195, 196, 226, 228, 242, 249, 257, 270
statistical analysis, 101
statistics, xxiii, 72, 123, 155, 270, 305
steel, 28, 60, 65, 93, 114, 164, 178
steel plate, 65
stiffness, 177
stimulus, 254
storage, 85, 104, 155, 187, 188, 189, 191, 220
Storm surge, 137, 145, 161, 172, 206, 213, 315
stormwater, 21, 165, 189
strain, xix, 268
strategic management, 238
strategies, 1, 6, 7, 10, 18, 19, 20, 21, 22, 23, 24, 74, 157, 215, 217, 223, 226, 227, 236, 238, 241, 242,

247, 248, 253, 255, 257, 258, 259, 263, 264, 266, 269, 270, 312
streams, 74, 75, 157, 165, 302, 308
strength, xix, xxii, 114, 118, 120, 127, 129, 140, 170, 209, 219, 246, 251, 254, 267, 292, 314, 317
stress, xxii, 12, 13, 26, 156, 178, 188, 208, 251, 254, 268, 308
stressors, 240, 245, 250, 251, 254, 257, 258, 317
stretching, 178
strikes, xix, 14, 125, 267
subjective, 254
submarines, 38
substances, 16, 307
substitution, 16
suburban, 50, 80, 88
sugar, 65
sulfur, 309
summaries, 246
summer, 12, 16, 97, 104, 106, 108, 113, 141, 142, 143, 144, 162, 178, 200, 265, 267
Sun, 142, 149, 152, 229
supplemental, 295
suppliers, 209
supply, 53, 54, 55, 63, 163, 192, 208, 210, 216, 222, 224
supply chain, 55, 163
surface area, xv, 77, 191
Surface Transportation Board, 92
surface water, 207, 307
surplus, 105, 289
surprise, 16
surveillance, 215
sustainability, 91, 97
switching, 92
Switzerland, 149, 150, 151, 271, 318
symbols, 56
synchronization, 17
synergistic, 258, 265
synthesis, xiv, 2, 23, 39, 138, 245

T

TAR, 131, 261
task force, 243
taxis, 311
taxonomy, xxii, 259
technological change, 85, 97, 193, 307
technological developments, 19, 313
temporal, 86, 144, 305
Tennessee, 68, 69, 149
tension, 178
terminals, xviii, 32, 59, 60, 62, 64, 66, 166, 173, 188, 189, 190, 192, 220, 221, 263, 273, 309
territory, 74
terrorist, 208
terrorist attack, 208
terrorists, 16
thawing, 7, 8, 11, 13, 15
thermal expansion, 4, 114, 131, 165
threat, 19, 209
threatened, 137, 266
thresholds, xix, 8, 17, 19, 146, 156, 246, 254, 256, 270
tides, 13, 32, 78, 137, 139, 265, 295
TIP, 236, 316
TIPS, 244
TMA, 235
tolerance, 163
topographic, 49, 95, 138
tornadoes, 4, 144, 162, 220
tracking, 124, 125, 129, 291
trade, 9, 62, 307
trading, 73, 207
traffic, xix, 14, 15, 27, 30, 34, 57, 58, 59, 60, 64, 65, 67, 70, 72, 164, 172, 177, 178, 191, 193, 200, 208, 211, 219, 220, 221, 223, 235, 243, 252, 253, 302, 303
training, 224
trajectory, xvii, xix, 146
transfer, 55, 62, 93, 187
transit ridership, 53
transition, 168, 206, 302, 308
transmission, 72, 204, 206, 207, 208, 210, 228
transnational, 7
transpiration, 308
transport, xv, xvi, xx, 1, 7, 10, 12, 49, 50, 53, 55, 75, 159, 163, 190, 192, 193, 214, 222, 225, 226, 268, 306, 308, 311, 313
travel, xv, xxi, 4, 10, 16, 25, 29, 35, 53, 58, 59, 66, 83, 92, 93, 170, 178, 238, 249, 261, 268, 302, 306, 317
tropical storms, xv, xix, 4, 49, 71, 76, 77, 90, 96, 118, 119, 123, 124, 125, 128, 139, 141, 144, 267
trucking, xvi, 15, 50, 54
trucks, 15, 59, 93, 159, 186, 187, 188, 220
tundra, 15
turbulence, 200
turbulent, 35

U

U.S. economy, 188
U.S. Geological Survey, xiv, 39, 74, 75, 77, 79, 90, 91, 92, 134, 147, 148, 149, 153, 230

uncertainty, xvii, xxi, 2, 6, 7, 17, 18, 21, 22, 40, 96, 101, 146, 156, 207, 233, 243, 250, 253, 257, 259, 307, 313
UNFCCC, 304, 317
uniform, 141, 159, 165, 246
United Kingdom, 9, 10, 40, 41, 42, 45, 46, 47, 92, 151, 153, 260, 261
United Nations, 304, 317
unmanned aerial vehicles, 215
updating, 76, 240, 241
urban areas, 7, 14, 53, 57, 83, 85, 88, 90, 223, 227, 241, 306
urban centers, 50, 317
urbanized, xx, 79, 80, 83, 235, 236, 242, 312
USDA, 90
USSR, 152
Utah, 46, 90

V

validation, 138
validity, 100
values, 97, 101, 104, 111, 114, 115, 132, 134, 142, 143, 145, 158, 159, 164, 265, 287, 288, 289, 317
vapor, 117, 142, 151, 200, 309, 310, 315, 316
variability, xiv, 1, 3, 22, 23, 75, 78, 86, 99, 100, 103, 104, 118, 123, 125, 131, 132, 145, 153, 158, 212, 238, 241, 242, 243, 244, 250, 251, 253, 255, 259, 264, 295, 302, 303, 304, 305, 307
variables, xvii, 10, 14, 78, 101, 105, 115, 144, 146, 156, 193, 231, 254, 269, 288, 303
variation, 81, 97, 128, 145, 158, 245, 251
vegetation, 11, 12, 27, 28, 96, 308
vehicles, xviii, 4, 16, 61, 62, 156, 162, 211, 212, 215, 223, 224, 267, 308, 309
ventilation, 14, 26
versatility, 198
vessels, 12, 33, 35, 65, 68, 163
Victoria, 116, 261
vision, xx, 236, 238, 241, 247, 312

W

walking, 242
warehousing, 4, 187
warrants, 269
Washington Post, 47
water evaporation, 308
water quality, 13, 21, 29, 144, 145
water table, 182, 206, 207, 210
water vapor, 117, 142, 151, 200, 310, 315, 316
watershed, 144, 274
waterways, xi, xiii, xvi, 7, 10, 11, 12, 15, 16, 26, 30, 50, 55, 68, 95, 145, 162, 188, 273
wavelengths, 309
WCR, 122
wear, 28
weathering, 308
wellbeing, 50
West Africa, 207
wetland restoration, 253
wetlands, 28, 77, 78, 79, 91, 142, 145, 148, 161, 162, 206, 242, 248, 264, 265, 292
wheat, 220
Wilma, 122, 147
wind, 4, 14, 18, 35, 77, 124, 128, 129, 137, 141, 149, 161, 162, 172, 177, 190, 191, 199, 203, 204, 215, 216, 221, 224, 226, 267, 291, 292, 303, 307, 310, 314, 316
wind speeds, 4, 14, 18, 129, 141, 226, 267
windows, 126
winter, 10, 11, 14, 15, 16, 20, 97, 104, 106, 141, 142, 200, 220
wireless, 215
wood, 64, 65
workers, 53, 84, 114, 178, 224

Y

yield, 249